SOCIETY OF GENERAL PHYSIOLOGISTS SERIES

Volume 37

Basic Biology of Muscles: A Comparative Approach

Society of General Physiologists Series

Published by Raven Press

SOCIETY OF GENERAL PHYSIOLOGISTS SERIES
Volume 37

Basic Biology of Muscles
A Comparative Approach

Editors

Betty M. Twarog, Ph.D.
Professor and Chairman
Department of Biology
Bryn Mawr College
Bryn Mawr, Pennsylvania

Rhea J. C. Levine, Ph.D.
Professor
Department of Anatomy
The Medical College of Pennsylvania
Philadelphia, Pennsylvania

Maynard M. Dewey, Ph.D.
Professor and Chairman
Department of Anatomical Sciences
State University of New York
Health Sciences Center
Stony Brook, New York

Raven Press ■ New York

Raven Press, 1140 Avenue of the Americas, New York, New York 10036

© 1982 by Raven Press Books, Ltd. All rights reserved. This book is protected
by copyright. No part of it may be reproduced, stored in a retrieval system, or
transmitted, in any form or by any means, electronic, mechanical, photocopying,
recording, or otherwise, without the prior written permission of the publisher.

Made in the United States of America

Library of Congress Cataloging in Publication Data

Basic biology of muscles.

    (Society of General Physiologists series ; v. 37)
    Includes bibliographical references and index.
    1. Muscles.   I. Twarog, Betty M.   II. Levine, Rhea J. C.
III. Dewey, Maynard Merle, 1932–   IV. Series.
QP321.B317   1982   591.1′852   82-47510
ISBN 0-89004-799-5

Great care has been taken to maintain the accuracy of the information contained
in the volume. However, Raven Press cannot be held responsible for errors or for
any consequences arising from the use of the information contained herein.

Dedication

This volume is dedicated to our honored guests, Emil Bozler and C. Ladd Prosser, who have inspired more than one generation of scientists to study muscle from a comparative point of view. It is my personal wish to mention yet another comparative physiologist in this dedication: John H. Welsh, an extraordinarily insightful comparative physiologist who completed a Ph.D. thesis with G. H. Parker at Harvard in 1929 in which he implicated contractile fibrils in the movement of distal pigment cells of the eyes of a shrimp. He remained interested in contractile mechanisms, although his subsequent research focussed on neurotransmitters and neurosecretion. He guided the preparation of my Ph.D. thesis on control of contraction in a molluscan catch muscle and I had the privilege of participating in and observing his studies on a molluscan heart, which revealed many principles of neurotransmitter action general to all animals. Interestingly, John Welsh's initial use of the molluscan heart stemmed from observations reported by C. Ladd Prosser. As one of John Welsh's numerous intellectual "offspring," I wish to express my gratitude for the inspiration of his teaching and research.

Betty M. Twarog

Preface

This volume presents the entire proceedings of the symposium held at Woods Hole, Massachusetts, September 10 to 13, 1981, under the sponsorship of the Society of General Physiologists and the National Science Foundation.

The theme of the symposium, a comparative approach to basic problems in muscle biology, aroused considerable enthusiasm among organizers and participants. As Andrew F. Huxley (1) has pointed out, for many decades the study of muscular contraction was profoundly influenced by the belief that all contractility is similar, yet our recent immense progress has come from studying highly specialized striated muscles in rabbit, frog, and insect. The importance of looking out for suitable experimental material when trying to solve a specific biological problem was clearly stated by August Krogh and restated by Hans Krebs (2), who wrote that a highly specialized function in a particular species can provide favorable material for the study of general questions, especially at the cellular level. Krebs quoted Pringle's description of the enormous variety of biological materials as the "treasure house of nature." The considerations which ran throughout this symposium and aroused our excitement—both because we shared new knowledge acquired from comparative studies and because we shared a vision of the rich rewards promised by future comparative studies—are expressed well by Graham Hoyle: "No two muscles or even muscle cells, from different species, have yet been found that cannot rather easily be distinguished from each other. It is a major task for the comparative muscle physiologist to understand the significance of these variations on the muscle theme. The variations provide an extremely rich set of natural experiments on which a fundamental muscle scientist may test concepts."

A basic unifying theme of this volume is the appreciation of similarities and differences in basic structure and function across the spectrum of muscle cell types. The assumption of underlying unity of the basic biology of muscle is the experimental rationale of many but not all participants. The comparative studies set forth herein exemplify three types of approach:

1) Those which focus on a detailed analysis of basic qualities of muscle structure and function. For these studies, the comparative approach facilitates selecting muscles in which a particular quality is best expressed or most conveniently studied.

2) Those which focus on specialized muscles per se to understand the unusual structures of functions they display. Novel principles are thereby revealed which later may be found to apply to muscles of diverse function.

3) Those which aim at understanding the evolutionary relationships of muscles by examining diverse muscle types throughout the animal phyla. This exploration of the "treasure house of nature" contributes to our "portfolio" of muscles suited for analysis as examples of specific principles.

Among the structural characteristics discussed in the first type of study is the molecular architecture of thick filaments. Is there a basic design of thick filaments? (Wray). What is the structure of the paramyosin core in quick-acting striated muscles and in smooth catch muscles? (A. Elliott and G. Elliott).

Pastra-Landis asks why there are two kinds of myosin light chain and explores possible functions of the alkali light chain. Many studies examine Ca^{2+} regulation and the role(s) of regulatory light chains (Kushmerick, Butler and Siegman, Perry, Kendrick-Jones, Adelstein, Ebashi). Perry summarizes our current understanding of regulation, pointing out that vertebrate striated muscle regulation is actin-linked, mediated by Ca^{2+} binding to a specific target protein, troponin, whereas in mollusc striated muscle, Ca^{2+} binds directly to myosin light chain. In smooth muscle, Ca^{2+}-induced phosphorylation of myosin light chain appears to trigger contraction, but this conclusion remains controversial: direct binding of Ca^{2+} to actin filaments or myosin has not been ruled out. Ebashi convincingly advocates the necessity of an open-minded approach to the mechanisms of Ca^{2+} regulation and emphasizes that in one known system (slime molds) Ca^{2+} is a repressor rather than an activator! Weber provides recent data suggesting that in actin-linked systems a regulatory mechanism exists in addition to Ca^{2+}-sensitive steric blocking.

Both Bozler and Rall seek to elucidate basic mechanisms that affect the capacity of muscles to work. The utility of the highly evolved striated muscles of frog, mammal, and insect for analysis of force generation in muscle and of the fundamental mechanisms of actomyosin cross-bridge reactions is exemplified in the work of Podolsky, Simmons, Kawai, and Tregear. In contrast, Fay's studies of isolated single smooth muscle cells aim to uncover the basic processes underlying cell motility, which are obscured in the very highly ordered and specialized striated muscles from which most recent concepts of contractile mechanisms derive.

The history and current status of the second type of study which attempts to delineate the mechanisms of seemingly unique structures or functions is detailed by Achazi. He sets forth evidence for the hypothesis that Ca^{2+}-dependent phosphorylation of paramyosin changes the properties of actomyosin-ATPase, resulting in a decreased velocity of cross-bridge cycling and increased stretch resistance in a mollusc muscle (catch). Understanding how phosphorylation of the paramyosin core of a thick filament in catch muscles affects the enzymic properties of the myosin molecules arrayed on its surface may help to explain the relationship between thick filament structure and mechanical properties, even in muscles lacking paramyosin. In a *Limulus* muscle, which operates over extreme ranges in length, the maintenance of functional overlap between thick and thin filaments at the longest lengths and ability to develop appreciable tension at the shortest lengths appear to be associated with thick filament shortening. Levine asks whether *Limulus* filament structure is related to function as well as to evolutionary affinities. Dewey, and Davidheiser and Davies, respectively investigate the structural and functional changes that occur during thick filament shortening and the energy expenditure associated with the shortening.

The third type of study starting from an evolutionary point of view is inherent in Levine's work, and is exemplified by Anderson's pioneering studies on myoepithelial cells of a marine worm which display a constellation of unique properties. Prosser's provocative discussion of the origins and diversity of muscles with respect to calcium regulatory mechanisms and the evolution of transversely aligned filaments (striation) presents a powerful argument for the evolutionary approach. In another vein, the contribution of Mykles can be seen as originating from evolutionary exploration: observations on molting in crustacea reveal that hormonally controlled claw muscle atrophy facilitates the withdrawal of claw muscles from the old exoskeleton. Mykles and Skinner have used atrophy in crustacean claw muscle (unique among other models because it is fully physiological rather than pathological) as a model for the study of protein metabolism and its manifestation in the structure of the sarcomere.

I, and my co-editors, Rhea Levine and Maynard Dewey, hope that the reader will share our enthusiasm for the varied explorations in the "treasure house of nature" presented here.

Betty M. Twarog

REFERENCES

1. Huxley, A. F. (1980): *Reflections on Muscle*. Princeton University Press, Princeton.
2. Krebs, H. A. (1975): The August Krogh Principle "For Many Problems There is an Animal on Which It Can Be Most Conveniently Studied". *J. Exp. Zool.*, 194:221–226.

Contents

xi

Unusual and Unfamiliar Muscles

Lectures by Honored Guests

Contributors

Rudolf K. Achazi
Free University
Institute of Animal Physiology and
  Applied Zoology
D-1000 Berlin 41, Germany

Robert S. Adelstein
Laboratory of Molecular Cardiology
National Heart, Lung, and Blood
  Institute
Bethesda, Maryland 20205

Norman R. Alpert
Department of Physiology and Biophysics
University of Vermont College of
  Medicine
Burlington, Vermont 05405

Margaret Anderson
Department of Biological Sciences
Smith College
Northampton, Massachusetts 01063

Toshiaki Arata
Laboratory of Physical Biology
National Institute of Arthritis, Diabetes,
  and Digestive and Kidney Diseases
National Institutes of Health
Bethesda, Maryland 20205

Pauline M. Bennett
King's College Biophysics Department
  and MRC Cell Biophysics Unit
London WC2B 5RL, England
  United Kingdom

J. Bordas
European Molecular Biology Laboratory
  Outstation
DESY
Hamburg, West Germany

Emil Bozler
The Ohio State University
Department of Physiology
Columbus, Ohio 43210

Peter Brink
Department of Anatomical Sciences
Health Sciences Center
State University of New York
Stony Brook, New York 11794

Thomas M. Butler
Department of Physiology
Jefferson Medical College
Thomas Jefferson University
Philadelphia, Pennsylvania 19107

M. L. Clarke
Institute of Animal Physiology
Babraham
Cambridge CB2 4AT, England
  United Kingdom

H. A. Cole
Department of Biochemistry
University of Birmingham
Birmingham B15 2TT, England
  United Kingdom

David Colflesh
Department of Anatomical Sciences
Health Sciences Center
State University of New York
Stony Brook, New York 11794

Mary Anne Conti
Laboratory of Molecular Cardiology
National Heart, Lung, and Blood
  Institute
Bethesda, Maryland 20205

Roger Craig
MRC Laboratory of Molecular Biology
Cambridge CB2 2QH, England
  United Kingdom

Michael T. Crow
Department of Medicine
Stanford University
Stanford, California 94305

Sandra Davidheiser
Department of Animal Biology
School of Veterinary Medicine
University of Pennsylvania
Philadelphia, Pennsylvania 19104

Robert E. Davies
Department of Animal Biology
School of Veterinary Medicine
University of Pennsylvania
Philadelphia, Pennsylvania 19104

Primal de Lanerolle
Laboratory of Molecular Cardiology
National Heart, Lung, and Blood
   Institute
Bethesda, Maryland 20205

Maynard M. Dewey
Department of Anatomical Sciences
Health Sciences Center
State University of New York
Stony Brook, New York 11794

Setsuro Ebashi
Department of Pharmacology
Faculty of Medicine
University of Tokyo
Hongo, Tokyo 113, Japan

Arthur Elliott
King's College Biophysics Department
   and MRC Cell Biophysics Unit
London WC2B 5RL, England
   United Kingdom

Gerald F. Elliott
Department of Biophysics Group
The Open University Research Unit
Oxford, England, United Kingdom

Shih-fang Fan
Shanghai Institute of Physiology
Academia Sinica
Shanghai, China

Fredric S. Fay
Department of Physiology
University of Massachusetts Medical
   School
Worcester, Massachusetts 01605

Michael A. Ferenczi
Department of Physiology
University College London
London WC1E 6BT, England
   United Kingdom

Kevin Fogarty
Department of Physiology
University of Massachusetts Medical
   School
Worcester, Massachusetts 01605

Keigi Fujiwara
Department of Anatomy
Harvard Medical School
Boston, Massachusetts 02115

Bruce Gaylinn
Department of Anatomical Sciences
Health Sciences Center
State University of New York
Stony Brook, New York 11794

R. J. A. Grand
Department of Biochemistry
University of Birmingham
Birmingham B15 2TT, England
   United Kingdom

N. Gural
Department of Anatomical Sciences
Health Sciences Center
State University of New York
Stony Brook, New York 11794

John C. Haselgrove
Johnson Research Foundation
University of Pennsylvania
Philadelphia, Pennsylvania 19104

Graham Hoyle
Biology Department
University of Oregon
Eugene, Oregon 97403

Ted W. Huiatt
Rosenstiel Basic Medical Sciences
   Research Center
Brandeis University
Waltham, Massachusetts 02254

Ross Jakes
MRC Laboratory of Molecular Biology
Cambridge CB2 2QH, England
   United Kingdom

Masataka Kawai
H. Houston Merritt Clinical Research
   Center for Muscular Dystrophy and
   Related Diseases
College of Physicians and Surgeons of
   Columbia University
New York, New York 10032

John Kendrick-Jones
MRC Laboratory of Molecular Biology
Cambridge CB2 2QH, England
   United Kingdom

Robert W. Kensler
Department of Anatomy
The Medical College of Pennsylvania
Philadelphia, Pennsylvania 19129

M. Koch
European Molecular Biology Laboratory
Outstation
DESY
Hamburg, West Germany

Martin J. Kushmerick
Department of Physiology
Harvard Medical School
Boston, Massachusetts 02115

B. A. Levine
Inorganic Chemistry Laboratory
University of Oxford
Oxford OX1 2QR, England
United Kingdom

Rhea J. C. Levine
Department of Anatomy
The Medical College of Pennsylvania
Philadelphia, Pennsylvania 19129

Susan Lowey
Rosenstiel Basic Medical Sciences
Research Center
Brandeis University
Waltham, Massachusetts 02254

Steven B. Marston
Cardiothoracic Institute
London W1N 2DX, England
United Kingdom

Susan U. Mooers
Department of Physiology
Jefferson Medical College
Thomas Jefferson University
Philadelphia, Pennsylvania 19107

Louis A. Mulieri
Department of Physiology and Biophysics
University of Vermont College of
Medicine
Burlington, Vermont 05405

Donald L. Mykles
Biology Division
Oak Ridge National Laboratory
Oak Ridge, Tennessee 37830

Geoffrey R. S. Naylor
Laboratory of Physical Biology
National Institute of Arthritis, Diabetes,
and Digestive and Kidney Diseases
National Institutes of Health
Bethesda, Maryland 20205

Styliani C. Pastra-Landis
Rosenstiel Basic Medical Sciences
Research Center
Brandeis University
Waltham, Massachusetts 02254

Mary D. Pato
Laboratory of Molecular Cardiology
National Heart, Lung, and Blood
Institute
Bethesda, Maryland 20205

Samuel V. Perry
Department of Biochemistry
University of Birmingham
Birmingham B15 2TT, England
United Kingdom

Richard J. Podolsky
Laboratory of Physical Biology
National Institute of Arthritis, Diabetes,
and Digestive and Kidney Diseases
National Institutes of Health
Bethesda, Maryland 20205

C. Ladd Prosser
Department of Physiology and Biophysics
University of Illinois
Urbana, Illinois 61801

Jack A. Rall
Department of Physiology
Ohio State University
Columbus, Ohio 43210

C. D. Rodger
Roehampton Institute of Higher
Education
London SW15, England, United Kingdom

Jonathan Scholey
MRC Laboratory of Molecular Biology
Cambridge CB2 2QH, England
United Kingdom

James R. Sellers
Laboratory of Molecular Cardiology
National Heart, Lung, and Blood
*   Institute*
Bethesda, Maryland 20205

Marion J. Siegman
Department of Physiology
Jefferson Medical College
Thomas Jefferson University
Philadelphia, Pennsylvania 19107

Robert M. Simmons
Department of Physiology
University College London
London WC1E 6BT, England
*   United Kingdom*

Dorothy M. Skinner
Biology Division
Oak Ridge National Laboratory
Oak Ridge, Tennessee 37830

John A. Sleep
Department of Physiology
University College London
London WC1E 6BT, England
*   United Kingdom*

Murray Stewart
MRC Laboratory of Molecular Biology
Cambridge CB2 2QH, England
*   United Kingdom*

Haruo Sugi
Department of Physiology
School of Medicine
Teikyo University
Itabashi-ku, Tokyo 173, Japan

Suechika Suzuki
Department of Physiology
School of Medicine
Teikyo University
Itabashi-ku, Tokyo 173, Japan

Phillip Tooth
MRC Laboratory of Molecular Biology
Cambridge CB2 2QH, England
*   United Kingdom*

Richard T. Tregear
Institute of Animal Physiology
Babraham
Cambridge CB2 4AT, England
*   United Kingdom*

Cynthia E. Trueblood
Department of Molecular, Cellular, and
*   Developmental Biology*
University of Colorado
Boulder, Colorado 80309

Richard Tuft
Worcester Polytechnic Institute
Department of Physics
Worcester, Massachusetts 01609

Terrence P. Walsh
Department of Biochemistry and
*   Biophysics*
University of Pennsylvania
Philadelphia, Pennsylvania 19104

Annemarie Weber
Department of Biochemistry and
*   Biophysics*
University of Pennsylvania
Philadelphia, Pennsylvania 19104

John S. Wray
Max-Planck Institute for Medical
*   Research*
Heidelberg, West Germany

Basic Biology of Muscles: A Comparative
Approach, edited by B. M. Twarog,
R. J. C. Levine, and M. M. Dewey.
Raven Press, New York © 1982.

Myosin Light Chain Isozymes: Assembly and Kinetic Properties

S. C. Pastra-Landis, T. W. Huiatt, and S. Lowey

*Rosenstiel Basic Medical Sciences Research Center, Brandeis University,
Waltham, Massachusetts 02254*

The isozymes of myosin, in particular those derived from the light chains, provide a means by which to address the question of the role of the light chains in the overall function of myosin. The hexameric myosin molecule is comprised of two heavy chains of molecular weight 200,000 and four smaller light chains of average molecular weight 20,000. It was a decade ago when chromatographic purification methods for myosin, the isotope dilution technique, and the availability of sodium dodecyl sulfate (SDS) polyacrylamide gel electrophoresis made possible the determination of the light chain stoichiometry (10,29). Each globular head, or S-1 region, of myosin contains two light chains—one member of each of two types. One class is known as "nonessential," as their removal with the sulfhydryl reagent, 5,5'-dithiobis (2-nitrobenzoic acid) (DTNB), does not affect the enzymatic activity. The other class, called the alkali light chains, became known as "essential" because their dissociation in alkali solutions was irreversible and caused complete loss of ATPase activity.

The earlier work of Bárány (1) demonstrated the polymorphic nature of myosin; differences in speed of contraction of fibers from physiologically diverse muscles were correlated with differences in ATPase activities of myosin isolated from these fibers. Myosins from different muscles also have correspondingly different light chain patterns (10,16), a result that led to the speculation that the light chains were involved either directly at the active site or in regulation of the enzymatic function of myosin.

Fast skeletal myosin exhibits an additional complexity: along with the DTNB light chain (LC2), there are two types of alkali light chains, A1 and A2 (otherwise called $LC1_f$ and $LC3_f$) of molecular weight 21,000 and 17,000 (6). These exist in nonintegral amounts and in different ratios among different species, or even among different muscles of the same species. For example, chicken pectoralis myosin contains almost equimolar amounts of A1 and A2, whereas the leg and PLD muscles have a smaller proportion of A2, as does rabbit skeletal muscle (10). In addition, the alkali light chain distribution changes systematically during development (3,20). The nonintegral stoichiometry suggested that there were myosin isozyme populations; subsequently, the existence of two homodimers, with either A1 or A2 light chain on each head, and a heterodimer containing both A1 and A2 has been demonstrated (4,7,9). Both alkali light chain variants have been localized within single isolated fast fibers (28) and in individual myofibrils (18). By immunoelectron microscopy, both isoforms were also found within a single myofilament (18). Thus, the isozymes coexist at every level of muscle organization.

Comparison of the sequences of the alkali light chains from rabbit myosin showed them to be highly homologous but phenotypically distinct, with identical amino-acid sequences over the C-terminal 140 residues (6). The A1 light chain contains an additional 41 residues at the N-terminus, known as the difference peptide, $\Delta 1$. The 8 residues between $\Delta 1$ and the long common sequence contain five amino-acid substitutions between A1 and A2, and constitute the $\Delta 2$ difference peptide. Thus, there must be two RNA coding sequences for these light chains in rabbit fast muscle.

The hypothesis that these light chains were associated with the regulation of ATPase activity received further support from experiments in which subfragment-1, the enzymatically active head portion of myosin obtained by chymotryptic proteolysis, was fractionated by ion exchange chromatography into two S-1 populations with either A1 or A2 light chains (30). The S-1 isozymes exhibited identical K^+-EDTA, Ca^{2+}, and Mg^{2+} ATPases, but were significantly different in the more physiological actin-activated kinetics. The V_{max} for A2 was double that of A1, and the K_{app} for actin binding was also much larger. These initial studies were carried out at low ionic strength (6 mM KCl) where the K_{app} of S-1 for actin is low. Parallel kinetic differences were also found between the isozymes of chymotryptic heavy meromyosin, which was fractionated on an ADP affinity column (23). These results established that differences in the actin-activated ATP hydrolysis between the A1 and A2 isozymes were not artifacts arising from the use of a single-headed proteolytic species, but were found with the two-headed subfragment as well.

Perhaps even more importantly, studies on hybrid S-1 molecules also suggested the involvement of light chains in controlling kinetic differences (27). Hybrids were obtained by alkali light chain exchange in high-salt (4.7 M NH_4Cl), conditions under which reversible dissociation of the subunits occurs. However, when the kinetic experiments on S-1 isozymes were performed at intermediate ionic strengths (25 mM KCl) and up to physiological salt conditions, the kinetic differences governed by distinct light subunits at 6 mM KCl disappeared: the V_{max} of A1 approached that of A2 and the K_{app} values also became very similar (26). With this experiment the entire question of the alkali light chain function was opened for reexamination. Why are there two kinds of light chain and what might their role be?

FILAMENTOUS MYOSIN

It became important to address the question using the whole myosin molecule, especially because developments in immunochemistry made possible the fractionation of the isozymes of native myosin (8,18). Studies using this material are free from complications due either to proteolysis or exposure to high-salt dissociating conditions. The fractionation was accomplished by immunoadsorption chromatography using rabbit antibodies specific to the difference peptides of A1 and A2, a method described in detail by Silberstein and Lowey (18). The purified preparations of homodimers contained approximately 15% of the heterologous light chain when analyzed by densitometry of SDS gels. This incomplete separation was due in part to the presence of heterodimer in the unfractionated myosin.

Using purified myosin isozymes, one can investigate not only the kinetic behavior of myosin, but also its structural properties. To examine the possibility that the isozymes differed in their assembly properties, synthetic filaments were formed from control (unfractionated) myosin and the isolated A1 and A2 isozymes under a variety of buffer conditions. An example of the synthetic filaments formed from A1 myosin is shown in Fig. 1; filaments formed from either A2 or control myosin under the same conditions were identical in appearance. Moreover, the length distribution of the filaments was identical for all three myosin samples (Fig. 2). Some variation in length of the synthetic filaments was seen when imidazole buffer was substituted for phosphate, or when $MgCl_2$ was omitted. However, the lengths of the filaments formed from the control, A1 and A2 myosins, were identical under the same buffer conditions, indicating that the particular alkali light chain present does not affect filament assembly.

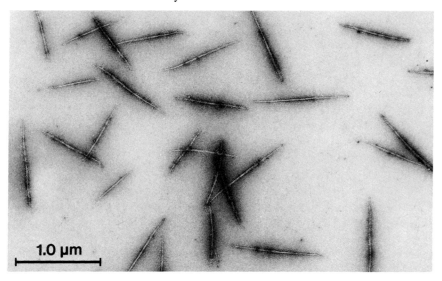

FIG. 1. Electron micrograph of synthetic filaments formed from purified A1 myosin homodimer. Filaments were formed by a procedure similar to that of Wachsberger and Pepe (22). A solution of A1 myosin (0.17 mg/ml) in 0.5 M KCl, 5 mM $MgCl_2$, 10 mM K-phosphate, pH 7.0, was diluted rapidly to 0.3 M KCl over a period of 1 min. The solution was allowed to stand on ice for 30 min and then diluted to 0.1 M KCl over 5 min. Filament samples were negatively stained with 2% aqueous uranyl acetate and examined in a Philips EM301 electron microscope operated at 80 kV. Some of these synthetic filaments do not exhibit a clearly defined bare zone, but they are all bipolar, as evidenced by the reversal of the polarity of the projections from the filament shaft (myosin heads) seen near the center of several of the filaments. ×23,000.

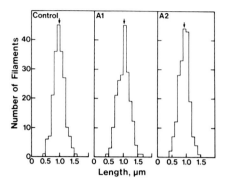

FIG. 2. Length distribution of synthetic myosin filaments formed from the myosin isozymes. Filaments were formed from control (unfractionated) myosin and the isolated A1 and A2 homodimers as described in the legend to Fig. 1. Lengths were measured with a ruler on prints at a final magnification of ×27,000. The ordinates were normalized to 200 filaments for the purpose of comparison. Average lengths (indicated by *arrows*) were: control myosin, 0.97 ± 0.19 μm; A1 myosin, 1.01 ± 0.21 μm; A2 myosin, 0.93 ± 0.18 μm.

TABLE 1. *Kinetic constants of myosin isozymes[a]*

	ATPase K^+ (EDTA)	ATPase Ca^{2+}	Actin-activated ATPase			
			25 mm KCl		55 mm KCl	
			V_{max}	K_m	V_{max}	K_m
Unfractionated myosin	9.8 ± 0.8	1.1 ± 0.1	3.7	7.6	2.8	31
A1 myosin	10.3 ± 1.0	1.5 ± 0.2	3.5	7.2	2.5	19
A2 myosin	10.2 ± 0.3	1.3 ± 0.2	2.7	8.1	1.8	19

[a]The activity is expressed in (mole Pi)/(mole protein active sites) (sec). Kinetic constants for the actin-activated data were obtained by use of double reciprocal plots. All assays were conducted by the pH-stat method at 25°C and pH 8.0. The K^+ ATPase was assayed in 0.66 m KCl, 1 mm EDTA, and 2.5 mm ATP. The Ca^{2+} ATPase was assayed in 0.25 m KCl, 2.5 mm $CaCl_2$, 2 mm Tris, and 2 mm ATP. The actin-activated assays were carried out in 25 or 55 mm KCl, as indicated, 2.5 mm ATP, and 3.75 mm $MgCl_2$. Actin concentrations in the range 5 to 50 μm were used. These actin-activated assays cannot be executed at significantly lower ionic strengths or at higher actin concentrations because the actomyosin solutions become exceedingly viscous and adequate stirring for a sensitive electrode response in the pH-stat is hindered.

Steady-state kinetics with the isolated native isozymes confirmed that the K^+-EDTA and Ca^{2+} ATPases were identical, and that the fractionated material retained the full enzymatic activity of unfractionated starting material (Table 1). The actin-activated kinetics were carried out at 25 and 55 mm KCl as well as at higher salt concentrations. At 55 mm KCl, the kinetic parameters were very similar (Table 1). At the lower ionic strength as well, only small differences could be discerned from one experiment to the other, and never were these differences in either V_{max} or K_{app} any larger than twofold, occasionally favoring a slightly higher V_{max} for the A1 isozyme. However, actin-activated kinetics of native myosin are difficult to reproduce because the myosin is present in an uncertain, heterogeneous state of aggregation at the ionic strengths used in the assays, resulting in some scatter of the data in double reciprocal plots. To overcome these experimental difficulties, additional kinetic studies on the isozymes were done, using the myosin minifilament system recently described by Reisler and co-workers (14,15).

MYOSIN MINIFILAMENTS

Minifilaments are highly homogeneous, short, bipolar filaments containing 16 to 18 myosin molecules (15). Rabbit myosin minifilaments behave kinetically like soluble subfragments; heavy meromyosin and minifilaments exhibit identical actin-activated ATPase activities (14). With rabbit myosin, minifilaments were formed from a monomeric solution, at 5 mg/ml in pyrophosphate buffer, by dialysis against 10 mM citrate-Tris, pH 8.0. Our results demonstrated that chicken myosin also assembled into minifilaments under these conditions. Figure 3a shows the Schlieren pattern obtained in the analytical ultracentrifuge at two different protein concentrations; the minifilaments sedimented as a hypersharp boundary ($s^0_{20,w}$ = 30.4S) with little evidence of myosin monomers. Assembly into minifilaments also took place at very dilute protein concentrations of around 0.2 mg/ml, such as those obtained from the immunofractionation (Fig. 3b). Minifilaments remained stable at these low protein concentrations even over a two-week period. Results similar to those in Fig. 3b were obtained with minifilaments formed from the isolated isozymes. Most importantly, the isozymes do not differ in their assembly behavior: the sedimentation coefficients are approximately the same as the $s^0_{20,w}$ value of 32.3S obtained by Reisler et al. (15) for rabbit myosin. An electron micrograph of a typical preparation of chicken myosin minifilaments is shown in Fig. 4.

Actin-activated kinetics on these homogeneous minifilaments do indeed show an improvement in the quality of data (Fig. 5). Assays done at comparable ionic strengths show noticeably less scatter in double reciprocal plots of the results. Kinetic parameters of the isozyme minifilament preparations are shown in Table 2. Once again, we found no differences due to distinct alkali light chains. In fact,

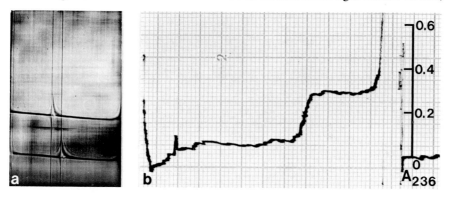

FIG. 3. Sedimentation velocity patterns obtained with chicken skeletal myosin minifilaments prepared according to the procedure of Reisler et al. (15). The solvent was 10 mM Citrate-35 mM Tris, pH 8.0. Rotor speed, 30,000 rpm; rotor temperature, 20°C. **a:** Schlieren patterns of minifilaments formed from myosin at an initial concentration of approximately 6 mg/ml. *Upper pattern*, 5.0 mg/ml, $s_{20,w}$ = 12.9S; *lower pattern*, 3.4 mg/ml, $s_{20,w}$ = 14.7S. Photograph taken 120 min after reaching speed; bar angle, 60°. **b:** Photoelectric scanner trace of the sedimentation profile of minifilaments formed from myosin at an initial concentration of 0.25 mg/ml. Scan initiated 68 min after reaching speed. Protein concentration in cell, 0.22 mg/ml; $s_{20,w}$ = 28.6S.

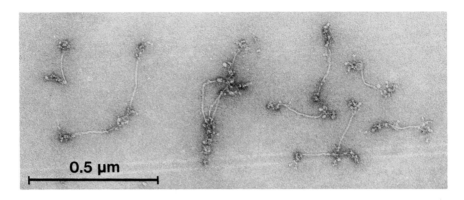

FIG. 4. Electron micrograph of negatively stained chicken myosin minifilaments. Minifilaments were fixed with 2.5% gluteraldehyde in 10 mM Citrate-35 mM Tris, pH 8.0, for 5 min at a protein concentration of 0.3 mg/ml, then diluted to 0.03 mg/ml and immediately applied to grids. The grids were washed with several drops of distilled H_2O and stained with 1% aqueous uranyl acetate. Fixation was necessary to prevent the disassembly of the minifilaments at the low concentrations used for staining. The minifilaments averaged 316 ± 33 nm in length and 7.4 ± 0.7 nm in diameter (n = 182). ×71,000.

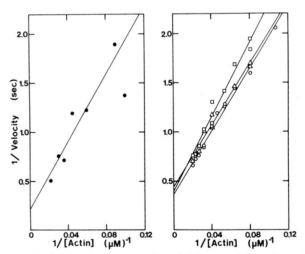

FIG. 5. Lineweaver-Burke plot of the actin-activated ATPase activities of myosin and myosin minifilaments. Myosin filaments *(closed circles)* were assayed at a myosin concentration of 92 nM in an assay medium consisting of 85 mM KCl, 2.5 mM ATP, and 3.75 mM $MgCl_2$. Minifilaments formed from unfractionated *(open circles)*, A1 *(squares)*, and A2 *(triangles)* myosin were assayed as described under Table 2.

the kinetic parameters of homogeneous minifilaments resembled those of the more heterogeneous myosin filaments formed at low ionic strength, suggesting that in both cases the steric hindrance due to assembly had approximately the same influence on the kinetic properties. This latter finding is in disagreement with the results of Reisler (14), who reported that the kinetic parameters of rabbit myosin minifilaments were the same as those of heavy meromyosin, namely higher than those of

TABLE 2. *Kinetic constants of chicken myosin minifilaments[a]*

Experiment	Stock myosin concentration (mg/ml)	V_{max} (sec^{-1})	K_m (μM)
Unfractionated	0.22	2.7	44
A1 isozyme	0.15	2.7	51
A2 isozyme	0.17	2.3	37
Unfractionated	0.22	1.8	29
A1 isozyme	0.12	2.5	33
A2 isozyme	0.18	1.6	32
Unfractionated	0.17	3.0	32
Unfractionated	2.7	3.1	45
Unfractionated	4.5	2.4	28

[a]Kinetic constants were obtained by use of double reciprocal plots (see Fig. 5). All assays were performed at 25°C in 10 mM KCl, 10 mM citrate, 35 mM Tris, pH 8.0, and 2.5 mM ATP, 3.75 mM MgCl$_2$. Actin concentrations in the range 8 to 62 μM were used and in the assay the myosin was 40 to 70 nM. Activity was determined from the levels of inorganic phosphate released, as measured by the method of Fiske and Subbarow (5).

myosin filaments. It is not yet certain whether the discrepancy between the two laboratories is a question of species, or whether there are some experimental differences, as yet not identified. The important aspect for the light chain question is that the isolated chicken myosin isozymes displayed the same kinetic behavior.

DISCUSSION AND CONCLUSIONS

At the level investigated, the alkali light chains in whole myosin do not regulate the ATPase activity or modulate filament formation and architecture. Indeed, a variety of results consistent with this conclusion has accumulated recently from a number of laboratories.

Pope et al. (13) have reported that myosin preparations enriched in either A1 or A2 light chains, prepared by light chain exchange in 4.7 M NH$_4$Cl, behaved in a similar manner. At a variety of ionic strengths, the maximum rates of actin-activated ATP hydrolysis by either the A1-enriched or A2-enriched myosins were nearly identical. Also by subunit exchange in ammonium chloride, Wagner (24) prepared myosin hybrids containing either heavy chains from ox ventricular myosin (a slow type) and light chains from rabbit fast muscle myosin, or, conversely, light chains from cardiac myosin and fast myosin heavy chains. The actin-activated ATPase activities of the hybrids were determined primarily by the heavy chain present, differing at most by one-third from the controls. Because the light chain was derived from a completely different muscle and species, these experiments again demonstrated that the particular light chain present had little influence on the kinetic properties of the myosin. In light of the sequence of the chicken myosin light chains

(11), this result may seem surprising, since the cardiac LC1 is only 30% homologous to the fast muscle LC1. On the other hand, the sequence indicates only conservative substitutions and rearranged sections, so that the subunits from cardiac and fast skeletal myosins have a very similar chemical nature.

The work of Sivaramakrishnan and Burke (19) on myosin subfragment-1 adds an interesting new aspect to the light chain issue. At elevated temperature, 37°C, and in 10 mM MgATP, the alkali light chains of S-1 can exchange freely, so that at physiological conditions of ionic strength and temperature, the interaction between subunits is extremely labile. However, it is not yet certain whether the same will hold true to the same extent for whole myosin, and, in particular, myosin arranged into thick filaments in myofibrils. A recent paper by Wagner and Giniger (25) is consistent with all the above indications that the light chain function does not follow traditional speculations. They isolated subfragment-1 heavy chains free from light subunits by a combination of dissociating conditions at high-salt and antibody-affinity chromatography, a procedure that removes the light chains. The subfragment-1 heavy chains, partly depleted of light subunits, bound F-actin reversibly and hydrolyzed ATP, sometimes with as much as 80% of native activity. On the other hand, heavy chains fully stripped of light subunits, retained only 30% of the ATPase activity. One can conclude that the active site and the actin-binding sites are primarily on the heavy chain. However, the light chain role is not altogether negligible: the denuded heavy chains have low activity and, at the very least, light chains are required for the stability of subfragment-1.

Several experiments argue strongly against simple dismissal of the light chains as "evolutionary vestiges" of earlier regulatory mechanisms as suggested by Chantler (2). From the earliest work it is known that different muscle fiber types carry myosins with different light chain patterns (10,16). Cross-reinnervation of mammalian fast-twitch and slow-twitch muscles has been shown to alter the light chain pattern along with hydrolytic activity of myosin and physiological response of the muscle (31). Therefore, light chain expression is regulated by the central nervous system. Developmental studies, both in the chicken and rat systems, have shown that the light chains appear in a particular temporal sequence (3,20). Recent work on chicken embryos in our laboratory (P. A. Benfield, D. D. LeBlanc, G. Waller, and S. Lowey, *manuscript in preparation*), as well as on the rat system (32), demonstrated that the appearance of a unique neonatal or early posthatch heavy chain is approximately synchronous and associated with the appearance of the A2 (or LC3) light chain. With this result even the fundamental question of whether there is specific association of light and heavy chain types *in vivo*, rather than the random assembly that is evidently possible *in vitro*, cannot yet be unambiguously answered in the negative.

What conclusions can be drawn? It has taken 10 years to disprove the earliest hypotheses—that the alkali subunits provide for differences in activity or filament architecture. The inadequacy of these simple theories should not, however, lead one to dismiss the presence of the light chains as adventitious. It has been shown that the LC2 light chains may be involved in modulating the length of synthetic skeletal muscle myosin filaments (12), and that phosphorylation of the analogous light chain of smooth muscle and nonmuscle myosins also regulates filament assembly (17,21). Although our results do not show similar large effects of the different alkali light chains on assembly, the answer may be more subtle. We have certainly not exhausted the search for differences that may be characterized as "fine tuning." Filament formation has only been tested under some buffer conditions; yet the recent work of Whalen et al. (32) on embryonic myosin has shown that the precise Mg^{2+} and ATP concentrations are important in determining small differences in assembly. Furthermore, only steady state kinetics have been examined with isolated A1 and A2 myosins. Although our results represent a necessary first step, the challenge of resolving the light chain question remains. To some extent, we must anticipate answers to their function before we can rationally design new experiments.

ACKNOWLEDGMENTS

This work was supported by grants from the National Institutes of Health (5 RO1 AM17350), the National Science Foundation (PCM 782 2710), and the Muscular Dystrophy Association to S.L. T.W.H. was the recipient of a National Research Service Award (5 F32 AMO6176–03) from the National Institutes of Health.

REFERENCES

1. Bárány, M. (1967): ATPase activity of myosin correlated with speed of muscle shortening. *J. Gen. Physiol.*, 50:197–218.
2. Chantler, P. D. (1981): Stripped for action. *Nature*, 292:581–582.
3. Chi, J. C. H., Rubenstein, N., Strahs, K., and Holtzer, H. (1975): Synthesis of myosin heavy and light chains in muscle cultures. *J. Cell Biol.*, 67:523–537.
4. d'Albis, A., Pantaloni, C., and Bechet, J. (1979): An electrophoretic study of native myosin isozymes and of their subunit content. *Eur. J. Biochem.*, 99:261–272.
5. Fiske, C. H., and Subbarow, Y. (1925): The colorimetric determination of phosphorus. *J. Biol. Chem.*, 66:375–400.
6. Frank, G., and Weeds, A. G. (1974): The amino-acid sequence of the alkali light chains of rabbit skeletal-muscle myosin. *Eur. J. Biochem.*, 44:317–334.
7. Hoh, J. F. Y. (1978): Light chain distribution of chicken skeletal muscle myosin isoenzymes. *FEBS Lett.*, 90:297–300.
8. Holt, J. C., and Lowey, S. (1977): Distribution of alkali light chains in myosin: Isolation of isoenzymes. *Biochemistry*, 16:4398–4402.
9. Lowey, S., Benfield, P. A., Silberstein, L., and Lang, L. M. (1979): Distribution of light chains in fast skeletal myosin. *Nature*, 282:522–524.
10. Lowey, S., and Risby, D. (1971): Light chains from fast and slow muscle myosins. *Nature*, 234:81–85.
11. Maita, T., Umegane, T., Kato, Y., and Matsuda, G. (1980): Amino-acid sequence of the L-1 light chain of chicken cardiac-muscle myosin. *Eur. J. Biochem.*, 107:565–575.
12. Pinset-Härström, I., and Whalen, R. G. (1979): Effect of aging of myosin on its ability to form synthetic filaments and on proteolysis of the LC2 light chain. *J. Mol. Biol.*, 134:189–197.
13. Pope, B., Wagner, P. D., and Weeds, A. G. (1981): Studies on the actomyosin ATPase and the role of the alkali light chains. *Eur. J. Biochem.*, 117:201–206.
14. Reisler, E. (1980): Kinetic studies with synthetic myosin minifilaments show the equivalence of actomyosin and acto-HMM ATPases. *J. Biol. Chem.*, 255:9541–9544.
15. Reisler, E., Smith, C., and Seegan, G. (1980): Myosin minifilaments. *J. Mol. Biol.*, 143:129–145.

16. Sarkar, S., Sréter, F. A., and Gergely, J. (1971): Light chains of myosins from white, red, and cardiac muscles. *Proc. Natl. Acad. Sci. USA*, 68:946–950.
17. Scholey, J. M., Taylor, K. A., and Kendrick-Jones, J. (1980): Regulation of non-muscle myosin assembly by calmodulin-dependent light chain kinase. *Nature*, 287:233–235.
18. Silberstein, L., and Lowey, S. (1981): Isolation and distribution of myosin isoenzymes in chicken pectoralis muscle. *J. Mol. Biol.*, 148:153–189.
19. Sivaramakrishnan, M., and Burke, M. (1981): Studies on the subunit interactions of skeletal muscle myosin subfragment 1. Evidence for subunit exchange between isozymes under physiological ionic strength and temperature. *J. Biol. Chem.*, 256:2607–2610.
20. Stockdale, F. E., Raman, N., and Baden, H. (1981): Myosin light chains and the developmental origin of fast muscle. *Proc. Natl. Acad. Sci. USA*, 78:931–935.
21. Suzuki, H., Onishi, H., Takahashi, K., and Watanabe, S. (1978): Structure and function of chicken gizzard myosin. *J. Biochem. (Tokyo)*, 84:1529–1542.
22. Wachsberger, P. R., and Pepe, F. A. (1980): Interaction between vertebrate skeletal and uterine muscle myosins and light meromyosins. *J. Cell Biol.*, 85:33–41.
23. Wagner, P. D. (1977): Fractionation of heavy meromyosin by affinity chromatography. *FEBS Lett.*, 81:81–85.
24. Wagner, P. D. (1981): Formation and characterization of myosin hybrids containing essential light chains and heavy chains from different muscle myosins. *J. Biol. Chem.*, 256:2493–2498.
25. Wagner, P. D., and Giniger, E. (1981): Hydrolysis of ATP and reversible binding to F-actin by myosin heavy chains free of all light chains. *Nature*, 292:500–562.
26. Wagner, P. D., Slater, C. S., Pope, B., and Weeds, A. G. (1979): Studies on the actin activation of myosin subfragment-1 isoenzymes and the role of myosin light chains. *Eur. J. Biochem.*, 99:385–394.
27. Wagner, P. D., and Weeds, A. G. (1977): Studies on the role of myosin alkali light chains. Recombination and hybridization of light chains and heavy chains in subfragment-1 preparations. *J. Mol. Biol.*, 109:455–473.
28. Weeds, A. G., Hall, R., and Spurway, N. C. S. (1975): Characterization of myosin light chains from histochemically identified fibres of rabbit psoas muscle. *FEBS Lett.*, 49:320–324.
29. Weeds, A. G., and Lowey, S. (1971): Substructure of the myosin molecule. II. The light chains of myosin. *J. Mol. Biol.*, 61:701–725.
30. Weeds, A. G., and Taylor, R. S. (1975): Separation of subfragment-1 isoenzymes from rabbit skeletal muscle myosin. *Nature*, 257:54–56.
31. Weeds, A. G., Trentham, D. R., Kean, C. J. C., and Buller, A. J. (1974): Myosin from cross-reinnervated cat muscles. *Nature*, 247:135–139.
32. Whalen, R. G., Sell, S. M., Butler-Browne, G. S., Schwartz, K., Bouveret, P., and Pinset-Härström, I. (1981): Three myosin heavy chain isozymes appear sequentially in rat muscle development. *Nature*, 292:805–809.

Basic Biology of Muscles: A Comparative
Approach, edited by B. M. Twarog,
R. J. C. Levine, and M. M. Dewey.
Raven Press, New York © 1982.

Structure of the Thick Filaments in Molluscan Adductor Muscle

Arthur Elliott and Pauline M. Bennett

*King's College Biophysics Department and MRC Cell Biophysics Unit,
London, WC2B 5RL, United Kingdom*

The opening and closing of the valves of molluscs is regulated by the adductor muscles, which usually consist of two readily distinguished parts, serving different functions. One part is frequently coloured and is relatively quick-acting, and may show striations when viewed in the polarising light microscope. In scallops, this muscle is cross-striated, evidence of regularity in its structure. In oysters and clams, the muscle appears obliquely striated and seems to be much less regular than in the scallops. The other muscle is usually white, opaque, and is a smooth muscle. It is slow acting but capable of developing much higher tension per unit cross section of muscle than most other muscles (26,28,29); in addition, it has the ability to maintain tension with a very small turnover of ATP and so can hold the shell closed for hours or even days. This, the "catch" property, is uncommon and its mechanism is not understood. In at least one species of mollusc, *Mytilus edulis*, the adductor appears to consist of a single muscle; it is not certain whether it contains two separate systems or whether the one muscle serves the two functions of closing the valves quickly and of holding them closed. The fast muscle in the oyster shares to some extent the catch property of the slow muscle (17).

Like vertebrate skeletal muscle, molluscan adductor muscles contain thick and thin filaments and are believed to contract by the sliding of one set of filaments over the other, through the operation of cross-bridges (17). The thick filaments contain, in relative amounts that vary widely from one muscle to another, both myosin and another protein paramyosin (21). It was suggested by Hanson and Lowy (17), and confirmed by Szent-Györgyi et al (35), that the myosin is on the surface and that the paramyosin forms a core.

For each type of muscle, the dimensions of the thick filaments vary with species, but greater differences are found in several species when filaments in the fast and the slow muscles are compared. Generally, the range as well as the magnitude of both length and maximum diameter in thick filaments in the fast muscle is smaller (sometimes much smaller) than those in the corresponding slow muscle. The scallop *Pecten maximus* furnishes an extreme example: it is uncertain whether thick filaments in the fast adductor vary at all (3,30). This muscle has much of the regularity of vertebrate striated muscles, with a thick filament length of about 1.76 μm and maximum diameter of 21 nm. In the slow muscle of *Pecten* we have observed diameters up to about 100 nm in the thick filaments, which are many microns long.

The difference between filament dimensions in the two parts of the oyster adductor muscle is shown in Fig. 1a and b. Thick filament length in the fast (obliquely, probably helically striated) muscle is not constant but the range is small (Fig. 1b), a length of about 5 μm being the most common (17). The maximum diameter is 60 nm. In striking contrast, the thick filaments of the oyster slow adductor are so long that it is difficult to separate them without breaks. In Fig. 1a, the longest filament (whose ends lie outside the limits of the electron micrograph) is more than 70 μm in length. Many much smaller thick filaments are found that are complete, as shown by the tapered ends. We have found filament diameters up to about 190

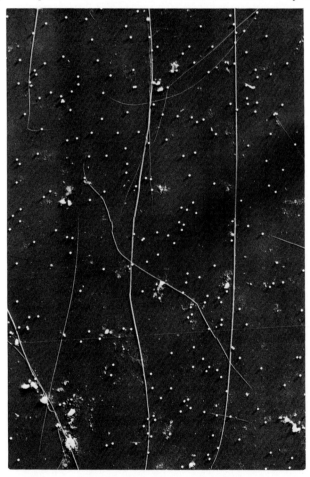

FIG. 1 a. Whole thick filaments from the white (slow) adductor muscle of the oyster *O. edulis*, dispersed without blending. Shadowed with tungsten, shadow angle 1 in 5. × 2,800 Polystyrene latex particles (for focusing).

nm in cross sections of the embedded slow adductor from *Ostrea edulis*. The most frequently occurring diameter is much less (60 nm in *Crassostrea angulata*) (15).

The difference in filament dimensions in the two components of the adductor in the clam *Mercenaria mercenaria* is less marked than in the oysters. The red adductor is obliquely striated and is presumably the fast muscle; the white adductor is smooth. The thick filaments of the red adductor vary in diameter from 20 to 25 nm, and those of the white from 50 to 150 nm (32). There is not much information concerning thick filament length. We have found filaments up to at least 15 μm length in the red adductor and it may be difficult to distinguish, by filament length, the red and the white parts since these long filaments are so easily broken in the preparation.

The thick filaments of both kinds of muscle in molluscs are thickest in the middle and taper to a very small diameter at each end. It has been shown by Szent-Györgyi et al. (35) that the thick filaments of the white adductor of *M. mercenaria* are bi-

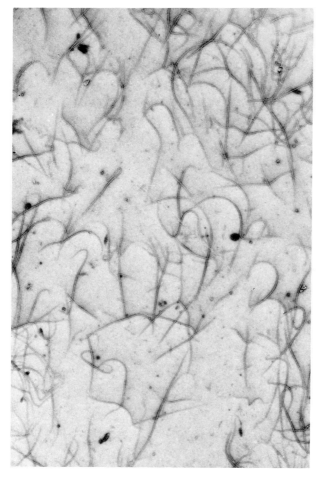

FIG. 1 b. Whole thick filaments from the yellow (fast) adductor muscle of the oyster *C. angulata*. Negatively stained with uranyl acetate (note that only the light "core" is the filament). ×2,800.

polar and we show and discuss later the region of polarity change in *O. edulis* and *C. angulata* thick filaments from the slow adductor (Fig. 3c and d). It is to be expected that this is a feature of the thick filaments of other adductor muscles but it is difficult to observe this when the diameters are small.

The thin filaments of the adductor muscles resemble those of vertebrate skeletal muscle, being composed of actin and tropomyosin, but they appear to lack troponin. The switching on and off of the cross-bridge mechanism is controlled by the action of Ca^{2+} on the light chains of the myosin on the thick filaments (22,36). The only molluscan adductor muscle known to have thin filaments that protrude from a well-developed Z-line is the fast muscle in *Pecten*. In other muscles examined, the thin filaments appear to come from "dense bodies" (17). In the obliquely (probably helically) striated fast adductor muscles, these may be Z-discs of restricted size, and there are recognisable sarcomeres. In the molluscan smooth adductor (slow) muscles, there are dense bodies, but no sarcomere can be recognized. The anterior byssus retractor muscle (ABRM) of *M. edulis*, though not an adductor, has many of the properties of slow adductor muscles (including the ability to go into "catch") (25), and Sobieszek (33) has suggested that sarcomeres can be recognized in that muscle. Since there are only three to four thick filaments per sarcomere, the term is being extended beyond its usual meaning.

The thin filaments in cross sections of embedded adductor muscles fixed in rigor may be seen to surround the thick filaments. In the most regular structure, the fast muscle of *Pecten*, there is a regular hexagonal lattice containing thick and thin filaments (3,30). In the oyster fast muscle, the thick filaments are arranged approximately hexagonally, and a number of thin filaments (probably 12) form a "rosette" round each thick filament in the region where the two types of filament overlap (17). In the obliquely striated red muscle of *M. mercenaria*, we have observed little sign of a regular arrangement. The slow adductors of *Pecten*, *Ostrea* (and *Crassostrea*) as well as of *Mercenaria* show no regularity in the thick filament arrangement. Rosettes of varying numbers of thin filaments are to be seen surrounding the thick filaments of these muscles, however, with dense-staining material between the thick filament and the rosette (see Fig. 5f).

STRUCTURE OF THE THICK FILAMENTS—HISTORICAL

The structure of the thick filaments has been a subject for speculation since Hall et al. (16) observed a regular two-dimensional pattern of spots on stained thick filaments from the adductor muscle of the clam *M. mercenaria*. It is now known that this pattern comes from the paramyosin core of the thick filaments, and it is the structure of this core that is our chief concern here, particularly in the smooth, slow adductor muscle. The pattern, like the X-ray diffraction pattern from the same muscle observed by Bear and Selby (2) could be equally well ascribed to a multi-stranded helical arrangement of molecules or to an essentially crystalline arrangement, each thick filament being a crystallite. The pattern ("Bear-Selby net") is shown in Fig. 2 and has been found in the smooth adductor muscles of oysters,

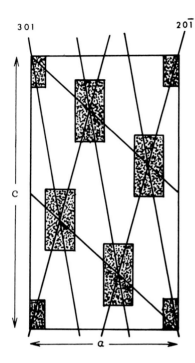

301   20$\bar{1}$

FIG. 2. Geometrical stain pattern (Bear-Selby net) on a paramyosin filament. The filament axis is along *c*. The 201 and 301 planes (normal to the plane of the diagram) are shown. They make small angles on either side of the filament axis.

scallops, and clams, as well as in the posterior adductor of the mussel *M. edulis* (above) and in another catch muscle, the ABRM in the same animal. The net is rectangular, having a repeat distance indicated by *c* along the fiber axis of 72 nm, which varies very little with species or with water content, and a transverse repeat indicated by *a*, which varies considerably with water content.

Elliott (14,15) carried out a series of observations on the slow (smooth) muscle of the oyster *C. angulata*, using X-ray diffraction and electron microscopy. He observed parallel striations in cross sections of some stained thick filaments, approximately 12 nm apart. These were most distinctly seen in sections from muscles that had been allowed to shorten to less than the normal minimum body length. He ascribed them to a layer structure in the filaments, and sought to correlate the spacing with that of an equatorial reflection in the X-ray diffraction pattern from pieces of muscle. This reflection was subsequently shown to come from the array of thin filaments (27) and the particular layer structure suggested by Elliott could not be accepted. As will be seen in the next section, however, these striations are indeed evidence of a layer structure.

Elliott and Lowy (12) attempted to use the concept of a helical as well as a layer structure by suggesting a model in which a surface bearing a two-dimensional Bear-Selby net was rolled up to form a helicoid, rather resembling rolled-up wire netting or mesh. The helical nature seemed to be confirmed by an electron micrograph of a thick filament from the smooth adductor of *P. maximus*, which had been shadowed with metal on both sides, and which showed a helix-like arrangement of strands on the surface (9). However, only one good example was found, and, in any case, the observation relates to the surface (presumably covered with myosin) and not to the core.

Striations similar to those observed by G. F. Elliott were recorded by Heumann (19) in cross sections of ABRM. He suggested that they might be planes of the Bear-Selby net in a layer structure, seen by chance tilting of the filaments from the orientation exactly transverse to the section cut, but did not suggest that they were evidence of a layer structure.

RECENT OBSERVATIONS ON THE STRUCTURE OF THE PARAMYOSIN CORE IN THICK FILAMENTS

One feature of the appearance of negatively stained thick filaments separated from smooth adductor muscles is that the Bear-Selby net is not seen on all filaments. Where it is absent, transverse lines 14.4 nm apart are seen. It has been suggested (35) that only those filaments from which the surface coating of myosin has been removed show the net, although Elliott (10) could not confirm this. Nonomura (31) also showed that some thick filaments from which the surface coating of myosin had been removed showed the net pattern whereas others showed only a transverse pattern of lines at 14.4 nm intervals. The explanation of the variable appearance has been revealed by tilting the filament-bearing grid on a goniometer stage so that the filament to be examined is rotated round its long axis (11). When this is done, the net disappears and is replaced by the pattern of transverse lines (Fig. 3a and b). This points clearly to a structure that is not helical, but in which the dark bars of negative stain that form the net run through the filament as shown in Fig. 4, or at least through a substantial part of it.

The axial lengths of the negatively stained regions that form the elements of the net are in all known cases somewhat greater than 14.4 nm (see below) and there are five such regions in one repeat length of 72 nm. Hence, when the filament is rotated so that the regions are viewed along the *a* direction in Fig. 4, dark lines separated by 14.4 nm are seen transverse to the fiber axis. If positive staining is also present, a light line is often seen adjacent to the dark one.

The above conclusions have been confirmed by a three-dimensional synthesis computed from a series of electron micrographs taken with a range of tilts (7,8). This shows the bars of stain running right through the filament.

Although the foregoing shows conclusively that filaments with a structure that is essentially crystalline are to be found in thick filaments from molluscan smooth muscles, it does not establish this as the common structure in such filaments in one muscle. To do this we have embedded and transversely sectioned muscles from several species (4). When examined on a goniometer stage, each filament, when suitably tilted, shows striations of spacing about 14.0 nm. These correspond to the 20$\bar{1}$ planes of the arrangement shown in Fig. 4. Figure 5d shows a cross section of the smooth adductor of the oyster *O. edulis* in which a number of these striations may be seen. In this, the tilting of the filament axis is accidental, a consequence

FIG. 3. Thick filaments from oyster slow adductor muscle, × 199,800. **a:** Filament rotated 5° round long axis (ammonium molybdate stain). **b:** The same filament rotated −40°. The net has almost disappeared, leaving a transverse series of lines at 14.4 nm intervals. **c:** Region showing the change in polarity (uranyl acetate stain). **d:** The structure of the region where the polarity changes is more clearly revealed by ammonium molybdate stain. (**a** and **b** are reproduced from Elliott, ref. 11, by courtesy of Academic Press.)

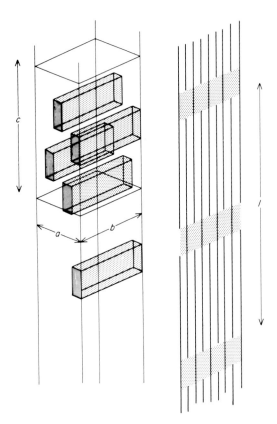

FIG. 4. Left: Three-dimensional perspective drawing showing regions of stain extending through a filament. The *b* direction is normal to the *a–c* plane. **Right:** Diagrammatic perspective representation showing a single layer of paramyosin molecules each 129.1 nm long with a displacement of *c* = 72 nm between neighbors (Cohen et al., ref. 6). Stain can lodge in bands 72 nm apart where there are gaps between the ends of the molecules. *l*, Molecular length. Neither the number of such layers nor their relation to neighboring layers within a region is known. (Reproduced from Elliott, ref. 11, courtesy of Academic Press.)

of imperfect orientation of the filaments. These are the striations observed by Elliott and by Heumann, who were able to observe them for the same reason. Figure 5 a–c also shows the same filament viewed with different angular tilts, the specimen having been rotated round an axis in the plane of the section, approximately parallel to the striations. The central tilt corresponds to accurate alignment of the central filament with the electron beam; no striations are to be seen, for the projection of the stain pattern of the Bear-Selby net is uniform in this view. The orientation is confirmed by the visibility of the thin filaments. On tilting 16° away from this position, fine striations corresponding to the 301 planes are visible. A larger angle (24°) in the opposite direction reveals clearly the $20\bar{1}$ planes. These planes are more easily seen than the 301 planes, not only because they are more widely spaced but because their contrast is greater, since they have more stained regions in a given

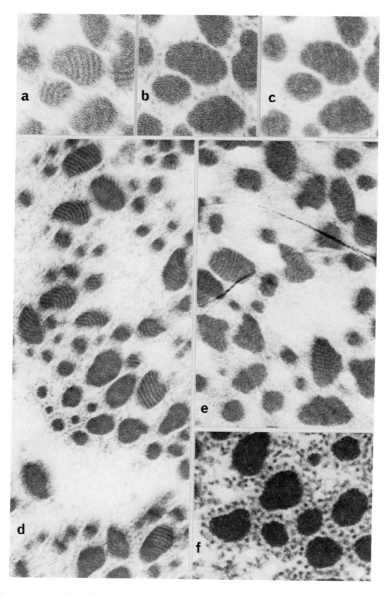

FIG. 5. Cross sections of embedded thick filaments from slow muscles. *a–e*, 4% uranyl acetate stain. **a:** From *O. edulis*, rotated 34° in grid plane. **b:** The same area, rotated +10°. **c:** The same, rotated −6°. All ×117,000. **d:** From *O. edulis*. Striations of varying degrees of regularity on different filaments, ×81,000. **e:** From *M. mercenaria*. As **d**. **f:** From *O. edulis*. 5% Phosphotungstic acid in ethylene glycol. ×126,000. (**a, b, c** and **e** from Bennett and Elliott, ref. 4, courtesy of Chapman and Hall Ltd.)

length in the line of sight. The measured ratio of spacings in the two sets is close to two-thirds, as expected for the planes in question, and both sets run in the same direction.

The *a* spacing of the net may be calculated from the striation spacing. This is around 27 nm for several smooth adductors of different species and is somewhat less than the value for *a* determined by X-ray diffraction from a piece of the embedded muscle (around 32 nm). The difference is only partly accounted for by shrinkage of the resin in the electron beam, and may indicate that the best-ordered filaments selected in the electron microscope have a somewhat smaller spacing than the average.

The results, which are reproduced in Table 1, show only a small variation between the smooth adductors of several species, and include one result for the *M. mercenaria* red adductor, which is obliquely striated. This contradicts an earlier reported variation in *a* with species (12). We are satisfied that the reported variation was caused by differences in the drying procedures used for different species. In all these muscles, the striations observed are usually somewhat curved, though straight forms can be found. Sometimes the striations may be seen to change direction by as much as 90° in one filament, and it is clear that the structure is only occasionally regular; it may be described as paracrystalline (4). Because of the generally smaller diameter of the filaments in the striated adductors, it is more difficult to observe striations in cross sections of these muscles. There are indications of this feature in the oyster obliquely striated adductor, which is to be expected since its X-ray diffraction pattern shows the Bear-Selby reflections (J. Lowy and B. M. Millman, *private communication*).

TABLE 1. *Striation spacing and Bear-Selby* a *spacing in molluscan smooth muscles*[a]

Muscle	Filaments measured	Mean 201 spacing (nm)	Mean *a* (nm)	Range (nm)	a From X-ray diffraction (embedded) (nm)
O. edulis (white)					
Standard embedding	11	13.4	27.3	25.8–29.8	32.0
20% acetone in media	25	13.8	28.0	26.1–30.2	32.4
C. gigas (white)	9	16.8	34.4	32.3–36.2	Not measured
M. mercenaria					
white	5	13.1	26.6	25.2–29.4	31.1
red	3	12.3	24.9	23.1–26.6	Not measured
M. edulis (ABRM)	12[b]	15.3	31.3	24.7–36.4	Not measured
P. maximus (white)	8	14.1	28.7	21.1–35.6	Not measured

From Bennett (3), courtesy of Chapman and Hall Ltd.

[a]The filaments of *O. edulis* and *M. mercenaria* have sufficient striations for the measurement error to be small compared with the real variations from one filament to another. With the other muscles this is not so and the range of spacings is significantly increased by the errors of measurement.

[b]Most of the filaments had only two measurable striations.

Heumann (20), on the basis of observations on tilted sections of pieces of embedded ABRM from *M. edulis*, has proposed a structure with concentric layers of cylindrical Bear-Selby nets. We are unable to see how this can be reconciled with both Heumann's and our observations of transverse sections.

STRUCTURE OF THE CORE OF MOLLUSCAN THICK FILAMENTS AT THE MOLECULAR LEVEL

The electron micrographs of sections of embedded thick filaments from slow molluscan adductor muscles show that the filaments resemble a crystal in their three-dimensional arrangement, but with a regularity considerably less than that in single crystals of many substances. This makes it more difficult to determine the "structure," by which we mean the ideally regular arrangement of molecules to which the filament cores approximate.

Paramyosin molecules have been shown by hydrodynamic and light-scattering data to be about 135 nm long with a molecular weight of 220,000 daltons (24) and constitute probably the whole of the thick filament core (35). Their rod-like nature is shown in Fig. 6 and they are believed to form coiled-coils (5). These are composed of two α-helical chains.

STUDY OF PARACRYSTALS

The examination of paracrystals formed from solutions of paramyosin separated from adductor muscles has contributed greatly to our understanding of the structure of the paramyosin core. Hodge (21) and Hanson et al. (18) examined such paracrystals and found an axial period 14.4 nm. A great advance was made when Cohen et al. (6) reported the use of divalent cations in the solutions from which the paracrystals were precipitated. This resulted in a number of new forms, most of which were elongated with an apparent axial period 72.0 nm, clearly related to the native paramyosin structure. The basic form, isolated alone as well as one component of more complicated structures, is the PI form. When negatively stained, this appears as an array of dark bands about 14.0 to 20.0 nm wide (depending on species) transverse to the paracrystal long axis, with the above axial period. When positively stained, the contrast is reversed and Cohen et al. (6) proposed the structure shown in perspective at the right-hand side of Fig. 4. Adjacent molecules are displaced by 72 nm, and, because paramyosin molecules are somewhat shorter than 144 nm, there are gaps between the ends of molecules that are mutually aligned. These gaps fill with stain (in negative staining) and give an apparent period of 72 nm, although the true molecular period is 144 nm. The period is the same in all species examined, and the variation in width of the dark band is taken as evidence of a small variation in molecular length in different species.

Two PI structures are frequently observed with a relative displacement of 2/5 × 72 nm. Cohen et al. (6) suggest that the Bear-Selby net is formed by a PI sheet, which develops without restriction in the axial direction, but whose width corre-

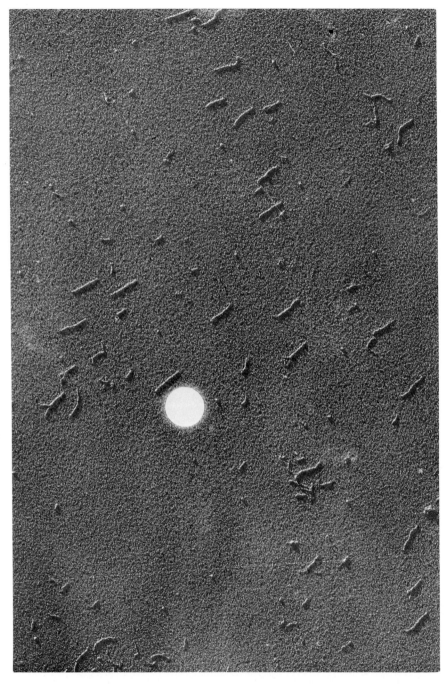

FIG. 6. Paramyosin molecules from the slow adductor of *C. angulata*, shadowed with platinum, shadow angle 1 in 10. ×70,000.

sponds to only a few intermolecular spacings, after which a $2/5 \times 72$ nm shear occurs. If this process continues, a Bear-Selby net is formed.

X-RAY DIFFRACTION

X-ray diffraction from native molluscan muscles yields little information about the molecular arrangement of the paramyosin core of the thick filaments, other than the Bear-Selby net and the fact that there are α-helices in the coiled-coil form, with their axes parallel to the filament axis (5). However, it has been known for some time that bathing the muscle in a variety of water-soluble organic liquids (e.g., acetone, dimethylformamide, dioxan) leads to an X-ray diffraction pattern that shows a small number of equatorial and near-equatorial reflections not otherwise seen. Their spacings (1.2 to 2.8 nm) suggest that they must be those of an intermolecular arrangement, which is better developed in the presence of these liquids than in aqueous solutions of salts (12,13). For a particular concentration of the organic liquid, the diffraction pattern is that corresponding to a body-centred tetragonal arrangement of scattering objects. It has been surmised that the paramyosin coiled-coil molecules would tend to pack in this way, which would give close-packing.

It is to be expected that the tetragonal unit would be part of a larger unit cell that would be closely related to the periodicities of the Bear-Selby net. The number of X-ray reflections is too small to allow a solution using X-ray data alone, so information from electron microscopy has to be used also. One of the difficulties of combining the data is that X-ray diffraction comes from a piece of muscle containing the whole range of filaments, of varying sizes and degrees of order. Information on molecular arrangement can be obtained from electron microscopy only by selecting the best-preserved and most-ordered filaments. Although we have made some progress in combining information from the two techniques, we have not yet a satisfactory solution.

BIPOLAR NATURE OF FILAMENTS

Molluscan thick filament cores can be seen to be bi-polar, as shown by Szent-Györgyi et al. (35), who recognized a reversal in the staining pattern of the Bear-Selby net. The stained areas that form this net when uranyl acetate is used as a negative stain are not rectangular but rather triangular and are sometimes called "daggers." The daggers are pointed toward the centre of the filament, and a region is sometimes seen where there is a change in the direction of the daggers on either side of a region where, clearly, the polarity changes. This is shown in Fig. 3c, but the detailed arrangement of daggers is not clear. That this region is not always seen in a filament showing the Bear-Selby net is doubtless a consequence of the fact that the usual method of preparation produces many fragmentary filaments.

Although uranyl acetate staining enables a polarity change to be seen, the accompanying positive stain is confusing, and there is good reason for thinking that the dagger shape is a staining artifact. Ammonium molybdate, especially when used with filaments prepared in the presence of 10 mM EDTA, gives nearly rectangular patches of negative stain, and little positive stain (11). In Fig. 3d, an example is shown where the polarity changes. One end of the rectangular patch of stain is sharp and the other is somewhat diffuse, which enables the polarity to be recognized. At the top of Fig. 3d most of the dark rectangles are sharp at their upper edges, and at the bottom most are sharp at their lower edges. In the middle of the filament the rectangles on the left are sharp at their upper edges, those on the right being sharp at the lower edge. By following the change carefully, it may be seen to occur along a line inclined at about 3.5° to the filament axis, in the clockwise sense. The region where the polarity changes is therefore very long, more than 1 μm in the thicker filaments. This is very different from the situation in thick filaments from vertebrate striated muscle. The paramyosin filaments may be compared to ropes joined by a long splice—even if the intermolecular forces between adjacent molecules of opposite polarity are weak, most of the filament retains its normal tensile strength.

One supposes that a filament grows by the formation first of a bi-polar region, and by the addition of molecules in a parallel (rather than antiparallel) manner for the rest of its length. There appears to be only one region of polarity change in each filament (35). Why the molecules should add so as to make a filament of (mainly) round shape and tapering ends is not clear.

DISCUSSION

The ratio of paramyosin to myosin in adductor muscles is known to vary widely. For example, in an investigation that included nonmolluscan muscles containing paramyosin, Levine et al. (23) report a molecular ratio 0.065 for the striated adductor of *Aequipecten* and 5.50 for *Mercenaria* white adductor. They find a linear relationship between thick filament length and paramyosin:myosin heavy chain molecular ratio. Although it is clear that great filament length and diameter go with a high paramyosin content (17,35), we doubt the significance of the linear relation in muscles where the filament dimensions are variable (see Fig. 1a). Squire (34) has suggested, very plausibly, that the myosin might form a layer of constant thickness (about 3.5 nm) on the paramyosin core. This would account for the high paramyosin:myosin ratio in very thick molluscan filaments. The distribution of filament length and diameter in most adductor muscles has not been determined; this, as well as a knowledge of paramyosin:myosin ratios, is needed before the significance of paramyosin:myosin ratios can be understood.

The arrangement of myosin on the surface of the paramyosin core is not known. It may be that part of the rod portion of myosin (perhaps the light meromyosin part) is added epitaxially, following the molecular arrangement of the core. The striations we observe on suitably tilted cross sections of embedded muscle appear to run right up to the edge of the filaments, but the myosin content of the thickest filaments (which show the striations most clearly) is probably so small that the myosin layer would not be distinguished if it had a different structure from the core.

It has been suggested that myosin might be epitaxially bound at first only on one side of the core that has the Bear-Selby net, and then perhaps continue round part or all of the filament in the same manner. This would give the surface layer something of the properties of a helical arrangement (11). In this way, it might be understood why a thick filament, shadowed on both sides, showed the characteristics of a helix (9). However, a solitary example does not serve to establish the structure, especially of a paracrystalline filament, and, in any case, it is not certain that myosin was present on the shadowed filament, although nothing had been done to remove it.

That there is myosin all round the core seems likely, for the thin filaments in rigor muscle (rosettes) appear roughly equi-spaced round each filament (Fig. 5f), and there is densely stained material between thick and thin filaments (presumably myosin heads).

It has been remarked (35) that apparently intact thick filaments of molluscan muscles appear smooth. Filaments from vertebrate skeletal muscle, on the other hand, appear rough or, if suitable preparative methods are used, show projections some of which resemble myosin heads (37). Nonomura (31) has investigated the relation between mode of preparation and the appearance of negatively stained molluscan thick filaments in the electron microscope. He has concluded that the interaction of myosin with paramyosin is so weak that myosin is easily detached from the paramyosin core, owing to its strong affinity for actin in the thin filaments. He found that homogenization of freshly dissected catch muscle in a relaxing medium containing ATP at a concentration greater than 10 mM produced thick and thin filaments. The thick filaments, when negatively stained, were rough in appearance, with globular parts projecting from the surface. These were presumably myosin heads. The surface projections were not seen if the homogenization of the adductor muscle was carried out with a concentration of added ATP less than 4 mM, and only smooth-surfaced filaments were then seen. Moreover, unless ATP was added continuously to a filament preparation in which the projections could be seen, thin filaments began to attach to the surface of the thick filaments and myosin apparently detached from them to leave smooth-surfaced thick filaments. Finally, all the thick filaments revealed smooth-surfaced structures. These observations are of great interest; we believe that more work is required before the conditions necessary for seeing projections are fully understood.

It is an interesting question whether all paramyosin cores in thick filaments have a single crystal type of structure. This structure has been found in the slow catch muscle in a number of species and it is a not unreasonable extrapolation to suppose that it is common to all such muscles. The fast adductor muscle of the oyster has a much smaller paramyosin content than the slow adductor; we have not been able to find good examples of the striations within the thick filaments in this fast muscle but we cannot exclude the possibility of their presence, which is the best indication of a crystalline arrangement. However, the fast adductor has an X-ray diffraction pattern that resembles that of the Bear-Selby net. Although not conclusive evidence for the same core structure as that in the slow adductor, it is now likely that the thick filaments in this obliquely striated muscle have crystal-like cores. The red muscle of *M. mercenaria* shows similar striations in the light microscope, and certainly has the crystalline structure illustrated in Fig. 4 (4). When the amount of paramyosin is very small, as in the scallop cross-striated muscle, no evidence bearing on the core structure can be cited. The possibility that the core in such muscles is crystal-like cannot be excluded at present.

We do not know whether these matters are relevant to the unsolved problem of the catch mechanism. The nature of catch and its control has recently been reviewed by Twarog (38), who considers that the evidence is strong that in catch the actin-myosin cross-bridges remain attached and cycle very slowly, if at all. This was suggested by Lowy et al. (26). However, the phosphorylation of paramyosin might be a factor affecting the actin-myosin dissociation so the structure of the core may have some relevance to the problem of catch (1).

SUMMARY

Molluscan adductor muscles are in general of two kinds—a slow-acting smooth muscle, which after contraction can go into a catch state, in which tension is maintained for many hours with a small turnover of ATP, and a quicker-acting muscle, usually to some extent striated, which has not the catch property in marked degree, if at all.

The thick filaments of both kinds of adductors contain myosin and paramyosin. Those from catch muscles are noteworthy for their high paramyosin content (up to about 90% by weight), which may be correlated with their great length and diameter. The paramyosin forms a core with a surface layer of myosin.

Recently, we have shown that the paramyosin core of catch muscles resembles a single crystal in its molecular arrangement, although the degree of regularity varies from one filament to another. The evidence is from electron microscopy both of individual filaments and of pieces of whole muscle embedded and sectioned.

Individual negatively stained filaments change greatly in appearance when rotated round their long axes, in a way not compatible with a helical structure. A three-dimensional reconstruction of the filament from micrographs of a rotated filament confirms this.

Transverse sections of smooth adductor muscles show no internal features within the thick filaments when viewed accurately along the filament axis. On tilting striations appear on the filaments that may be correlated with the planes of the Bear-Selby net. They are a general feature of the filaments and have been seen in the smooth adductors of *P. maximus*, *O. edulis*, *Crassostrea gigas*, *M. mercenaria*, of the anterior byssus retractor of *M. edulis* and in the red (obliquely striated) adductor of *M. mercenaria*.

The arrangement of the myosin on the paramyosin core is not known.

ACKNOWLEDGMENTS

We wish to express our thanks to Professor Jack Lowy for allowing us to use the electron micrographs shown in Figs. 1b and 3c. These are unpublished micrographs taken by Professor Lowy and the late Professor Jean Hanson. Figure 6 is from unpublished work by one of the authors (A. E.) and Dr. Gerald Offer, whose part in this we are glad to acknowledge.

REFERENCES

1. Achazi, R. K., (1979): Phosphorylation of Molluscan Paramyosin. *Pfluegers Arch.*, 379:197–201.
2. Bear, R., and Selby, C. C. (1956): The structure of paramyosin fibrils according to X-ray diffraction. *J. Biophys. Biochem. Cytol.*, 2:55–69.
3. Bennett, P. M. (1977): *Structural Studies of Muscle Thick Filaments and Their Constituents.* Ph.D. Thesis, University of London.
4. Bennett, P. M., and Elliott, A. (1981): The structure of the paramyosin core in molluscan thick filaments. *J. Muscle Res. Cell Motil.*, 2:65–81.
5. Cohen, C., and Holmes, K. C., (1963): X-ray diffraction evidence for α-helical coiled-coils in native muscle. *J. Mol. Biol.*, 6:423–432.
6. Cohen, C., Szent-Györgyi, A. G., and Kendrick-Jones, J. (1971): Paramyosin and the filaments of molluscan "catch" muscles. I. Paramyosin: Structure and assembly. *J. Mol. Biol.*, 56:223–237.
7. Dover, S. D., and Elliott, A. (1979): Three-dimensional reconstruction of a paramyosin filament. *J. Mol. Biol.*, 132:340–341.
8. Dover, S. D., Elliott, A., and Kernaghan, A. K. (1981): Three-dimensional reconstruction from images of tilted specimens: The paramyosin filament. *J. Micros.* 122:23–33.
9. Elliott, A. (1971): Direct demonstration of the helical nature of paramyosin filaments. *Philos. Trans. R. Soc. Lond. [Biol.]*, 261:197–199.
10. Elliott, A. (1974): The arrangement of myosin on the surface of paramyosin filaments in the white adductor muscle of *Crassostrea angulata*. *Proc. R. Soc. Lond. [Biol.]*, 186:53–66.
11. Elliott, A. (1979): Structure of molluscan thick filaments: A common origin for diverse appearances. *J. Mol. Biol.*, 132:323–341.
12. Elliott, A., and Lowy, J. (1970): A model for the coarse structure of paramyosin filaments. *J. Mol. Biol.*, 53:181–203.
13. Elliott, A., Lowy, J., Parry, D. A. D., and Vibert, P. J. (1968): Puzzle of the coiled coils in the α-protein paramyosin. *Nature*, 218:656–659.
14. Elliott, G. F. (1960): *Electron Microscope and X-ray Diffraction Studies of Invertebrate Muscle Fibres.* Ph.D. Thesis, University of London.
15. Elliott, G. F. (1964): Electron microscope studies of the structure of the filaments in the opaque adductor muscle of the oyster *Crassostrea angulata*. *J. Mol. Biol.*, 10:89–104.
16. Hall, C. E., Jakus, M. A., and Schmidt, F. O. (1945): The structure of certain muscle fibrils as revealed by the use of electron stains. *J. Appl. Physics*, 16:459–465.
17. Hanson, J., and Lowy, J. (1961): The structure of the muscle fibres in the translucent part of the adductor of the oyster *Crassostrea angulata*. *Proc. R. Soc. Lond. [Biol.]*, 154:173–196.
18. Hanson, J., Lowy, J., Huxley, H. E., Bailey, K., Kay, C. M., and Rüegg, J. C. (1957): Structure of molluscan tropomyosin. *Nature*, 180:1134–1135.
19. Heumann, H. G. (1973): Substructure of paramyosin filaments prepared by freeze-substitution techniques. *Experientia*, 29:469–471.
20. Heumann, H. G. (1980): Paramyosin structures in the thick filaments of the anterior byssus retractor muscle of *Mytilus edulis*. *Eur. J. Cell. Biol.*, 22:780–788.
21. Hodge, A. J. (1952): A new type of periodic structure obtained by reconstruction of paramyosin from acid solution. *Proc. Natl. Acad. Sci. USA*, 38:850–855.
22. Kendrick-Jones, J., Lehman, W., and Szent-Györgyi, A. G. (1970): Regulation in molluscan muscles. *J. Mol. Biol.*, 54:313–326.
23. Levine, R. J. C., Elfvin, M., Dewey, M. M., and Walcott, B. (1976): Paramyosin in invertebrate muscles. II. Content in relation to structure and function. *J. Cell. Biol.*, 71:273–279.
24. Lowey, S., Kucera, J., and Holtzer, A. (1963): On the structure of the paramyosin molecule. *J. Mol. Biol.*, 7:234–244.
25. Lowy, J., and Millman, B. M. (1963): The contractile mechanism of the anterior byssus retractor muscle of *Mytilus edulis*. *Philos. Trans. R. Soc. Lond. [Biol.]*, 246:105–148.
26. Lowy, J., Millman, B. M., and Hanson, J. (1964): Structure and function in smooth tonic muscles of lamellibranch molluscs. *Proc. R. Soc. Lond. [Biol.]*, 160:525–536.

27. Lowy, J., and Vibert, P. J. (1967): Structure and organisation of actin in a molluscan smooth muscle. *Nature*, 215:1254–1255.
28. Millman, B. M. (1963): *The Mechanical Properties of Molluscan Smooth Muscle*. Ph.D. Thesis, University of London.
29. Millman, B. M. (1964): Contraction in the opaque part of the adductor muscle of the oyster *(Crassostrea angulata). J. Physiol.*, 173:238–262.
30. Millman, B. M., and Bennett, P. M. (1976): Structure of the cross-striated adductor muscle of the scallop. *J. Mol. Biol.*, 103:439–467.
31. Nonomura, Y. (1974): Fine structure of the thick filament in molluscan catch muscle. *J. Mol. Biol.*, 88:445–455.
32. Philpott, D. E., Kahlbrock, M., and Szent-Györgyi, A. G. (1960): Filamentous organisation of molluscan muscles. *J. Ultrastruct. Res.*, 3:254–269.
33. Sobieszek, A. (1973): The fine structure of the contractile apparatus of the anterior byssus retractor muscle of *Mytilus edulis. J. Ultrastruct. Res.*, 43:313–343.
34. Squire, J. M. (1971): General model for the structure of all myosin-containing filaments. *Nature*, 233:457–462.
35. Szent-Györgyi, A. G., Cohen, C., and Kendrick-Jones, J. (1971): Paramyosin and the filaments of molluscan "catch" muscles. II Native filaments: Isolation and characterization. *J. Mol. Biol.*, 56:239–258.
36. Szent-Györgyi, A. G., Szentkiralyi, E. M., and Kendrick-Jones, J. (1973): The light chains of scallop myosin as regulatory subunits. *J. Mol. Biol.*, 74:179–203.
37. Trinick, J., and Elliott, A. (1979): Electron microscope studies of thick filaments from vertebrate skeletal muscle. *J. Mol. Biol.*, 131:133–136.
38. Twarog, B. (1979): The nature of catch and its control. Motility in Cell Function. In: *Proceedings of the First John M. Marshall Symposium in Cell Biology*, edited by F. Pepe. pp. 231–241. Academic Press, New York.

Basic Biology of Muscles: A Comparative Approach, edited by B. M. Twarog, R. J. C. Levine, and M. M. Dewey. Raven Press, New York © 1982.

Organization of Myosin in Invertebrate Thick Filaments

J. S. Wray

Max-Planck Institute for Medical Research, Heidelberg, West Germany

The comparative study of muscle leads to important insights into the relation between its structure and function (for review, see 8). I discuss here the diversity of muscle structure specifically at the level of filament architecture, in relation to molecular events in contraction. The specializations of particular muscles appear to be associated with their thick (myosin-containing) filaments, rather than with the actin filaments which are conservative in structure. Therefore, this chapter will be concerned with the structure of myosin filaments alone.

The thick filaments of molluscan smooth muscles, reviewed by A. Elliott *(this volume)*, are a valuable source of information about paramyosin. But they tell us little about myosin, which is located on the surface of the large paramyosin core and forms only a small part of the filament mass. To understand the arrangement of myosin, we turn to striated muscles, where it is the predominant component of the thick filaments, although paramyosin may still be present in smaller amounts in the core. Striated muscles are employed where greater speed of shortening is required at the expense of contractile force. In parallel with the reduction in the amount of paramyosin, the myosin filaments have smaller diameters. Both they and the thin filaments are also shorter, and are organized into sarcomeres whose often remarkable three-dimensional order is advantageous in structural studies. The results described here have been obtained principally by X-ray diffraction, as reviewed in greater detail elsewhere (17). For technical reasons, X-ray studies have hitherto been largely confined to three groups of animals, the vertebrates, arthropods, and mollusks, where suitably large muscles are common.

The structure of the myosin filaments is most naturally studied when the muscle is in the relaxed state. The myosin cross-bridges are detached from actin, and in their regular arrangement round the myosin filament backbone they diffract X-rays strongly and characterize the filament architecture. A conserved feature of all myosin filaments that have been studied is the axial spacing (close to 145 Å) between levels of cross-bridges. A key parameter of the structure is the helical repeat of the cross-bridge arrangement, which is a measure of the screw symmetry or "twist" between successive levels. This repeat is easily obtained from the X-ray pattern by direct measurement. Another parameter, the number of myosin molecules at each level, is less easily deduced from the X-ray pattern; if it is known, however, the radial position of the cross-bridges also can be directly obtained from the intensity distribution on the layer lines (18).

DIVERSITY OF CROSS-BRIDGE ARRANGEMENTS

In a preliminary study by Wray et al. (19), the X-ray patterns from three invertebrate striated muscles in the relaxed state were compared. These muscle types serve as representatives of three classes of myosin filament structure, distinguished by the following characteristics.

First, scallop myosin filaments (the filaments of the *striated* adductor muscles of scallops) give an X-ray pattern showing characteristic strong layer lines, which show the helical repeat of the surface lattice to be 480 Å (see also 12), and the number of myosin molecules per level to be six or seven. From the profiles of the layer lines, conclusions could be drawn about the cross-bridge configuration, especially that the myosin heads must depart significantly from the perpendicular to the filament axis. The profiles showed directly that the bridges are crowded unusually close with their neighbors within the same level. These conclusions have now been strikingly confirmed by electron microscope studies (R. Craig and P. Vibert, *personal communication*). The structure seems to be characteristic of scallop muscles, and bears no apparent relation to the structure of the smooth adductor of this or other mollusks. Although paramyosin is present in these muscles, the diffraction pattern shows no sign of the layer lines typical of paramyosin in smooth muscles.

Second, in *Limulus* (horseshoe crab) muscles, the comparably detailed but quite distinct X-ray pattern indicates almost the same degree of cross-bridge order. The helical repeat is 435 Å, exactly three times the axial spacing of 145 Å. As in the scallop, the bridges appear to be markedly tilted, and R. J. C. Levine et al. *(this volume)* have confirmed this from electron microscopy and have shown that there are, in fact, four myosin molecules at each level. I have found recently that the leg muscles of tarantula spiders give an almost identical X-ray pattern. (Horseshoe crabs are not really crabs: they and spiders belong to the same class of arthropods, the chelicerates, but have been separated evolutionarily for at least 400 million years.) A more surprising affinity is with vertebrate striated muscle, which appears to fall into the same class. The screw symmetry of frog muscle filaments (7) is again the same as in *Limulus*, and the layer line intensities are closely similar, especially after stretching (19). The pattern from frog muscle shows, however, that the filaments are substantially specialized and may not be perfectly uniform in structure along their length. In contrast to scallop and *Limulus*, the actin filaments are regularly arranged in the filament lattice. The X-ray pattern, together with the now widely accepted assumption that there are three myosin molecules at each level, shows that the bridges are centered at 135 Å from the axis of the myosin filament.

The third class of myosin filaments is not so easily delimited; it comprises certain crustacean and insect muscles, and will be termed the "lobster class." The myosin layer lines are weak, as first described by Miller and Tregear (11) for *Lethocerus* flight muscle, showing that the bridges are very disorderd compared to those of scallop and *Limulus* filaments. The helical repeat is not fixed in this class: layer lines of essentially the same appearance are obtained from a variety of different muscles, each yielding a characteristic value (15). Thus, the helical repeat can range from 300 to 320 Å in fast muscles of lobsters and crayfish, through values of about 330 Å for crab muscles (Y. Maéda, *personal communication*), 350 to 360 Å for lobster slow muscles and 360 Å for dragonfly flight muscles, to 385 Å for asynchronous insect flight muscles. Despite the evidence for bridge disorder and for variability of the helical repeat, there is nothing imprecise about the filament architecture. The layer lines, although weak, are very sharp, meaning that in each case the helical repeat is very accurately maintained along the filament length and over the millions of filaments within the fiber under examination. No treatment has been found that can induce cross-bridge order comparable to that found in scallop and *Limulus* filaments. The disorder thus appears to be characteristic of these muscles, and makes it difficult to study the conformation of the relaxed cross-bridges. Presumably the disorder is not static, but represents random movement of the bridges about their average positions. The correlation between X-ray diffraction and electron microscopy extends to this class of filaments also, in that negatively stained lobster and insect filaments, unlike those from *Limulus* and scallop muscles, have hitherto shown no well-defined surface lattice (M. K. Reedy, *personal communication*). The muscles in the lobster class are diverse not only in helical repeat but also in filament diameter, presumably reflecting variation in the number of myosin molecules per level. Insect flight muscles and the fast muscles of lobsters and crayfish seem to form a series in which there are four molecules per level. Assuming this rotational symmetry, the cross-bridges must be centered at 200 Å from the axis of the myosin filament.

Figure 1 summarizes the interrelationships of myosin filament backbone, cross-bridges, and actin filaments in the latter two classes of myosin filaments, using frog and insect muscles as representatives since they have especially regular filament lattices. It is interesting to note, from consideration of Figure 1, that the rotational symmetries of frog and insect muscles are unlikely to be larger than those assumed (namely three- and fourfold, respectively), since proportionately larger distances of the cross-bridges from the myosin filament axis would be required for consistency with the X-ray data and would hardly be accommodated within the known filament lattices.

MYOSIN FILAMENT BACKBONE

Since the three classes of filaments described represent assemblies of essentially similar myosin molecules, it is natural to seek some relationship between them. It would be especially valuable to know more about the structure of the filament

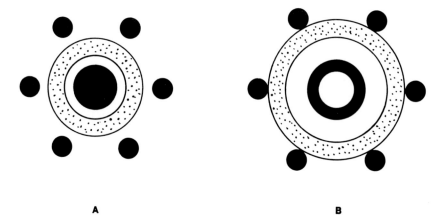

FIG. 1. Diagram representing the location of cross-bridges *(stippled ring)* in relation to the myosin backbone and surrounding actin filaments *(solid)* in the relaxed state of frog muscle **(A)** and insect flight muscle **(B)**. Outer radius of backbone, center of cross-bridge density, and actin-myosin separation are 75, 135, and 225 Å for frog and 100, 200, and 265 Å for insect.

backbone. The most information on this question has come from the lobster class of filaments, where the diffraction pattern contains additional features due not to the cross-bridges but to the backbone (15). From electron microscopy, both crustacean and insect thick filaments are often found to be hollow or to have a less densely staining core, and the diffraction from the backbone likewise suggests a tubular structure formed from subfilaments running in a specific direction through the surface lattice. The different helical repeats observed for the cross-bridge arrangement correspond naturally to different rates of subfilament twisting. The spacing between subfilaments is about 42 Å, and the presence and exact form of a reflection near 40 Å in the near-equatorial pattern constitute the crucial evidence on filament organization. These conclusions refer specifically to the fast and slow muscles of lobsters and crayfish and related species, having diverse screw and rotational symmetries. Crab muscles show apparently identical features, and the diffraction from flight muscles of dragonfly and bumble bee (15) and water-bug (1) shows sufficient similarities to suggest that all these muscles fall into the same class on the basis of backbone structure as well as type of cross-bridge organization.

Despite their greater cross-bridge order, *Limulus* and frog muscles give less information about their backbone structure from X-ray diffraction. It has not been possible to single out any diffraction feature that is unambiguously due to the backbone. The patterns from tarantula muscles do show at least a prominent 40 Å equatorial reflection, perhaps indicating a homologous subfilament structure, for some reason more clearly visible than in *Limulus*. The screw symmetry observed for this class of filaments would be generated by merely untwisting a filament of the lobster class so that the subfilaments run parallel to the axis (15). If the classes are indeed so related, a natural question is whether the simplicity of the untwisted structure is somehow related to the greater cross-bridge order. The backbone structure of the scallop myosin filament is completely unknown. Our present information is thus inadequate as a basis for comparing backbone structures, and further studies are needed.

Although paramyosin is present in all of the invertebrate muscles mentioned, the X-ray patterns show no sign of it either in the form of the Bear-Selby net familiar from smooth molluscan muscles or as an additional component of the structure congruent with the myosin surface lattice. The latter might be expected if the function of paramyosin in noncatch muscles is to increase the mechanical strength of long filaments or to assist in filament formation if myosin alone cannot form long assemblies (3). The arrangement of paramyosin in striated muscles is thus a continuing mystery. Perhaps further work will reveal a muscle type in which myosin and paramyosin form a more equal partnership, from which their interrelationship can be discerned.

RELATION TO MUSCLE FUNCTION

A striking instance where filament architecture can be related to the activity of cross-bridges is that of asynchronous insect flight muscles, whose particular filament geometry may contribute to the marked stretch-activation that underlies their myogenic rhythmicity (16). The symmetry of the cross-bridge array when the helical repeat is exactly 385 Å may be such that the bridges are systematically located relative to the envelope of target areas on the surrounding actin filaments. The myosin periodicity is then no longer in a vernier relationship to that of the envelope of actin filaments; hence changes of muscle length affect the number of bridges that can attach to actin, and stretch-activation results. These muscles exhibit legendary structural regularity (13), which presumably ensures that imposed length changes are translated into uniform molecular effects throughout the cell. It is surprising to note that the myosin filaments themselves belong to the class characterized by minimal cross-bridge order: but the important point is the precision of the average locations (or long-range order) of the cross-bridges in relation to the actin lattice, not the exact uniformity of bridge conformation. This mechanism presupposes the known staggered arrangement of the thin filaments in three dimension. Descherevsky (4) proposed a related mechanism based on the coincidence of spacing between individual myosin and actin filaments and requiring the actin filaments to be in register.

In other muscles, the specific function, if any, of the long-range order is not known, and only local geometrical relationships need be considered. Movements of the cross-bridges away from their positions in relaxed muscle are expected from the concept of the contractile cycle, raising important questions about the relation to force generation: whether, and in what conformations, attachment to and detachment from actin occur, and what changes of conformation intervene. Observations of equatorial and axial intensity changes from frog muscle can be interpreted in terms of cross-bridge movements, but the axial, azimuthal, and radial components of these movements are still unresolved. It is natural to suppose both the state of the bridges in relaxed muscle and that in rigor to be part of the cycle, implying radial movements toward the actin filaments as well as axial movements accompanying force-generation. Figure 1 emphasize a difference among the classes of myosin filament. The cross-bridges are further from the actin filaments in frog muscle than in insect muscle, and a similar relation will apply in other members of the same classes where the rotational symmetry is larger and the lattice dimension likewise. Thus, in general there will be differences in the cross-bridge movements accompanying activation, or attachment to actin in contraction or rigor. Such differences are, in fact, already apparent from measurements of equatorial intensities. These indicate a much larger mass movement between relaxed and rigor states in frog muscle than in insect muscle, and in *Limulus* than in crab muscle (11,18,20), which clearly cannot be taken merely as indicating differences in the numbers of cross-bridges involved. The differences in geometry may be related also to the differences in cross-bridge order. Thus, in frog and *Limulus* muscles, the bridges may be held closer to the ordered surface lattice of the backbone. This order can be reduced, for example, by stretching frog muscle to nonoverlap lengths (5,6), which causes a large decrease in lattice spacing; it is restored by immersion in more dilute Ringer or merely by waiting (5,6). Since either of these causes the filament lattice to expand (10), the original loss of order may be due to interaction with neighboring filaments. In the lobster, the bridges seem to be permitted considerable freedom of movement, and possess only limited order that apparently cannot be improved. In a sense, one may say that lobster and insect muscles have no counterpart of the relaxed state of frog muscle. Such differences between classes of filaments presumably have their basis in the architecture of the myosin molecule. Interestingly, lobster myosin may indeed be more flexible than vertebrate muscle in the S2 region (14).

The diffraction from cross-bridges in patterns from frog muscle is greatly weakened or lost also in contraction (7); X-ray information on other muscles in contraction is not yet available. This loss might be explained if only an undetectably small number of cross-bridges have returned to their ordered positions close to the backbone at any one time. Alternatively, it could reflect an activation of the thick filaments—"preparatory reaction" (9)—implying that the state of cross-bridges in the relaxed muscle is not part of the contractile cycle. Further dynamic X-ray measurements may allow these possibilities to be distinguished, and may suggest the reason why certain muscles possess a relaxed state in which the cross-bridges are sequestered away from the actin filaments. The reason does not seem to be related to the absence of myosin-linked regulation in vertebrate skeletal muscle, since *Limulus* exhibits a comparable arrangement of relaxed cross-bridges and double regulation. The scallop muscle, which lacks actin-linked regulation, provides a further important example. It is tempting to speculate that the unusual proximity of the relaxed cross-bridges with their neighbors at the same level is related to the cooperative interactions between molecules that occur in the calcium-activation of the myosin filament in this muscle (2).

CONCLUSION

This short survey leaves structural, functional, and evolutionary puzzles. One cannot yet point to a "basic" design of myosin filament comparable to the seemingly conservative actin filament. Instead, the limited range of striated muscles that have been examined establish certain classes of highly symmetric structures. The observed diversity prompts the question of what molecular interactions promote the formation of particular packing arrangements. Thus, the helical repeats in different lobster muscles are presumably specified by small differences in the sequence of the myosin molecule. The myosin filaments of the lobster class illustrate a versatility in design that allows diverse structures even within a single animal. The purpose of such diversification is not clear: perhaps it achieves some control over the dynamics of the muscle akin to, but more subtle than, the dramatic effects observed in asynchronous insect flight muscles. On the other hand, the *Limulus* class shows that a single design can be retained by animals as different in physiology as the horseshoe crab and the spider, over the lapse of hundreds of millions of years, and the apparently related design found for frog muscles probably characterizes all vertebrate striated muscles as well. Myosin filament designs do not seem to be significantly constrained by phylogeny, despite the striking instances of evolutionary conservatism. Instead, one must try to understand every type of myosin filament as adapted to its function. The particular virtue of, for example, the *Limulus* design remains to be discerned.

The different designs are likely to have their most marked consequences in the S2 and "neck" regions of the myosin molecule, determining especially how far the myosin heads are specifically bound to positions on the filament surface in the relaxed state or, alternatively, are in more random motion further from the backbone. Perhaps the diversity of filament designs contains clues about the function of that region. Although myosin molecules themselves are generally similar, the known differences between them often concern light chains located in the same region. We still know little about the conformational changes of cross-bridges, or even the site of the structural change accompanying energy transduction, and variations in the molecular environment at the myosin filament surface may yet prove to reflect diversity in the mechanism of contraction as well as of regulation.

ACKNOWLEDGMENTS

The ideas in this paper have grown out of a collaboration over a number of years with Carolyn Cohen and Peter Vibert, and I thank them for much inspiration.

REFERENCES

1. Barrington-Leigh, J., Goody, R. S., Hofmann, W., Holmes, K., Mannherz, H. G., Rosenbaum, G., and Tregear, R. T. (1977): The interpretation of X-ray diffraction from glycerinated flight muscle fibre bundles: New theoretical and experimental approaches. In: *Insect Flight Muscle*, edited by R. T. Tregear, pp. 137–146. North-Holland, Amsterdam.
2. Chantler, P. D., Sellers, J. R., and Szent-Györgyi, A. G. (1981): Cooperativity in scallop myosin. *Biochemistry*, 20:210–216.

3. Cohen, C., and Szent-Györgyi, A. G. (1971): Assembly of myosin filaments and the structure of molluscan catch muscles. In: *Contractility of Muscle Cells and Related Processes*, edited by R. J. Podolsky, pp. 23–36. Prentice-Hall, Englewood Cliffs, N.J.
4. Descherevsky, V. I. (1971): A kinetic theory of muscle contraction. *Biorheology*, 7:147–170.
5. Haselgrove, J. C. (1975): X-ray evidence for conformational changes in the myosin filaments of vertebrate striated muscle. *J. Mol. Biol.*, 92:113–143.
6. Huxley, H. E. (1972): Structural changes in the actin- and myosin-containing filaments during contraction. *Cold Spring Harbor Symp. Quant. Biol.*, 37:361–376.
7. Huxley, H. E., and Brown, W. (1967): The low-angle X-ray diagram of vertebrate striated muscle and its behaviour during contraction and rigor. *J. Mol. Biol.*, 30:383–434.
8. Josephson, R. K. (1975): Extensive and intensive factors determining the performance of striated muscle. *J. Exp. Zool.*, 194:135–154.
9. Lowy, J. (1972): X-ray diffraction studies of striated and smooth muscles. *Boll. Zool.*, 39:119–138.
10. Matsubara, I., and Elliott, G. F. (1972): X-ray diffraction studies on skinned single fibres of frog skeletal muscle. *J. Mol. Biol.*, 72:657–669.
11. Miller, A., and Tregear, R. T. (1972): Structure of insect fibrillar flight muscle in the presence and absence of ATP. *J. Mol. Biol.*, 70:85–104.
12. Millman, B. M., and Bennett, P. M. (1976): Structure of the cross-striated adductor muscle of the scallop. *J. Mol. Biol.*, 103:439–467.
13. Reedy, M. K. (1968): Ultrastructure of insect flight muscle. *J. Mol. Biol.*, 31:155–176.
14. Siemankowski, R. F., Zobel, C. R., and Manuel, H. (1980): Comparative studies on the structure and aggregative properties of the myosin molecule. II. The substructure of the lobster myosin molecule. *Biochim. Biophys. Acta*, 622:25–35.
15. Wray, J. S. (1979): Structure of the backbone in myosin filaments of muscle. *Nature*, 277:37–40.
16. Wray, J. S. (1979): Filament geometry and the activation of insect flight muscles. *Nature*, 280:325–326.
17. Wray, J. S., and Holmes, K. C. (1981): X-ray diffraction studies of muscle. *Ann. Rev. Physiol.*, 43:553–565.
18. Wray, J. S., Vibert, P. J., and Cohen, C. (1974): Cross-bridge arrangements in *Limulus* muscle. *J. Mol. Biol.*, 88:343–348.
19. Wray, J. S., Vibert, P. J., and Cohen, C. (1975): Diversity of cross-bridge configurations in invertebrate muscles. *Nature*, 257:561–564.
20. Yagi, N., and Matsubara, I. (1977): The equatorial X-ray diffraction patterns of crustacean striated muscles. *J. Mol. Biol.*, 117:797–803.

Basic Biology of Muscles: A Comparative Approach, edited by B. M. Twarog, R. J. C. Levine, and M. M. Dewey. Raven Press, New York © 1982.

Molecular Organization of *Limulus* Thick Filaments

*Rhea J. C. Levine, *Robert W. Kensler, **Murray Stewart, and †John C. Haselgrove

*Department of Anatomy, The Medical College of Pennsylvania, Philadelphia, Pennsylvania 19129; **C.S.I.R.O. Division of Computing Research, P.O. Box 1800, Canberra City, A.C.T. 2601, Australia; and †Johnson Research Foundation, University of Pennsylvania, Philadelphia, Pennsylvania 19104*

The skeletal or body-wall muscles of invertebrates exhibit a broad spectrum of functional and structural diversity. One example of such diversity is related to extreme ranges in length over which many of these muscles operate *in situ*. In different species, the muscles that express this property have solved the problems associated with: (a) maintenance of functional overlap between thick and thin filaments at the long extreme of muscle length, and (b) the ability to develop appreciable active tension at the shortest *in situ* muscle length, by a variety of mechanisms superimposed on the basic mechanism of sliding filaments. Thus, in smooth and obliquely striated muscle cells of coelenterates, molluscs, nematodes, and annelids, the thick filaments that are completely misaligned at long muscle lengths either shear past each other to lie in better register (especially in obliquely striated muscles) (21) or crumple and become disarrayed (20) at short muscle lengths.

True striated muscles faced with similar problems must overcome the constraint of sarcomeric organization as well, and exhibit a variety of adaptive structural alterations, to achieve similar results. At one extreme are the one myofibril thick, very short-sarcomere, retractor muscles of the bryozoan, *Membranipora serrilamella*, which undergo a 10-fold change in length in the intact animal. When fully extended, these muscles have well ordered striations with ~1.0 μm sarcomeres and ~0.5 μm A bands. When retracted, or fully shortened, however, the muscles show complete loss of striations (A. Wagner, *unpublished observations*). Thus, in

---

Dr. Stewart's present address is MRC Laboratory of Molecular Biology, Hills Road, Cambridge CB2 2QH, United Kingdom.

this species, an enormous change in muscle length is accommodated by reversible disruption and reordering of sarcomeric organization. A somewhat different mechanism has evolved in the scutal depressor muscles of the giant barnacle, *Balanus nubilis*, which operate over a 3-fold change in length. At very short muscle lengths, the >5.0 μm long thick filaments have been observed to penetrate the Z-discs, which are perforated in these muscles, and thick filaments of adjacent sarcomeres overlap with each other (12). Still another mechanism to overcome the sarcomeric restraints has been evolved by the telson levator muscles of the horseshoe crab, *Limulus polyphemus*, and it is these muscles we have studied.

LENGTH CHANGES IN *LIMULUS* TELSON MUSCLES

The telson (fused segmental caudal appendage) of *Limulus* is used as a lever to right the animal when it is overturned and to balance swimming movements (22). This appendage has greater than 180° of freedom of azimuthal movement, and the antagonistic pairs of muscles that act to raise and lower the telson, the telson levators and depressors, undergo a greater than twofold change in length with change in telson position (5–7). The fibers of *Limulus* telson muscles are uniform with respect to sarcomere length at a given muscle length, and *in situ* sarcomere lengths range from >10.0 μm when the muscles are fully extended to ~4.0 μm when the muscles are fully shortened (5,7,9,16). DeVillafranca (3,4) first reported changes in length of A-bands of *Limulus* skeletal muscle that paralleled changes in sarcomere length, and this has been extended and confirmed in both our laboratory and that of Dewey and co-workers (5,7,9,16). We have resolved the overall change in A-band length (from >6.0 μm in 10.0 μm sarcomeres to ~3.0 μm in 4.0 μm sarcomeres) into two separate components. As sarcomeres shorten from their longest *in situ* length down to rest length (~7.0 μm), the concomitant decrease in A-band length is due to the realignment of out-of-register ~4.0 μm long thick filaments, reminiscent of the situation in obliquely striated muscle. Abundant evidence has accumulated that the further decrease in A-band length observed as sarcomeres shorten below rest length is the result of a conformational change of the thick filaments themselves, which decrease in length by as much as 40% with an apparent corresponding increase in diameter. A structural change of this magnitude undoubtedly requires modification in the molecular organization of the thick filaments. To elucidate this phenomenon, our recent efforts have been devoted to an examination of the packing of the component protein molecules in *Limulus* thick filaments.

LIMULUS THICK FILAMENTS

We began our investigation with the following background information in hand. First, unlike the 1.6 μm long thick filaments of vertebrate striated muscles, in which myosin constitutes the single major structural protein, longer invertebrate thick filaments contain paramyosin as an additional structural protein. Paramyosin forms the filament core and is surrounded by a cortical layer of myosin (1,2,16, 23,24,26). The paramyosin:myosin ratio correlates directly with filament length, and *Limulus* thick filaments contain twice as many myosin as paramyosin molecules (17). Second, our studies on isolated thick filaments, in collaboration with Dewey and co-workers (6,7), demonstrated that stimulation is necessary to effect filament shortening. Thus, long filaments ($\geq$4.0 μm) are separated from unstimulated, relaxed muscle, whereas short filaments ($\sim$3.0 μm) are obtained from relaxed muscle that was previously stimulated to contract isotonically or (in recent studies in collaboration with S. Davidheiser and R. E. Davies, *unpublished data*) isometrically, at lengths beyond overlap. Therefore, filaments can be isolated and examined preferentially in either conformation, depending on the treatment of the intact muscle prior to tissue homogenization. Third, although the X-ray diffraction studies of both Millman et al. (19) and Wray et al. (29,30) showed no change in the 1/14.6 nm^{-1} meridional reflection between *Limulus* muscles glycerinated at long or short sarcomere lengths, both groups found that there was a great degree of order to the structure of the thick filaments in the lengthened fibers, as revealed by the appearance of the myosin reflections. Further, Wray et al. (29) interpreted some of their data: off-meridional position of the 1/14.6 nm^{-1} reflection in lengthened muscle and lack of change of intensity of thin filament reflections at different sarcomere lengths as consistent with stagger of thick filaments within the A-bands, in the first case, and no change in length of the overlap region over a range of sarcomere lengths, in the second. Both of these findings fit our interpretation of the overall change in A-band length with sarcomere length.

In the case of *Limulus* muscle, one drawback to comparative X-ray analysis of lengthened and shortened fibers to determine differences in filament structure is the inability to know the precise length of the filaments in the tissue being examined. Electron microscopy, although potentially less powerful than X-ray diffraction, has the advantage of permitting the examination, measurement, and interpretation of the actual structure under analysis. Indeed, when combined with other techniques such as optical diffraction (18) and computer image analysis (27), electron microscopy provides a useful alternative technique for the analysis of periodic structures. We have, therefore, employed the latter approach to examine the structure of *Limulus* filaments, isolated under conditions that minimize damage to them.

Our studies to date have focused largely on the structure of *Limulus* thick filaments in their lengthened conformation. This approach is predicated by the rationale that in order to discern differences between long and short filaments, it is necessary to establish base-line parameters of the structure of one of these. With this in mind, our goals were to determine the organization of the myosin molecules on the surfaces of the long filaments, including the hand of the cross-bridge helix, and also to examine the structure of the paramyosin core. The results of these studies would be compared with those obtained from subsequent analysis of shortened filaments.

NATIVE STRUCTURE OF LENGTHENED FILAMENTS

Filament Preparation

By rapid and gentle separation of thick filaments from fresh, unstimulated muscle, we consistently obtain filament populations that are uniform in length (4.1 $\pm$ 0.2

μm SD, N = 100) and show minimal breakage. Visualization of structural features is maximized by use of a very thin (5.0 to 7.5 nm) carbon film over a carbon-stabilized perforated Formvar® support as the substrate to which the filaments are adsorbed. Damage to the filaments by the electron beam is minimized by focusing on an area adjacent to the filaments and moving the grid prior to photography (13,14).

Cross-Bridge Arrangement

Electron microscopic images of negatively stained *Limulus* thick filaments pre-pared in this manner show well-ordered periodic structure. At low magnification the filaments appear to have a regular periodicity, interrupted only by a central bare zone (Fig. 1a). At higher magnifications this axial periodicity is increasingly ap-parent and is seen to arise from the helical array of cross-bridges, which project from the filament backbone. The helices generated by the cross-bridges give the filament a surface pattern resembling a chain of angulated "loops" that repeat at ~43 nm intervals. The filament shows bilateral symmetry across a plane parallel to its long axis with a paired arrangement of cross-bridges at each 14.6 nm level. Frequently the positions of the paired cross-bridges within a single repeat is such that the overall appearance of the loop is reminiscent of a six-pointed star (Fig. 1B). Since no direct attempt is made to rid the thick filament preparations completely of thins, we frequently observe the situation where a thin filament runs alongside a thick for a considerable distance. The parallel periodic array of cross-bridges projecting from the surface of the thick filament and attaching it to the actin monomers of the neighboring thin filament are readily visible.

This appearance is largely absent from filaments we isolate from relaxed gly-cerinated muscle, which seem to have lost most of their surface projections. This finding suggests that there may be sufficient interaction between the thick and thin filaments still present in our preparations of relaxed glycerinated muscle to pull either whole myosin molecules or the cross-bridge portions of them off the thick filament surface during homogenization. The difference between our present results and previous results of preliminary studies of the structure of isolated thick filaments (6,15) may be due, in part, to the fact that the earlier studies were done using filaments isolated mainly from glycerinated muscle.

Optical diffraction patterns obtained from the electron micrograph negatives of the negatively stained filaments are highly rewarding. They are strong and sharp with excellent separation of the intensity of the filament reflections from the minimal background noise, and, like the electron micrographs themselves, frequently show resolution of periodic information out to ≤4.0 nm. Both our optical diffraction patterns and the transforms computed from digitized densities on electron micro-graph images of the filaments (25) display a series of layer lines indexing on orders of $1/43.8$ nm^{-1}, in agreement with values obtained from X-ray analysis of relaxed, glycerinated *Limulus* muscle (29,30). Both types of transform show meridional reflections at $1/14.6$ and $1/7.3$ nm^{-1}, and on patterns from the best specimens a

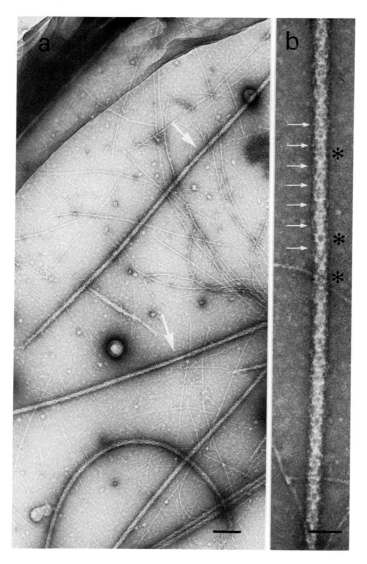

FIG. 1. Electron micrographs of negatively stained *Limulus* thick filaments. Thin filament fragments are visible in the backgrounds. **a:** Low magnification electron micrograph. Note the periodic appearance of the thick filaments except in the regions where bare zones *(arrows)* are visible. Thin filaments lie in close association with the curved thick filament. Bar = 0.2 μm. **b:** Higher magnification electron micrograph, showing the "angulated loop" appearance of the cross-bridge array at the filament surface; 43 nm repeats are indicated by *arrows*, and the star-like appearance of the helical array by *asterisks*. Bar = 75 nm.

reflection at $1/4.9$ nm^{-1} is also visible. These meridional reflections index on orders of $1/14.6$ nm^{-1}, which represents the axial spacing along the filament of successive levels of cross-bridge projections. The first, third, fourth, and sixth layer lines are strong, whereas the second and fifth are weak, also in agreement with the X-ray diagram (Fig. 2b and c). The similarity between the diffraction patterns we obtain and the X-ray results indicates that we achieve optimal preservation of the filament structure, and that the images and optical transforms describe the actual organization of molecules on the filament surface. The major diffraction maxima of the optical transform index on reciprocal nets, which correspond both to the upper and lower surfaces of the helical cross-bridge array, thus indicating that the electron micro-scopic image represents the superposition of detail from the helical array on both the upper and lower surfaces of the filament (Fig. 2c). This array is described by the equation: $L = n/N + 3m$, where L = layer line, n = order of Bessel function predicted for that L, N = number of strands in the helix, and m = an integer.

One of the important structural parameters to determine is N, the number of strands in the helix, or the number of cross-bridges that project from the filament surface (and give rise to the helical pattern) every 14.6 nm along the long axis of the filament. N can be estimated from the position of the primary maxima on the first layer line of the diffraction pattern if other parameters, including the radius, from the center of the filament, at which the greatest mass of the cross-bridges is centered, are known.

The radius at which the greatest mass of the cross-bridges is centered is calculated as 15.5 nm from the radial position of the satellite maxima on the third layer line (meridional $1/14.6$ nm^{-1} reflection). This radius is consistent with measurements of filament diameter on electron micrographs, which give a minimum of 23 nm in the region of the bare zone, and a maximum of ~36 nm at the furthest discernable projection of the paired cross-bridges. Using 15.5 nm = r, and the radial position of the primary maxima on the first layer line (17.8) = R, an argument: $(2 \pi Rr) = 5.5$ is obtained for the Bessel function that best describes the position of the reflections on the first layer line; 5.5 is closest to the expected maximum of a J_4 Bessel function and corresponds to an N = 4, or four cross-bridges/14.6 nm. The information retrieved from the computer transform essentially confirms these calculations and further enables more precise definition of the number of strands in the helix, since this analytic technique has the ability to retain the phase information lost in the optical transform and thus allows determination of whether the number of helical strands is even or odd. The primary maxima corresponding to the near and far sides of the helix, on both the first and fourth layer lines, are in phase, respectively, consistent with an even-order Bessel function. The individual subunits or cross-bridges can be visualized by computerized reconstruction, using data from both filament surfaces or from just one surface. In the first instance, the result is re-markably similar to the appearance of the filaments on electron micrographs, whereas the one-sided image clearly reveals a four-stranded structure consistent with four cross-bridges at each level (25) (Fig. 3a and b).

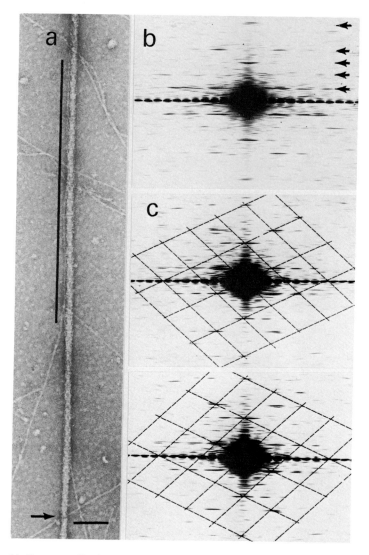

FIG. 2. Medium magnification electron micrograph of a region of a negatively stained thick filament and optical diffraction patterns. **a:** Electron micrograph of a thick filament. Bar along side image of filament delineates the region from which the optical diffraction pattern in **b** was obtained. Note the thin filaments that run alongside the thick filament. The bare zone is visible at the *arrow.* Bar = 0.1 μm. **b:** Optical diffraction pattern obtained from the region in *a* alongside the dark bar. Illustrates the detailed nature of patterns from *Limulus* thick filaments, with layer-lines at 1/43.8, 1/21.9, 1/14.6, 1/10.9, and 1/7.3 nm⁻¹ *(arrows),* and the satellite maxima on the 1/14.6 nm⁻¹ layer-line. **c:** An optical diffraction pattern indicating the tentative indexing of the major diffraction maxima on reciprocal nets *(dashed lines)* corresponding nominally to the near *(top)* and far *(bottom)* sides of the filament. The reciprocal nets are consistent with a screw symmetry of three. There is good similarity of the corresponding reflections from the two filament surfaces with respect to both their radial distances from the meridian and their intensities, which demonstrates good preservation and staining of both filament surfaces.

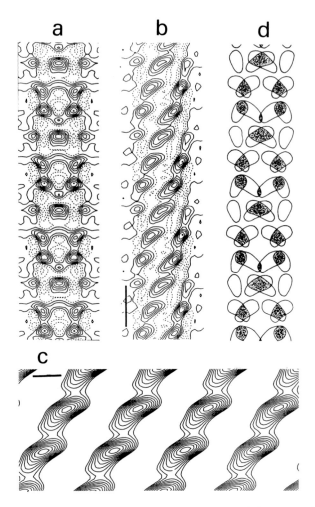

FIG. 3. Computer-derived image reconstructions of *Limulus* thick filament surface structure and a filament model derived from optical transforms and electron micrographs. **a** and **b**: Contour plots of Fourier-filtered images (data averaged from electron micrographs of 7 filament areas). *Full contour lines* represent positive density (protein), *broken contour lines* represent negative density (stain). **a:** Image obtained using all layer line data. **b:** One sided image produced by appropriate masking of the transform according to Wray et al. (29). The images are oriented so the bare zone is toward the bottom of the figure. Bar = 25 nm. **c:** Circumferential section computed at a radius of 16 nm through a three-dimensional reconstruction of the thick filament. Only positive densities (protein) are shown. The reconstruction is oriented so that the bare zone direction is toward the bottom of the page. The individual subunits (cross-bridges) are bent and tilted approximately 30° to the direction of the helix. In each subunit, the end furthest from the bare zone appears denser than the other end. Bar = 10 nm. **d:** Model of filament surface structure, computed from optical transform data and measurements from electron micrographs. Each cross-bridge is shown as a fat, comma-shaped structure. The cross-bridges are axially tilted at 35° and azimuthally at 80° (11,30). The appearance generated by this model closely resembles the computer reconstruction of both filament surfaces *(a)*, and each cross-bridge resembles, in shape, the computed subunits seen in the circumferential reconstruction *(c)*. *(a, b,* and *c,* from Stewart et al., ref. 25, with permission.)

The results of these studies are, therefore, in good agreement with those of the X-ray studies on lengthened, relaxed glycerinated *Limulus* fibers, with respect to the screw symmetry of the cross-bridge array and the axial spacing between sequential levels of bridges. Further, our results determine that there are four cross-bridges/14.6 nm, which was suggested, but not proven, by the X-ray analysis. Thus, the surface of lengthened *Limulus* thick filaments is covered by myosin molecules arranged in four helical strands with a pitch of 175 nm (12/4 helix), having an axial rise of 14.6 nm and threefold screw symmetry. The diameter of the filament shaft is ~23.5 nm and, although cross-bridges may project to a radius of >18 nm, their centers of mass lie about 4 nm from the filament surface and 16 nm from its center.

SHAPE AND ORIENTATION OF CROSS-BRIDGES

The individual repeating units that give rise to the helical structure of the filament surface (the cross-bridges) are composed of the head regions of myosin molecules. Cross-bridges can extend away from the filament surface, interact with actin monomers on the thin filaments, and possess ATPase activity.

Using layer line data averaged from seven different areas of different filaments, three dimensional reconstructions of the filament are computed, which provide information in greater detail about the shape of the cross-bridges than can be observed directly (25). Similar to measurements from electron micrographs and calculations from the optical transforms, in the computed reconstruction the center of mass of the bridges is located at a radius of ~16 nm and individual bridges have diameters of ~8 nm. In computed cylindrical sections, the bent-elongated shape (~20 nm in length) of the entire cross-bridge can be seen, along with its ~30° inclination to the direction of the helix around the filament and the location of its greatest mass in the region of the bridge farthest from the bare zone of the filament (Fig. 3c). Individual heads cannot be resolved. Since preliminary estimates from quantitative gel analysis give the number of myosin molecules/14.6 nm of thick filaments as 4.1, we interpret each cross-bridge to be composed of a single myosin molecule, having two heads. The shape of the cross-bridge revealed in the reconstruction is similar to that proposed from analyses of decorated thin filaments (27).

MODELS OF FILAMENT STRUCTURE

To confirm further that the proposed arrangement of myosin is consistent with the appearance of the *Limulus* filaments in electron micrographs, we have modeled the expected appearance of such a filament, using a minicomputer program. In the computations, the cross-bridge is modeled as a bent rod composed of three overlapping spheres with diameters of 8, 8, and 6 nm (Fig. 3d). The middle sphere is centered 16.0 nm from the filament axis. The axial and azimuthal tilt angles of the cross-bridges are defined as described by Wray et al. (30) and Haselgrove (11), and models with cross-bridges at various axial tilt and azimuthal angles are computed. We note a strong resemblance between the computed models and images of the real filaments both in the original micrographs (Figs. 1 and 3d) and in the computer-filtered images (Fig. 3a and b). The model accounts for the stain excluding regions, which appear to occur centrally every third level of bridges along the filament axis and the frequently seen pairs of white dots (Fig. 1b, white arrows), which occur at successive levels along the length of the filament. As illustrated in the model, these dots probably represent the regions at which the cross-bridges at the back and front of the helix appear to overlap in the projected view. These results thus suggest that the electron microscopic appearance of *Limulus* thick filaments is consistent with the proposed helical arrangement of the myosin cross-bridges.

SHADOWED THICK FILAMENTS: THE HAND OF THE CROSS-BRIDGE HELIX

Although the negatively stained images delineate most features of the helical array of cross-bridges, these images do not allow a determination of whether the helix is right- or left-handed. This is because the negatively stained image represents a superposition of details of the helix on both the upper and lower surfaces of the filament without distinguishing which portion of the image derives from either surface. To approach this problem, we have employed heavy metal shadowing of isolated filaments adsorbed to grids, since this technique delineates only the features on the upper surface of the filament. In these studies, filaments were deposited on carbon-coated grids, lightly stained with uranyl acetate, then rinsed with buffer and shadowed at ~30° with platinum or platinum-carbon. Filaments not exposed to uranyl acetate tend to collapse during air drying and produce poor images. This may indicate that uranyl acetate may act as a fixative for the filament or a mordant for the vaporized metal.

Filaments oriented with their long axes perpendicular to the direction of shadowing show a striking right-handed helical structure. This rope-like structure extends uniformly, in the same direction, across both arms of the filament, but is absent along the entire length of the bare zone. The strands of the helix appear to run diagonally across the filament surface at regular intervals of 43 nm. Filaments oriented with their long axes nearly parallel to the direction of shadowing have a more complex surface structure in which individual cross-bridges are delineated. The image fits the expected one-sided image of a four-stranded helix that repeats every 43.8 nm (Fig. 4a).

Filaments of either appearance described above give very strong optical diffraction patterns that are dominated by reflections arising from only one side of the surface helix. Thus, the layer lines observed in transforms of negatively stained filaments are still present. The maxima on the first and fourth layer lines appear in the same two quadrants of the pattern, whereas those on the second layer line, which are stronger than when obtained from negatively stained filaments, occur in the two opposite quadrants (Fig. 4b).

The inclusion of the outer maxima of the first layer line in the pattern from one surface of the shadowed filament indicates that these are secondary peaks of the

FIG. 4. Electron micrograph of a shadowed *Limulus* thick filament and the optical diffraction pattern corresponding to the electron microscope image. **a:** Electron micrograph of *Limulus* thick filaments fixed briefly with uranyl acetate, washed and shadowed with platinum carbon. *White arrow* indicates direction of shadowing. Note the different appearance of the surface of filaments, depending on their position in relation to the direction of shadowing. In some, a right-handed helix is clearly visible on both sides of the bare zone *(black arrow)*, whereas in those oriented differently, individual subunit structure is visible on the surface. *Bar* alongside filament indicates region from which the optical diffraction pattern seen in *b* was obtained. Bar = 0.15 μ. **b:** Optical diffraction pattern obtained from the region of the shadowed thick filament indicated by the dark bar in *a*. Note the persistence of all the layer lines seen in diffraction patterns of negatively stained filaments, but their restriction to only two opposite quadrants, here. The first and fourth layer line occur in the same quadrants; the second does not.

Bessel function N (N = 4), rather than 2N. This result agrees with the phase information of the transform computed from the image of the negatively stained filament, which also suggests that these are subsidiary maxima arising from the same filament surface as the primary maxima.

Thus, by shadowing, we have determined that the myosin cross-bridge helix on the surface of *Limulus* thick filaments is right-handed. In addition, the optical diffraction patterns arising from only one surface of the filament in the shadowed preparations clearly confirm our indexing (Fig. 2a) of the diffraction patterns from the negatively stained filaments. Differences between the optical diffraction patterns of shadowed filaments and negatively stained images, such as the stronger second layer line in patterns from the shadowed filaments, undoubtedly arise from the relatively poorer resolution in the shadowed images. In the shadowed preparations, the structure and orientation of individual cross-bridges probably contribute little to the diffraction pattern.

STRUCTURE OF THE PARAMYOSIN CORES OF LENGTHENED FILAMENTS

In the very large (up to 150 nm diameter, 30 μm long) thick filaments of molluscan smooth muscles, the core paramyosin molecules are organized into highly ordered arrays (2,10). These arrays are easily visualized by electron microscopy and appear to be side-to-side associations between organized aggregates of paramyosin, each 10 to 20 nm wide (10). Even though the diameter of the shaft of *Limulus* thick filaments is only 23.5 nm, it is of interest to examine the nature of the organization of core paramyosin molecules in order both to determine whether the core structure is related to and/or determines the arrangement of surface myosin molecules, and to be able to compare the molecular packing of the core (as well as of the cortex) between lengthened and shortened filaments. Paramyosin is presumably distributed throughout the interiors of *Limulus* thick filaments, since they appear solid in electron micrographs of cross-sectioned muscle fibers, unlike many other invertebrate thick filaments with dimensions similar to those of *Limulus* muscle, which appear hollow.

Paramyosin cores are obtained by incubation of separated *Limulus* thick filaments in 0.4 M KCl at pH 6.0 for 10 min, either in suspension or after collection of the separated filaments on the carbon-coated electron microscope grids. The cores, deposited on or still adsorbed to the grids, are negatively stained in the same way as are intact filaments.

Unlike the highly ordered structure of the molluscan thick filament cores, those obtained from lengthened *Limulus* thick filaments show little axial periodicity in negatively stained images. Longitudinally oriented subunits are visible and in some regions transverse banding appears at intervals of 70 to 75 nm. There is no evidence of a Bear-Selby net and there is little indication of a 14.6 nm axial periodicity (Fig. 5). The surfaces of the cores are smooth along their entire lengths and their diameters are 19 to 20 nm. The difference between the diameter of the core and that of the

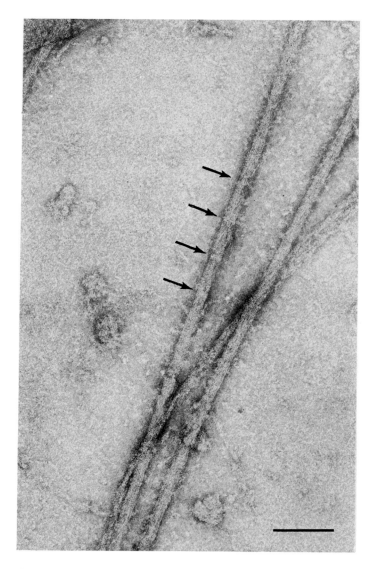

FIG. 5. Moderately high magnification electron micrograph of the paramyosin cores obtained after stripping the myosin off of the surfaces of *Limulus* thick filaments. The cores appear relatively smooth. Longitudinal striations are visible. Cross-striations appear at periods of 70 to 75 nm in some regions *(arrows)*. Bar = 0.1 μ.

intact filament shaft (measured in the region of the bare zone) is consistent with removal of the entire cortical layer of myosin.

Optical transforms of the electron micrograph images of the cores are generally as disappointing as the images themselves, and show no evidence of the presence of major periodic structures.

It is possible that the cores of *Limulus* thick filaments, which are much thinner than those of molluscan thick filaments, are less stable than the molluscan cores, and even than the intact *Limulus* filaments, and thus undergo molecular rearrangement during the preparative procedures. Such a rearrangement might lead to a loss of repeating structure and result in the rather featureless appearance that we observe. It is also possible, however, that the diameter of the *Limulus* thick filament core is too small ($\sim$ 20 nm) to permit lateral organization of sufficient numbers of basic paramyosin aggregates to produce a recognizably ordered structure. Finally, even though paramyosins from widely divergent phyla appear to be fairly well conserved, evolutionarily (8), it is also possible that the variations that do exist among paramyosins from molluscs and those from arthropods are sufficient to produce differences in their aggregating properties. Further work is underway to differentiate among these possibilities.

RELATIONSHIP TO OTHER THICK FILAMENTS

It would indeed be helpful to everyone interested in thick filament structure if many, if not all, thick filaments were organized in a similar fashion to those of *Limulus* skeletal muscle. That hope is of course unreasonable, in view of the tremendous disparity in size of many thick filaments, but it turns out that it is even unreasonable to hope that thick filaments of the approximate dimensions and paramyosin : myosin molecular ratios as *Limulus* thick filaments are structurally similar to the latter.

There has been some indication of the tremendous structural differences among thick filaments from muscles of different animals from the comparative X-ray studies of Wray and his colleagues (28,30). More recently, using the techniques described here to investigate thick filament structure, we and others have found great dissimilarities among thick filaments from molluscan fast adductors, insect flight and jumping muscles, and crustacean slow and fast muscles. Some of these differences may be related to the number of strands in the helix, others to the organization of cross-bridges into nonintegral surface helices. The thick filaments from other invertebrate muscles may also differ from those of *Limulus* with respect to the orderliness of the arrangement of their cross-bridges, either in the intact system or when the filaments are separated from the rest of the contractile apparatus. Structures that exhibit this type of disorder will be less amenable to analysis by the techniques we have used for *Limulus* thick filaments. Very preliminary observations in our laboratory suggest that in the arthropods, the thick filaments of insect leg and crustacean fast and slow abdominal and slow claw muscles may be more similar to each other than to those of other fast (molluscan) or slow *(Limulus)* muscles. Thus far, the only thick filaments examined that appear to be structurally similar to those of *Limulus* skeletal muscle are those of tarantula leg muscle (M. Reedy and M. Reedy, *unpublished observation*). Interestingly, except for scorpions, the spiders constitute the group most closely related to the horseshoe crab, phylogenetically. This may mean that similarity among thick filaments is dependent on evolutionary affinities as well as functional parameters.

ACKNOWLEDGMENTS

We thank K. McGlynn for photographic assistance and Dolores Wells and Kathy Golden for their secretarial help. Many thanks also to M. and M. Reedy, P. Vibert, R. Craig, and J. Wray, for suggestions and discussion. This work was supported by U.S.P.H.S. grants: HL-15835 to the Pennsylvania Muscle Institute and NRSA GM 07475 to R. W. Kensler.

REFERENCES

1. Bullard, B., Luke, B., and Winkelman, L. (1973): The paramyosin of insect flight muscle. *J. Mol. Biol.*, 75:359–367.
2. Cohen, C., Szent-Györgyi, A. G., and Kendrick-Jones, J. (1971): Paramyosin and the filaments of molluscan "catch" muscles. I. Paramyosin: Structure and assembly. *J. Mol. Biol.*, 56:223–237.
3. deVillafranca, G. W. (1961): The A and I band lengths in stretched or contracted horseshoe crab skeletal muscle. *J. Ultrastruct. Res.*, 5:109–115.
4. deVillafranca, G. W., and Marschhaus, C. M. (1963): Contraction of the A-band. *J. Ultrastruct. Res.*, 9:156–165.
5. Dewey, M. M., Levine, R. J. C., and Colflesh, D. E. (1973): Structure of *Limulus* striated muscle. The contractile apparatus at different sarcomere lengths. *J. Cell Biol.*, 58:574–593.
6. Dewey, M. M., Levine, R. J. C., Colflesh, D., Walcott, B., Brann, L., Baldwin, A., and Brink, P. (1979): Structural changes in thick filaments during sarcomere shortening in *Limulus* striated muscle. In: *Crossbridge Mechanism in Muscle Contraction*, edited by H. Sugi and G. H. Pollack, pp. 3–22. University of Tokyo Press, Tokyo.
7. Dewey, M. M., Walcott, B., Colflesh, D. E., Terry, H., and Levine, R. J. C. (1977): Changes in thick filament length in *Limulus* striated muscle. *J. Cell Biol.*, 75:366–380.
8. Elfvin, M. J., Levine, R. J. C., and Dewey, M. M. (1976): Paramyosin in invertebrate muscles. I. Identification and localization. *J. Cell Biol.*, 71:261–272.
9. Elfvin, M. J., Levine, R. J. C., and King, H. A. (1979): Uniformity of fiber type in *Limulus* telson levator muscles. *Biophys. J.*, 31:112a.
10. Elliott, A. (1979): Structure of molluscan thick filaments: A common origin for diverse appearances (with appendix by Dover, D., and Elliott, A.: Three-dimensional reconstruction of a paramyosin filament). *J. Mol. Biol.*, 132:323–341.
11. Haselgrove, J. C. (1980): A model of myosin cross-bridge structure consistent with the low-angle X-ray diffraction pattern of vertebrate muscle. *J. Mus. Res. Cell Motil.*, 1:177–191.
12. Hoyle, G., McAlear, J. H., and Selverston, A. (1965): Mechanism of super-contraction in a striated muscle. *J. Cell Biol.*, 26:621–640.
13. Kensler, R. W., and Levine, R. J. C. (1981): Structure of *Limulus* thick filaments. *Biophys. J.*, 33:242a.
14. Kensler, R. W., and Levine, R. J. C. (1982): An electron microscopic and optical diffraction analysis of the structure of *Limulus* telson muscle thick filaments. *J. Cell Biol.*, 92:443–451.
15. Levine, R. J. C., and Dewey, M. M. (1979): Changes in molecular packing during shortening of *Limulus* thick filaments. In: *Motility in Cell Function*, edited by F. A. Pepe, J. Sanger, and V. Nachmias, pp. 341–345. Academic Press, New York.
16. Levine, R. J. C., Dewey, M. M., and deVillafranca, G. W. (1972): Immunohistochemical localization of contractile proteins in *Limulus* striated muscle. *J. Cell Biol.*, 55:221–236.
17. Levine, R. J. C., Elfvin, M. J., Dewey, M. M., and Walcott, B. (1976): Paramyosin in invertebrate muscles. II. Content in relation to structure and function. *J. Cell Biol.*, 71:273–278.
18. Millman, B. M., and Bennett, P. M. (1976): The structure of a cross-striated molluscan muscle: The adductor muscle of the scallop. *J. Mol. Biol.*, 103:439–467.
19. Millman, B. M., Warden, W. J., Colflesh, D. E., and Dewey, M. M. (1974): X-ray diffraction patterns from glycerol-extracted *Limulus* muscle. *Fed. Proc.*, 33:1333.
20. Perkins, F. O., Ramsey, R. W., and Street, S. F. (1971): The ultrastructure of fishing tentacle muscle in the jellyfish *Chrysaora quinquecirrha*: A comparison of contracted and relaxed states. *J. Ultrastruct. Res.*, 35:431–450.

21. Rosenbluth, J. (1965): Ultrastructural organization of obliquely striated muscle fibers in *Ascaris lumbricoides. J. Cell Biol.*, 25:495–516.
22. Silvey, G. E. (1973): Motor control of tailspine rotation of the horseshoe crab, *Limulus polyphemus. J. Exp. Biol.*, 58:599–625.
23. Squire, J. M. (1971): General model for the structure of all myosin-containing filaments. *Nature (Lond.)*, 283:457–462.
24. Squire, J. M. (1973): General model of myosin filament structure. III. Molecular packing arrangements in myosin filaments. *J. Mol. Biol.*, 77:291–323.
25. Stewart, M., Kensler, R. W., and Levine, R. J. C. (1982): The structure of *Limulus* telson muscle thick filaments. *J. Mol. Biol.*, 153:781–790.
26. Szent-Györgyi, A. G., Cohen, C., and Kendrick-Jones, J. (1971): Paramyosin and the filaments of molluscan "catch" muscles. II. Native filaments: Isolation and characterization. *J. Mol. Biol.*, 56:239–258.
27. Taylor, K. A., and Amos, L. A. (1981): A new model for the geometry of the binding of myosin crossbridges to muscle thin filaments. *J. Mol. Biol.*, 147:297–324.
28. Wray, J. S. (1979): Structure of the backbone in myosin filaments of muscle. *Nature (Lond.)*, 277:37–40.
29. Wray, J. S., Vibert, P. J., and Cohen, C. (1974): Cross-bridge arrangements in *Limulus* muscle. *J. Mol. Biol.*, 88:343–348.
30. Wray, J. S., Vibert, P. J., and Cohen, C. (1975): Diversity of cross-bridge arrangements in invertebrate muscles. *Nature (Lond.)*, 257:561–564.

Basic Biology of Muscles: A Comparative
Approach, edited by B. M. Twarog,
R. J. C. Levine, and M. M. Dewey.
Raven Press, New York © 1982.

Structural, Functional, and Chemical Changes in the Contractile Apparatus of *Limulus* Striated Muscle as a Function of Sarcomere Shortening and Tension Development

*M. M. Dewey, *D. Colflesh, *P. Brink, **Shih-fang Fan,
*B. Gaylinn, and *N. Gural

*Department of Anatomical Sciences, Health Sciences Center, State University of
New York, Stony Brook, New York 11794; and **Shanghai Institute of Physiology,
Academia Sinica, Shanghai, China

Since its first scientific description in seventeenth century France (5) the horseshoe crab, *Tachypleus polyphemus*,[1] has intrigued zoologists because of its nearly unique systematic and lonely position in the world of invertebrates. Its restricted global distribution along the eastern coast from Maine to Mexico, on the one hand, and that of the two closely related forms from Japan to India, on the other hand, is of particular evolutionary interest. Near the turn of the last century two prominent zoologists with differing theories both suggested that *Limulus* lies close to the ancestral tree of the vertebrates. Today the "living fossil" is of renewed interest because of its hemolymph (clinical diagnostic value), eyes (photochemoelectrical transduction), and striated muscle (chemomechanical transduction). As the currently held model of vertebrate striated muscle (16,17) was being firmly established over a period of a decade and a half, deVillafranca (8) and deVillafranca and Marschaus (10) described shortening of A-bands in glycerinated muscle fibers induced to shorten by the addition of ATP, Mg^{2+}, and Ca^{2+}. For the forthrightness of this observation we believe George deVillafranca's name need be especially noted. Further, deVillafranca and Leitner (9) isolated paramyosin from *Limulus* striated muscle. Paramyosin at that time was thought to be restricted to annelids and mollusks, but has subsequently been demonstrated to occur in most invertebrate muscles

---

[1]We have used the American horseshoe crab supplied us from Woods Hole Marine Biology Laboratory or collected on the shore of the Long Island Sound in Belle Terre or Mount Sinai, Long Island, New York. Revision in nomenclature now designates this species as *Tachypleus polyphemus*. We will continue to use the name *Limulus* as a common name.

that have been studied. It has been suggested that paramyosin's role in the core structure of thick filaments—where it is found (4,6,21,30)—is to increase thick filament length. There is a good positive correlation between filament length and paramyosin content in invertebrate striated and smooth muscles (22).

Intrigued with deVillafranca's observation and Levine's subsequent confirmation (19), we began collaborative studies with Dr. Rhea J. C. Levine about 15 years ago to analyze structural correlates and, more recently, the functional parameters involved in A-band and thick filament shortening in *Limulus* striated muscle. In this chapter we describe a "laboratory notebook" illustrating approaches to the analysis of the discrepancies between vertebrate and an invertebrate striated muscle in terms of the descriptions of sarcomeric changes during shortening. We describe also analytic approaches at the molecular level.

STRUCTURAL AND IMMUNOCYTOCHEMICAL CHANGES IN SARCOMERES OF *LIMULUS* STRIATED MUSCLE AT VARIOUS SARCOMERE LENGTHS

Using phase optics, fluorescent microscopy of immunocytochemically stained myofibrils (anti-paramyosin and anti-myosin), and transmission electron microscopy of sectioned material, we demonstrated that there is a positive linear relationship between A-band length and sarcomere length in sarcomeres from ~4.0 to ~12.0 μm. The A-band increases from ~3.0 to ~7.5 μm. In addition, at the suggestion of Professor J. K. Blasie, we have calculated Fourier transformations from relative intensities of layer lines in optical transforms of muscle fibers that had been glycerinated or chemically fixed at various sarcomere lengths. Two additional conditions were analyzed: glycerinated fibers were shortened incrementally with optical transforms obtained at each step, or living fibers were electrically stimulated and allowed to shorten to various lengths where they were not allowed to shorten further, and all possible transforms calculated at each step (11,13). Phases that gave transforms matching interference micrographs of fibers in the various conditions were selected. Again these data showed a linear increase in A-band length with increase in sarcomere length. An additional aspect of these data was that in reconstructions of sarcomeres above ~7.1 μm shoulders developed on either side of the A-band. These shoulders increased with increasing sarcomere length. We interpreted this to be the result of skewing of thick filaments relative to each other in the A-band as the sarcomere lengthened.

In further analysis of electron micrographs of longitudinally sectioned muscle, total A-band lengths were measured as well as individual thick filament lengths (12). We were cognizant of the difficulty in accurately measuring thick filament length in sectioned material, but statistical analysis of the data showed that as the sarcomere increased in length, the A-band increased linearly in length. The individual thick filaments also increased in length from 2.9 to about 4.9 μm in sarcomeres with lengths from 4.0 to about 7.2 μm. In muscles with sarcomeres ranging from about 7.2 to 11.0 μm, thick filaments remained constant in length. Both the optical diffraction data and the electron microscopic data were consistent with the view that in *Limulus* muscle as the sarcomere shortens from long lengths to rest length, thick filaments realign in the A-band, and as the sarcomere shortens below rest length, 7.0 μm, both the thick filaments and the A-bands shorten.

We described an interesting correlation with these structural changes using fluorescene labeled anti-*Limulus* myosin and anti-*Limulus* paramyosin (21). Bundles of muscle fibers were glycerinated at various sarcomere lengths and fibrils were isolated and stained. At long sarcomere lengths (from 7 to 11 μm) anti-myosin stained the entire A-band whereas anti-paramyosin stained only the lateral margins of the A-band. In short sarcomeres (from 3 to 7 μm), the staining pattern was reversed. Anti-myosin stained only the lateral margins of the A-band and anti-paramyosin stained the entire A-band. These studies as stated above confirmed A-band shortening and were interpreted in the conventional manner (24,25). We suggested that, in long sarcomeres, myosin heads were arrayed on the surface of the thick filament and accessible to anti-myosin, but sterically hindered anti-paramyosin binding except at the ends of the thick filaments where paramyosin might extend beyond myosin in the tapered ends of filaments or myosin heads might be lifted off the thick filaments interacting with actin in the region of overlap, thus exposing the paramyosin core. At short sarcomere lengths, we argued that actin-myosin interaction leads to steric hinderance or covering of the antigenic site in the binding of anti-myosin and that this same interaction lifted myosin heads off the thick filament exposing the core paramyosin for the binding of anti-paramyosin. Some of us believe we have never adequately explained the lateral or marginal binding of anti-myosin in the A-bands of fibrils with short sarcomeres.

Although great skepticism was expressed over these results, we and others generated at least three models of how thick filaments could appear to decrease in length during sarcomere shortening. The three mechanisms were envisioned to be:

1. Actual shortening by rearrangement of constituent thick filament proteins, i.e., subfilaments sliding past each other within the thick filaments.

2. Alignment and skewing of consistently short thick filaments with end-to-end interaction at long sarcomeres.

3. Dissolving of thick filaments at short sarcomere lengths.

One might have assumed that an easy solution to the problem would have been an X-ray diffraction analysis of the muscle at various sarcomere lengths. Such studies, however, were disappointing to us because no changes were observed at short sarcomere lengths in the myosin or paramyosin meridional reflections (23,32). We reported similar findings from optical diffraction of isolated, negatively stained long and short thick filaments (20). Changes might have been predicted if the filaments shorten in a nonuniform way. It is of interest, however, that Wray et al. (32) reported that there was no significant variation in the intensity of the decorated actin layer lines at various sarcomere lengths. This would suggest that the degree of overlap of thick and thin filaments remains rather constant over various sarcomere lengths.

Several approaches were used in an attempt to distinguish between the three alternatives.

FILAMENT DIAMETER AS A FUNCTION OF SARCOMERE LENGTH

Bundles of muscle were fixed at various lengths, and sarcomere lengths were determined by optical diffraction. The fibers were then embedded and transversely sectioned. Diameters of thick filaments in A-I overlap regions were measured (13). The filaments are elliptical in profile and both dimensions were measured on tracings of their profiles made from projected images ($\sim 2 \times 10^6 \times$) of negatives. It was also determined that the elliptical profiles were randomly oriented within the sarcomere. As can be seen in Table 1 and Fig. 1, thick filament diameter remains constant at sarcomere lengths in which thick filament length does not change (~ 7.0 to 9.4 μm). A highly significant increase in diameter occurs at those sarcomere lengths in which the filament shortens (note the standard deviations). This would suggest that the mass of the thick filament is not decreased during shortening but rather structural rearrangements within the filament occur, leading to shortening.

ISOLATION OF THICK FILAMENTS FROM MUSCLE FIBERS WITH DIFFERENT SARCOMERE LENGTHS

Thick filaments were isolated from glycerinated muscle with long, intermediate, and short sarcomeres and living muscle treated with EGTA or living muscle electrically stimulated or shortened in solutions containing high levels of potassium (450 mM) (10). As can be seen in Table 2, thick filaments isolated from muscle with long sarcomeres, either glycerinated or living, are long (~ 4.0 to 4.4 μm), whereas filaments isolated from muscle bundles, either from glycerinated muscle or living muscle electrically or K^+ stimulated, are short (~ 2.9 μm). Special attention should be given to the magnitude of standard deviations. They do not overlap nor were the populations of isolated filaments modal or skewed. These observations were interpreted to exclude the second model of shortening: alignment, skewing, end-to-end interaction of consistently short thick filaments. The first and third models were not excluded by these data.

TABLE 1. *Thick filament dimensions at different sarcomere lengths*[a]

Sarcomere length (μm)	Thick filament diameter (nm)	Thick filament cross-sectional area (nm²)	No. of observations
9.4	25.1 ± 2.5	398 ± 6.1	746
7.4	24.6 ± 1.6	425 ± 2.1	111
6.4	26.6 ± 3.5	459 ± 7.7	1793
5.9	33.0 ± 3.2	673 ± 8.8	526
4.4	36.6 ± 4.6	841 ± 14.4	543

[a]Mean ± SD.

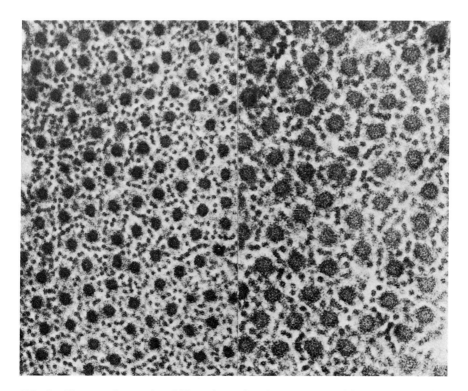

FIG. 1. Electron micrographs of AI overlap regions in sarcomeres of *Limulus* muscle. Note both were taken at the same magnification (magnification printed by the microscope) and contact printed. Micrograph on *left* was taken from a muscle fiber with sarcomere lengths of 9.4 μm (determined by optical diffraction). Micrograph on *right* is from a fiber with sarcomeres of 4.4 μm. Note differences in filament diameter. Fiber with long sarcomere *(left)* was fixed in a relaxed state. Fiber with short sarcomere *(right)* was fixed while being stimulated with high K⁺. × 90,000.

MASS DETERMINATION OF ISOLATED THICK FILAMENTS

We have expended considerable effort the past years attempting to isolate thick filaments from *Limulus* striated muscle, which are increasingly pure of actin filaments. Using thoroughly skinned, living, extensively relaxed muscle bundles and glycerol gradients, we now can isolate thick filaments 85% free of actin as determined by quantitating SDS gels of the isolated thick filaments. We have used these filaments to make mass measurements similar to those described by Reedy et al. (26).

We have measured the mass of isolated thick and thin filaments using the dedicated STEM Biotechnology Resource at Brookhaven National Laboratory in collaboration with Drs. J. Wall and B. Panessa-Warren. The mass is measured by comparing the electron scattering signal per unit length of unstained filaments with

TABLE 2. *Relationship between sarcomere, A-band, and thick filament lengths under various conditions[a]*

	Sarcomere length (μm)	A-band width (μm)	Isolated thick filament length (μm)
	8.1 ± 1.2 (230)	4.5 ± 1.9 (230)	4.4 ± 0.5 (95)
	6.4 ± 1.0 (208)	3.4 ± 0.6 (208)	2.9 ± 0.5 (90)
	7.7 ± 1.2 (273)	4.0 ± 0.6 (273)	4.0 ± 0.7 (127)
Living (stimulated) electrically or high K^+	5.0[b]		2.9 ± 0.3 (100)

[a]Mean ± SD (number of observations in parentheses).
[b]Determined by laser diffraction.

that from tobacco mosaic virus particles in the same image. Our results are most preliminary. Thick filaments were isolated from muscle bundles with long sarcomeres and were long (3.9 to 4.3 μm). Their mass was determined to be 16,154 daltons/Å with a standard deviation (SD) of 964 daltons/Å, n = 18. Filaments isolated long were treated with calcium and ATP to shorten them (2.9 to 3.2 μm). Mass of these shortened thick filaments was determined to be 20,374 daltons/Å (SD = 6571, n = 75). These data suggest that the total mass of a long thick filament is in the range of 600 Mdal and does not change as the filament shortens. If so, this would point to the model of molecular rearrangement for shortening as opposed to a dissolving model. As yet, however, the data are too preliminary as indicated by the high standard deviation of the shortened filaments. This deviation may be due to nonuniform shortening along the filament. Additional position-determined mass measurements need to be performed to determine whether such nonuniformity of shortening exists. We feel that it is premature to calculate the number of myosin heads per crown in either long or short filaments from these data since we believe our original estimate (22) of myosin heavy chain to paramyosin ratio is inaccurate. Recent SDS gel work indicates that the paramyosin content probably was overestimated originally.

SARCOMERIC AND FILAMENT STRUCTURE IN RELATION TO THE LENGTH TENSION CURVE

Figure 2 illustrates the relationship between A-band length, thick filament length and diameter, and I-band length to the length tension curve of *Limulus* striated muscle (31). Figure 3 graphically illustrates the gross structural changes in the sarcomere at various lengths. Several points are of interest. As the sarcomere shortens from ~10 μm to l_0 (~7 μm), the total amount (counting all thick and thin filaments per sarcomere) of overlap of thick and thin filaments increases and probably reaches a maximum at sarcomere lengths slightly greater than l_0. Because of the realignment of the thick filaments during this amount of shortening, the apparent I-band length does not appear to change significantly. The increase in degree of overall overlap is reflected in the rising arm (increased tension development) of the

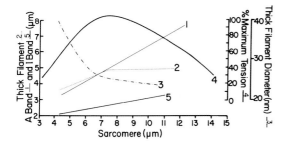

FIG. 2. Length tension curve of *Limulus* striated muscle in relation to structural changes in the sarcomere. (Data from refs. 11–14, 21, and 31; 1 is in part data from Fourier reconstructions of optical diffraction patterns of glycerinated; glycerinated and incrementally shortened with ATP, Ca^{2+}, and Mg^{2+}; fresh; fixed; and living muscle bundles.)

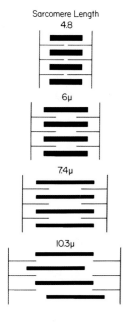

FIG. 3. Schematic diagram of structural changes that occur in sarcomeres at various lengths.

length tension curve. It is also important to note that 90% of maximum tension is developed at a sarcomere of ~10 μm. Based on this, one might not expect to see a significant change in the decorated actin layer lines in X-ray diagrams at sarcomere lengths 10 μm and below. At sarcomere lengths below l_0, thick filaments shorten and increase in diameter. This is associated with a downward slope of the length tension curve. The fall in tension might be due to steric hindrance of myosin bridges as the thick filaments increase in diameter. However, also note that the relative degree of overlap of thick and thin filaments does not change as the sarcomere shortens below l_0.[2] This suggests to us that there are two machines operative in *Limulus* striated muscle: sliding filaments and shortening thick filaments. The short-

---

[2]This has recently been confirmed by analysis of videotapes of glycerinated sarcomeres during shortening induced by treatment with ATP, Mg^{2+}, and Ca^{2+} (M. M. Dewey and D. Colflesh, *unpublished data*).

ening thick filaments must generate a force equal to that generated by thick and thin filaments near l_0. The decrease in tension generation of the thick filament shortening at very short lengths must be equal to the fall in tension due to steric hindrance between thick and thin filaments at sarcomeres below l_0. Alternatively, below l_0, the interaction between thick and thin filaments may go into a kind of "catch" state and active tension is generated solely by thick filament shortening. This process might be less efficient in energetic terms. Critical mechanical studies have not yet been performed to determine whether two machines are present. Is thick filament shortening tension generating?

REQUIREMENTS FOR *IN VITRO* SHORTENING OF ISOLATED THICK FILAMENTS

Thick filaments for these experiments (13) were isolated from briefly glycerinated muscle bundles with long sarcomeres, as determined by optical diffraction. Activation was minimized by doing the isolation in a relaxing solution containing 5 mM ATP and 1 mM EGTA. The effect of Ca^{2+} was determined by adding Ca^{2+} to suspensions of long thick filaments. Lengths of thick filaments were determined from negatively stained preparations by electron microscopic observations. Only filaments with a clear bare zone and tapered ends were measured. By varying the Ca^{2+} concentration while maintaining constant concentrations of EGTA, ATP, and Mg^{2+}, a calcium activation curve was determined (Fig. 4). The activation curve is sigmoidal with the maximum rate of shortening at a EGTA/Ca^{2+} ratio of 1 (estimated pCa of 5.5). It is of particular interest to note that intermediate Ca^{2+} concentrations produced filaments of intermediate lengths. This would suggest that there are multiple binding sites for Ca^{2+} and that Ca^{2+} is not simply "triggering" filament shortening. That Mg^{2+} in the presence of ATP does not shorten the filaments is seen at high EGTA/Ca^{2+} ratios since 2 mM MgCl was always present. However, Mg^{2+} appears to be essential not only for actomyosin enzymatic activity but also for filament shortening since filaments did not shorten in the presence of 5 mM EGTA with Ca^{2+} and ATP present in activating concentrations.

An ATP activation curve was similarly determined by keeping the EGTA/Ca^{2+} ratio constant and varying the ATP concentration. Concentrations of 1 mM ATP and higher gave maximal shortening. Further, we demonstrated that ATP hydrolysis was essential for filament shortening by employing adenyl imidodiphosphate (AMP-

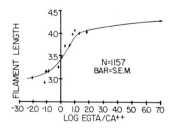

FIG. 4. Lengths of isolated thick filament determined by negative stain as a function of the EGTA/Ca^{2+} ratio.

PNP), a nonhydrolysable ATP analog (3). Long thick filaments treated with AMP-PNP and sufficient Ca^{2+} did not shorten.

Thus, it seems that the requirements for isolated thick filament shortening include the presence of Ca^{2+} (pCa^{2+} ~5 to 6), Mg^{2+}, and ATP (Figs. 5–8). These results are consistent with those reported for "physiological" levels for shortening and

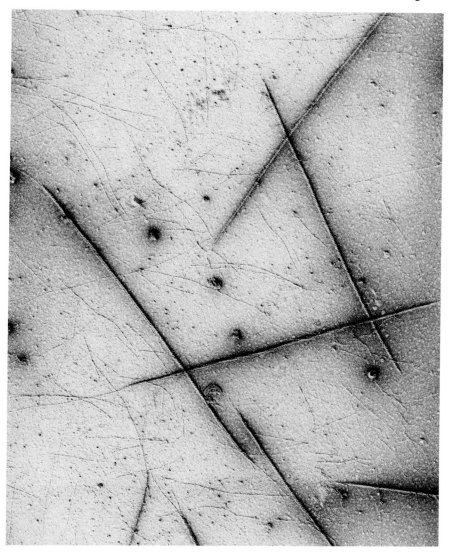

FIG. 5. Electron micrograph of isolated thick filaments isolated in relaxing solution from muscle bundles with long sarcomeres (~8.5 μm). Filaments have been negatively stained and shadowed (angle) with platinum (1 nm, 32 degree angle). Filaments, 4.2 μm in length.

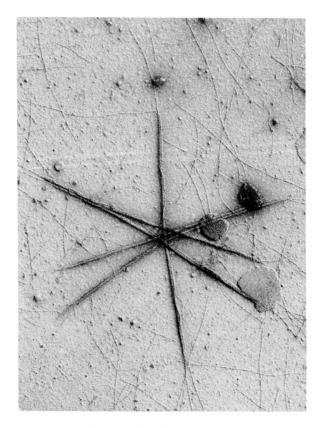

FIG. 6. Electron micrograph of isolated thick filaments from preparation in Fig. 5. This aliquot of filaments had been treated with EGTA/Ca^{2+} = 0.7 in the presence of ATP. Filaments, 2.2 μm in length negatively staining and shadowed with platinum (1.0 nm, 32 degree angle).

tension development in other muscles (15,18,29). Further, purified thick filaments shorten so that actin is not essential for shortening.

Since ATP hydrolysis appeared necessary for shortening, we hypothesized that phosphorylation of constituent filament protein or proteins was instrumental in filament shortening. We then attempted to relengthen shortened filaments by treating with a nonspecific alkaline phosphatase (3). Filaments from living muscle were isolated long, 4.3 ± 0.3 μm (n = 150) and shortened by adding Ca^{2+}, Mg^{2+}, and ATP, 3.4 ± 0.6 μm (n = 123) with a statistical significance $p < 0.001$. The shortened filaments were then treated with alkaline phosphatase (Worthington) and the filaments relengthened, 3.7 μm ± 0.4 (n = 88). The relengthening was statistically significant, $p < 0.001$, when compared to control filaments. Shortening and relengthening could be repeated several times on the same preparation. Since the filaments were pelleted between each cycle of shortening and relengthening, we felt that this suggests that the filaments did not shorten by dissolution. The

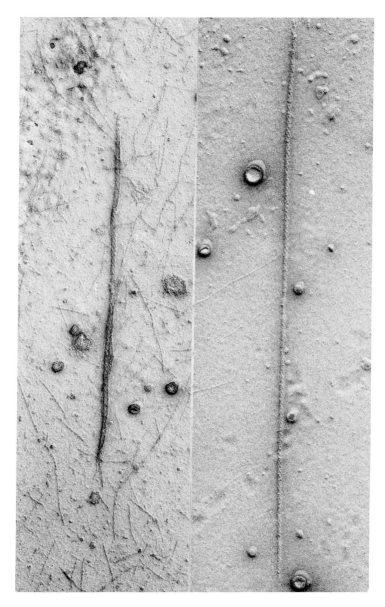

FIG. 7. Electron micrographs at higher magnification of thick filaments taken from preparations in Figs. 5 and 6. It is not clear why in negatively stained preparations of short thick filaments the increased diameter is not as apparent as in fixed and sectioned material. In fact, optical diffraction patterns of these filaments do show a diameter increase in the short filament. Long filament, 4.8 μm in length. Short filament, 2.8 μm in length. Negatively stained shadowed with platinum (1.0 nm, 32 degree angle).

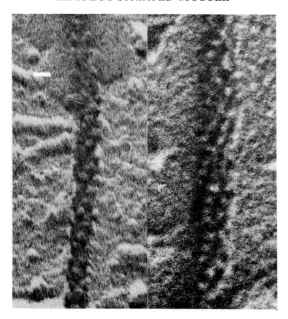

FIG. 8. Electron micrographs at higher magnification of long *(left)* and shortened thick filaments. Negatively stained and shadow casted with platinum. Note arrangement of myosin heads and filament diameters. ×218,000.

dissolved myosin and paramyosin from shortened filaments would have been discarded with the supernatant.

These observations led us to look at phosphorylation and dephosphorylation of contractile proteins in *Limulus* muscle. At the onset we believed we would see, at a minimum, phosphorylation of myosin light chains and paramyosin. As often happens, our predictions were somewhat overzealous.

PHOSPHORYLATION OF *LIMULUS* CONTRACTILE PROTEINS

We have found that glycerinated muscle retains a complex system for phosphorylation and dephosphorylation of several proteins that is controlled by calcium and may be involved in thick filament shortening as well as myosin ATPase activity.

Radioactive γ-^{32}P-ATP was used to identify proteins phosphorylated in glycerinated muscle under various conditions. Muscle bundles were soaked in a relaxing solution (100 mM KCl, 5 mM MgCl, 5 mM Tris, 2 mM ATP, 2 mM EGTA, pH 7.4) to remove the glycerol. These bundles were then either allowed to shorten in ^{32}P-labeled activating solution (relaxing solution containing 4 mM EGTA, 8 mM calcium) or soaked in ^{32}P-labeled relaxing solution. The muscles were then quickly frozen, denatured, and homogenized, electrophoresed on SDS polyacrylamide gels and autoradiographed.

Sellers has reported (27) that isolated *Limulus* myosin can be phosphorylated on two of its three light chains by a calcium dependent kinase. Figure 9 demonstrates

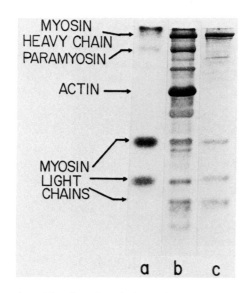

MYOSIN
HEAVY CHAIN
PARAMYOSIN

ACTIN →

MYOSIN
LIGHT
CHAINS

a   b   c

FIG. 9. SDS slab gel electrophoresis, 10 to 22½% polyacrylamide gradient. Columns *a* and *b* show respectively, the radiograph and corresponding Coomassie blue staining from Ca²⁺ activated, ³²P-labeled, glycerinated *Limulus* muscle. Column *c* shows the stained gel of column purified *Limulus* myosin.

that this phosphorylation of myosin light chains also occurs when glycerinated muscle bundles shorten.

When glycerinated muscle bundles were incubated in labeled relaxing solution in the absence of calcium, the myosin light chains were not phosphorylated, but other proteins were (see Fig. 10, column c). The proteins that incorporate phosphate preferentially in the absence of calcium include one of approximately 100 kdal chain weight and one of about 35 kdal chain weight. The 100 kdal band seemed likely to be paramyosin as this protein has been found to be phosphorylated in molluscan muscle protein preparations (1,7). Closer examination (Fig. 11) shows that on lower percentage gels the 100 kdal band does not co-electrophorese with purified paramyosin. The mobility of this band phosphorylated or dephosphorylated remains different from paramyosin, and it does not co-purify with paramyosin.

When muscle that has been ³²P-labeled in the absence of calcium is then allowed to shorten in a solution containing calcium and γ-³²P-ATP, myosin light chains are phosphorylated, the 100 kdal band remains phosphorylated, and the 35 kdal band is specifically dephosphorylated (Fig. 10, column d).

To examine how these phosphorylations and dephosphorylations might relate to thick filament shortening, we examined samples where the A-bands had already been shortened before the muscle was allowed to shorten in length. These samples were prepared by potassium contracting living muscle or ATP Mg²⁺ Ca²⁺ activating glycerinated muscle that was held isometrically. This produced muscle with long sarcomeres and short A-bands, which remained short during glycerination or short A-bands with long sarcomeres if glycerinated muscle was held isometric during activation (Fig. 12). The ³²P-labeling of this kind of glycerinated muscle is shown in Fig. 10, columns b', c', and d' (as compared to b, c, and d for muscle not potassium contracted). In the short A-band muscle, the 35 kdal band does not

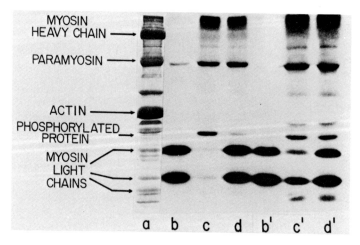

FIG. 10. ³²P-labeled glycerinated *Limulus* muscle, SDS slab gel electrophoresis, and autoradiography on a 10 to 18% polyacrylamide gradient. Column *a* shows the Coomassie blue staining pattern of this muscle that was radiographed to produce column *b*. Columns *b* to *d* are radiographs of relaxed long sarcomere, long A-band muscle samples. Columns *b'* to *d'* are radiographs of K⁺-contracted, long sarcomere, short A-band muscle samples. Columns *b* and *b'* were labeled in activating solution, *c* and *c'* were labeled in relax solution, and *d* and *d'* were labeled in activating solution after previous labeling in relax solution (see text for composition of solutions).

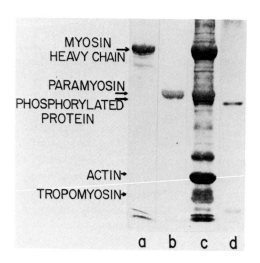

FIG. 11. SDS slab gel electrophoresis, 7.5% polyacrylamide. The various columns contain *a* column purified *Limulus* myosin, *b* purified *Limulus* paramyosin, *c* glycerinated *Limulus* muscle ³²P-labeled in relaxing solution, *d* radiograph of column *c*.

dephosphorylate as it does in the long A-band muscle, and thus this dephosphorylation may correlate with thick filament shortening. If so, it would suggest the hypothesis that thick filament shortening and myosin activation are controlled by independent phosphorylation systems and that a specific dephosphorylation is associated with thick filament shortening.

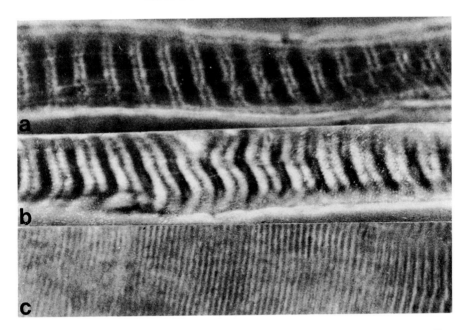

FIG. 12. **a:** A single muscle fiber glycerinated at a long sarcomere length. **b:** The same fiber held isometrically and treated with ATP, Ca^{2+}, and Mg^{2+}. Note shortened A-band. **c:** The same fiber relaxed and allowed to shorten isotonically. Dark band is A-band, as determined on videotapes. All ×320.

DONNAN POTENTIALS IN *LIMULUS* STRIATED MUSCLE

The measurement of Donnan potentials in glycerinated muscle fibers is an indicator of fixed charge density under appropriate ionic conditions (2,28). Here we wish to thank Professor G. F. Elliott for suggesting we use this method in our analysis of *Limulus* muscle.

Glycerinated muscle bundles (20 to 100 μm in diameter) were used. Potentials were measured by placement of 3 M KCl microelectrodes (5 to 20 MΩ) into the muscle matrix and measuring the potential differences between it and a silver-silver chloride reference electrode in the bathing media surrounding a muscle bundle. The muscle fiber and electrodes were visualized by use of a Nikon® inverted microscope (400×). Microelectrodes were easily placed in known regions of the sarcomere (Fig. 13). Increasing the ionic strength of the bathing solution masked the Donnan potential as expected (Fig. 14).

By changing the pH of the bathing media, the isoelectric point of the A-band could be measured. In all cases the solutions in which the potential was measured contained 50 mM KCl, 10 mM buffer, and 50 μm EGTA. The pH was adjusted by addition of HCl or KOH. The amplitude of the potential, as expected, varied with pH. At pH 7.4 in Tris buffer A-band potentials of long sarcomeres (8.5 to 10 μm) ranged from negative 3 to 10 mV (Fig. 15). The fiber was then activated and allowed to shorten. Activation was accomplished by exchanging the bathing solution

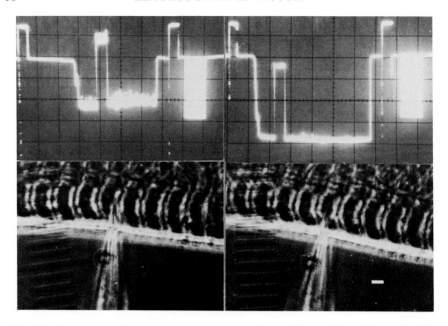

FIG. 13. Donnan potential measurements made in 50 mM KCl and 10 mM Tris at pH 7. The positive deflections are 1.0 nA pulse passed through the microelectrode to measure electrode resistance when it was out of and within the muscle matrix. The solid negative diffractions are 10 mV calibration pulses. The *lower frames* show the placement of the microelectrode within the muscle. The first is in I-band, the second is in A-band.

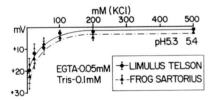

FIG. 14. The Donnan potential was measured at pH = 5.3 to 5.4, with increasing ionic strength to demonstrate concentration's effect on the potential measurement. As predicted by Donnan potential theory, high ionic strength masks the potentials.

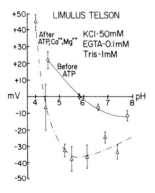

FIG. 15. Donnan potentials measured in a single bundle with long and short sarcomeres of *Limulus* muscle in 50 mM KCl at various pH values using Tris buffers.

with one containing 5 mM ATP, 1 mM MgCl$_2$, and 1 mM CaCl$_2$. Following shortening the fiber was thoroughly washed in buffer without ATP, Mg^{2+}, or Ca^{2+}. Following activation and shortening (sarcomeres <4 μm), the A-band potential ranged from negative 20 to 40 mV. Of equal importance the isoelectric point of the A-band shifted from ∼ pH 6 to pH 5. Similar experiments were performed with glycerinated sartorius muscle of frog *(Rana pipiens)* (Fig. 16). Shortening of sarcomeres in the sartorius resulted in a slight increase in the A-band potential but little or no change in the isoelectric point. Some bundles of live *Limulus* muscle were exposed to a skinning solution (100 mM KCl, 1 mM EGTA, 10 mM Tris, and 1% Triton) for 24 hr. Measurements of A-band potentials were similar to those of glycerinated fibers. Thus, exposure to glycerol did not alter the Donnan potential.

The similar experiments with *Limulus* muscle were done with citrate as the buffer and showed an alkaline shift in pH following shortening (Fig. 17) pH 5.5 (long sarcomeres) to pH 6.6 (shortened sarcomeres). Acetate buffer gave results similar to citrate buffer, whereas results in glycine buffer were like those in Tris buffer. When muscle activated in Tris buffer (isoelectric point pH 5.0) was then placed in acetate buffer, the isoelectric point shifted to pH 6.5. The fact that in *Limulus* muscle cationic (Tris) or zwitterionic (glycine) molecules shift the isoelectric point to a more acid range and anionic buffers cause a shift to a more alkaline point demonstrates that the buffers bind with selected fixed charge groups depending on their specific characteristics. Presumably, Tris is masking the more alkaline charge

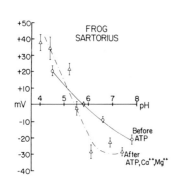

FIG. 16. Donnan potentials measured in long and short sarcomeres of a single bundle of frog sartorius muscle in 50 mM KCl at various pH values using Tris buffer.

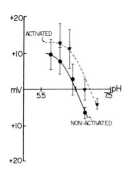

FIG. 17. *Limulus* striated muscle in 50 mM KCl, 10 mM K$^+$ citrate. Donnan potentials measured in a single bundle with long and shortened sarcomeres in citrate buffer. Note alkaline shift in isoelectric point. Alkaline shift is more pronounced.

groups allowing acid groups to dominate and hence causing the more acid isoelectric point. Additionally, the amplitude of the potentials for shortened *Limulus* muscles (Fig. 15) was greater for short sarcomere than in frog sartorius. Frog muscle showed no significant increase in potential with shortening. Thus, one conclusion might be that *Limulus* bundles, on shortening, acquire or expose more fixed charge groups observed as larger potentials with a shifted isoelectric point.

The alkaline shift observed in anionic buffers probably is an unmasking of fixed charge groups that are weakly acid because removal of strong acid groups would cause the amplitude of the potential to decrease. The potential would remain unchanged if only a small number of groups were involved, an interpretation consistent with the actual acetate data.

The data indicate a possible change in fixed or absorbed charge density in *Limulus* striated muscle with shortening of thick filaments. It is tempting to speculate that the changes in potential are caused by changes in charge groups that might ultimately be responsible for shortening thick filaments. In both the frog and *Limulus* striated muscle, Tris and glycine acted similarly. Citrate and acetate buffers caused alkaline shifts in shortened striated muscle of frog no greater than 0.2 pH units whereas in *Limulus* the shift was larger than 0.5 pH units. The potential changes seen in activated versus nonactivated fibers were presumably due to changes on thick filaments. To test this further, thick filaments were isolated as described and dialyzed against acetate buffer. After 24 hr the dialysate was pelleted and Donnan potential measurements were made. The values of the Donnan potentials for preshortened thick filaments were -13 mV $\pm$ 1.3 mV (n = 15) and for long thick filaments were -5.1 mV $\pm$ 1.4 mV (n = 11), pH = 7. The results are consistent with the data for intact sarcomeres and show a much greater increase in potential amplitude for shortened filaments. In fact, the magnitude of the Donnan potential in isolated, shortened filaments in acetate buffer was greater than potentials measured in shortened A-bands of fibers in acetate buffer. But this effect may be concentration related.

Further analysis of the Donnan potentials by binding reactive groups with blocking reagents may provide evidence as to whether charge changes are involved in filament shortening.

CONCLUSIONS

We believe that adequate data are now available to conclude that thick filaments in *Limulus* striated muscle shorten below l_0. Further, the data are strongly suggestive that the muscle works by both a sliding filament system and a thick filament shortening system. The former occurs above rest length and the latter occurs below rest length. The thick filament shortening system is correlated with phosphorylation changes and charge accumulation in the thick filaments. The shortening of thick filaments is most likely due to rearrangement of constituent proteins of the thick filaments. Such rearrangement, however, apparently does not alter the axial repeat of myosin heads along the filament.

ACKNOWLEDGMENTS

We wish to thank Dr. Gerald Elliott, Dr. Susan Gilbert, and Dr. Barry Millman for many useful discussions. Special thanks go to Eileen Petite for her help in preparing the manuscript. This work was supported by U.S.P.H.S. grants AM30053 and GM26392.

REFERENCES

1. Achazi, R. K. (1979): Phosphorylation of molluscan paramyosin. *Pflugers Arch.*, 379:197–201.
2. Bartels, E. M., and Elliott, G. F. (1981): Donnan potentials from the A- and I-band of skeletal muscle, relaxed and in rigor. *J. Physiol. (Lond.)*, 317:85P–87P.
3. Brann, L., Dewey, M. M., Baldwin, E. A., Brink, P., and Walcott, B. (1979): Requirements for *in vitro* shortening and lengthening of isolated thick filaments of *Limulus* striated muscle. *Nature*, 279:256–257.
4. Bullard, B., Luke, B., and Winkelman, L. (1973): The paramyosin of insect flight muscle. *J. Mol. Biol.*, 75:359–367.
5. Clusi, C. (Charles de L'Ecluse) (1605): Cancer mollucanus. Capitulo XIV. In: *Exohcorum Libi VI*, edited by Chez B. Leyde, and A. Elzeviro, pp. 128–129. Antwerp.
6. Cohen, C., Szent-Györgyi, A. G., and Kendrick-Jones, J. (1971): Paramyosin and the filaments of molluscan "catch" muscles. I. Paramyosin: Structure and assembly. *J. Mol. Biol.*, 56:223–237.
7. Cooley, L. B., Johnson, W. H., and Krause, S. (1979): Phosphorylation of paramyosin and its possible role in the catch mechanism. *J. Biol. Chem.*, 254:2195–2198.
8. deVillafranca, G. W. (1961): The A- and I-band lengths in stretched or contracted horseshoe crab skeletal muscle. *J. Ultrastruct. Res.*, 5:109–115.
9. deVillafranca, G. W., and Leitner, V. E. (1967): Contractile proteins from horseshoe crab muscle. *J. Gen. Physiol.*, 50:2495.
10. deVillafranca, G. W., and Marschaus, C. (1963): Contraction of the A-band. *J. Ultrastruct. Res.*, 9:156–165.
11. Dewey, M. M., Blasie, J. K., Levine, R. J. C., and Colflesh, D. E. (1972): Changes in A-band structure during shortening of a paramyosin containing striated muscle. *Biophys. Soc. Annu. Meet.*, 12:82a.
12. Dewey, M. M., Levine, R. J. C., and Colflesh, D. E. (1973): Structure of *Limulus* striated muscle. The contractile apparatus at various sarcomere lengths. *J. Cell Biol.*, 58:574–593.
13. Dewey, M. M., Levine, R. J. C., Colflesh, D., Walcott, B., Brann, L., Baldwin, A., and Brink, P. (1979): Structural changes in thick filaments during sarcomere shortening in *Limulus* striated muscle. In: *Cross-bridge Mechanism in Muscle Contraction*, edited by H. Sugi and G. H. Pollack, pp. 3–22. University of Tokyo Press, Tokyo.
14. Dewey, M. M., Walcott, B., Colflesh, D. E., Terry, H., and Levine, R. J. C. (1977): Changes in thick filament length in *Limulus* striated muscle. *J. Cell Biol.*, 75:366–380.
15. Fabiato, A., and Fabiato, F. (1975): Effects of Mg^{++} on contraction activation of skinned cardiac cells. *J. Physiol. (Lond.)*, 249:497–517.
16. Huxley, A. F., and Neidergerke, R. (1954): Structural changes in muscle during contraction. Interference microscopy of living muscle fibers. *Nature*, 173:971–973.
17. Huxley, H., and Hanson, J. (1954): Changes in the cross-striations of muscle during contraction and stretch and their structural interpretation. *Nature*, 173:973–976.
18. Kushmerick, M. J., and Davies, R. C. (1969): The chemical energetics of muscle contraction. II. The chemistry, efficiency and power of maximally working sartorius muscles. *Proc. R. Soc. Lond.* [Biol.], 174:315–353.
19. Levine, R. J. C. (1966): Intrasarcomeric localization of Limulus myosin B by the direct fluorescent antibody technique. Doctoral dissertation, New York University.
20. Levine, R. J. C., and Dewey, M. M. (1979): Changes in molecular packing during shortening of *Limulus* thick filaments. In: *Motility in Cell Function*, edited by F. A. Pepe, J. Sanger, and V. Nachmias, pp. 341–345. Academic Press, New York.
21. Levine, R. J. C., Dewey, M. M., and deVillafranca, G. W. (1972): Immunohistochemical localization of contractile proteins in *Limulus* striated muscle. *J. Cell Biol.*, 55:221–235.

22. Levine, R. J. C., Elfvin, M., Dewey, M. M., and Walcott, B. (1976): Paramyosin in invertebrate muscles. II. Content in relation to structure and function. *J. Cell Biol.*, 71:273–279.
23. Millman, B. M., Warden, W. J., Colflesh, D. E., and Dewey, M. M. (1972): X-ray diffraction from glycerol-extracted *Limulus* muscle. *Biophys. Soc. Annu. Meet. Fed. Proc.*, 33:1333.
24. Pepe, F. A. (1966): Some aspects of the structural organization of the myofibril as revealed by antibody-staining methods. *J. Cell Biol.*, 28:505–525.
25. Pepe, F. A. (1967): The myosin filament. II. Interaction between myosin and actin filaments observed. *J. Mol. Biol.*, 27:227–236.
26. Reedy, M. K., Leonard, K. R., Freeman, R., and Arad, T. (1981): Thick myofilament mass determination by electron scattering measurements with the scanning transmission electron microscope. *J. Mus. Res. Cell Motil.*, 2:45–64.
27. Sellers, J. R. (1981): Phosphorylation-dependent regulation of *Limulus* myosin. *J. Biol. Chem.*, 256:9274–9278.
28. Stephenson, D. G., Wendt, I. R., and Forrest, Q. G. (1981): Non-uniform ion distributions and electrical potentials in sarcoplasmic regions of skeletal muscle fibers. *Nature*, 289:690–692.
29. Storer, A. C., and Cornish-Bowden, A. (1976): Concentration of Mg-ATP and other ions in solution. *Biochem. J.*, 159:1–5.
30. Szent-Györgyi, A. G., Cohen, C., and Kendrick-Jones, J. (1971): Paramyosin and the filaments of molluscan "catch" muscles. II. Native filaments: Isolation and characterization. *J. Mol. Biol.*, 56:239–258.
31. Walcott, B., and Dewey, M. M. (1980): Length-tension relation in *Limulus* striated muscle. *J. Cell Biol.*, 87:204–208.
32. Wray, J. S., Vibert, P. J., and Cohen, C. (1974): Cross-bridge arrangements in *Limulus* muscle. *J. Mol. Biol.*, 88:823–830.

Basic Biology of Muscles: A Comparative
Approach, edited by B. M. Twarog,
R. J. C. Levine, and M. M. Dewey.
Raven Press, New York © 1982.

Structure of Paramyosin Filaments

Gerald F. Elliott

*Biophysics Group, The Open University Research Unit, Boars Hill,
Oxford, United Kingdom*

It is hardly necessary to say how delighted I am that Arthur Elliott, my long-time friend and former colleague at King's College, London, has developed a model for the paramyosin filament (4), from negative-staining and sectioning studies, that is in broad agreement with the "stack-of-layers" model I put forward from my own electron-microscope and X-ray diffraction studies of oyster adductor muscle (6–9). My conclusions were largely based on the wide variety of band patterns seen in longitudinal sections, which clearly showed a structure that did not have cylindrical symmetry. As I wrote at the time "the band patterns could also be explained by a filament made of a single paramyosin net of thick rods, perpendicular to the net plane, extending through the whole thickness of the filament... such rods might give the structure seen in transverse sections" (8). I discussed this "rod" possibility as an origin for the transverse structure (6) in terms very similar to Arthur Elliott but then preferred a different possible origin for the transverse structure, layers in the plane of the paramyosin net (and, thus, at right angles to the hypothetical rods). This preference arose largely because of some equatorial X-ray evidence (10), which was later shown not to be due to paramyosin but to the actin packing (12). The same X-ray evidence made me then prefer a loose interlayer distance. I have now no doubt that Arthur Elliott is correct in the very tight interlayer packing implied in his model; given this tighter packing, the relative positions of the molecules in the two models are identical.

IMPLICATIONS FOR PARAMYOSIN FUNCTION

Paramyosin is a highly alpha-helical protein, and the Elliott-Elliott structure must be essentially due to the packing of alpha helices. I am intrigued, now as 20 years ago, by the fact that the structure shows two different intermolecular arrangements, in directions in the plane of the Bear-Selby net (1) and at right angles to it. These two arrangements must have molecules staggered in the plane of the net and in register in the plane at right angles to the net. Presumably, this implies different intermolecular bonds in these two directions and perhaps makes it possible that interlayer movement might be much easier in one plane than in the other. Since the time of my thesis (6), I have speculated that paramyosin filaments might be a sort of biological graphite, able to shorten or extend by interlayer slippage. This could be a factor in "catch," and also in the thick-filament shortening observed by Dewey et al. *(this volume)*.

The long, oblique, line along which the Bear-Selby pattern sense changes at the center of the filament, which Bennett and Elliott have noted *(this volume)*, is also interesting in this connection. If the filaments do change length by interlayer slippage, toward the centre of the filament, a diagonal glide plane would be a necessary feature. It was previously pointed out (5) that the side-chain packing is very likely to affect the X-ray pattern, as well as the structure, in paramyosin filaments. Perhaps the particular side-chain arrangement and interaction is responsible for the two different orthogonal planes in the filaments.

In this connection, we should remember that the electron-microscope measurements and the low-angle X-ray patterns of paramyosin show a wide spacing variation in the *a* direction (the spacing across the filament in the plane of the Bear-Selby net, see Fig. 1). Values in the literature range from 19.3 nm (electron microscope, dry) (11) to 42 nm (X-ray, wet) (7). These values may depend both on the species of the animal from which the muscle is taken and on the degree of hydration of the specimen examined, although Bennett and Elliott (2) have recently reexamined this question and have concluded that the species variation is not a major factor, and that when the conditions are the same, the *a* spacing is approximately constant from one muscle species to another. This leaves hydration as the probable major factor in this large *a* spacing variation (Fig. 1) and it would be of great interest to see a study comparing the effect of filament hydration on the *a* spacing and, simultaneously, on the intermolecular spacings in the high-angle pattern, which were observed by Elliott et al. (5).

X-RAY INTENSITY VARIATION WITH HYDRATION AND THE PARAMYOSIN FOURIER SYNTHESIS

As well as the spacing variation that has been mentioned, the low-angle X-ray pattern from paramyosin undergoes profound intensity changes between dried and living muscle. X-ray intensity data from the paramyosin pattern has been collected for over two decades, from muscles native (hydrated) and dried, living and fixed, stained and unstained (1,6,13–15). Two-dimensional Patterson plots of unstained material suggest a centered structure at least as a first approximation and the phases (O or π) may then be determined by a low-angle adaptation of the "heavy-atom" method (7). The resultant Fourier synthesis gives about 1.2 nm resolution in the fiber-axis direction, but only about 30.0 nm resolution across the fiber axis. (In the paramyosin pattern layer, lines up to $l = 50$ are observed, but only the row lines $h = 0$, ± 1, and a very few reflections on $h = \pm 2$, are seen.) The data from the PTA-stained specimens also give values for the phases; to a first approximation the position of the PTA stain is known from electron-microscope observations. The phases thus determined are the same as those given by the hypothetical centered structure (7).

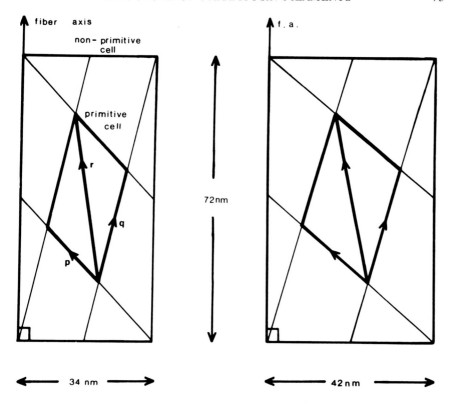

FIG. 1. The paramyosin net in the opaque adductor muscle of the oyster *Crassostrea angulata*. The nonprimitive (Bear-Selby) net is orthogonal, with a fibril axis period of 72.0 ± 0.5 nm and a cross-fibril period *(a)* = 34 ± 2 nm (dry) and 42 ± 2 nm (wet). The vectors *p* and *q* mark the edges of a primitive unit cell. In the direction *r* (= *p* + *q*) the distance between net points is close to 43.5 nm, with threefold symmetry. This might be related to myosin packing on the surface of paramyosin filaments. (Data from Elliott, ref. 6.)

The Fourier synthesis for dry paramyosin is shown in Fig. 2, and for wet, native, paramyosin in Fig. 3. The data for both these figures is for the Portuguese oyster, *Crassostrea angulata* (6). Both these Fouriers show large peaks at the origins of the primitive unit cell; it is these peaks that dominate the X-ray pattern. Similar Fourier syntheses, using extra intensity data, have been calculated for both wet and dry paramyosin filaments and for filaments stained with silver nitrate and PTA in my laboratory and in B. M. Millman's laboratory at Guelph, Canada. Intensity data is available from physiologically active and contracting muscle (14) and it would be easy to calculate a synthesis for this also.

The interpretation of these Fouriers, however, poses a problem. Because of Babinet's principle, it is not clear whether the origin peaks are regions of high electron density, or are holes in an otherwise rather uniform filament core. For this reason, the Fourier maps are not as useful as they might otherwise be in deciding between various tentative models for the molecular packing (2,3,7,8). If the peaks are regions of high density, they might be overlap between molecules, or some

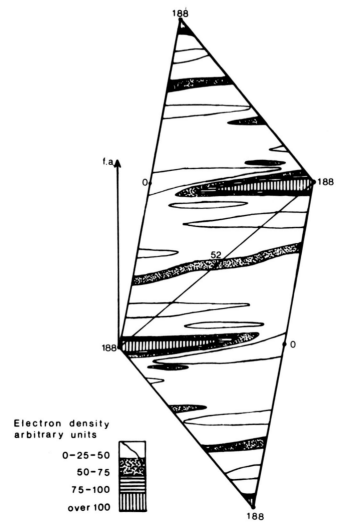

FIG. 2. Fourier synthesis for the electron density in the primitive unit cell of paramyosin, in the same muscle as Fig. 1. The electron-density scale is arbitrary, and has been zeroed on the lowest trough. This is for dry muscles, muscles dried *in situ* (between the shells) in a dessicator.

component that has escaped biochemical detection, or regions of high ion-binding. If they are low density, they might be gaps in the molecular packing (cf., collagen); this is the solution favored by Cohen et al. (3), from electron microscope and biochemical evidence. Against this it does seem strange that gaps should become more pronounced, and their electron contrast greater, as the filament hydrates from dry (see Figs. 2 and 3).

An intensity-matching study with neutrons, collecting intensity data from specimens equilibrated in D_2O/H_2O mixtures of differing ratios, might give enough data

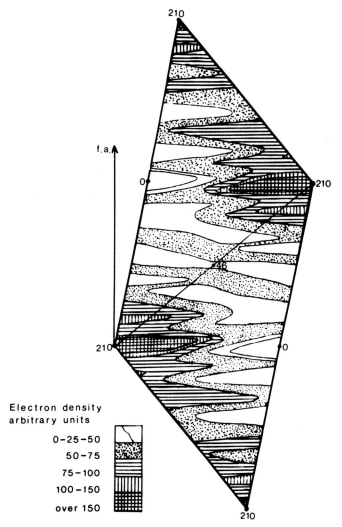

FIG. 3. As for Fig. 2, but for wet muscle. This includes formaldehyde-fixed muscles and living, resting, muscles examined in appropriate Ringers solution, without the addition of electron stains.

to decide between the high-density or low-density alternatives. This information would be of great value in increasing our understanding of the structure and function of these most interesting filaments.

ACKNOWLEDGMENTS

My interest in the structure of paramyosin filaments has been stimulated over a long period by many discussions with Arthur Elliott, J. Lowy, B. M. Millman, and the late Jean Hanson, and, more recently, by interaction with M. M. Dewey and members of his laboratory.

REFERENCES

1. Bear, R. S., and Selby, C. C. (1956): The structure of paramyosin fibrils according to X-ray diffraction. *J. Biophys. Biochem. Cytol.*, 2:56–69.
2. Bennett, P. M., and Elliott, A. (1981): The structure of the paramyosin core in molluscan thick filaments. *J. Muscle Res.*, 2:65–81.
3. Cohen, C., Szent-Gyorgyi, A. G., and Kendrick Jones, J. (1971): Paramyosin and the filaments of molluscan "catch" muscles. *J. Mol. Biol.*, 56:223–237.
4. Elliott, A. (1979): The structure of molluscan thick filaments: A common origin for diverse appearances. *J. Mol. Biol.*, 132:323–341.
5. Elliott, A., Lowy, J., Parry, D. A. D., and Vibert, P. G. (1968): The puzzle of the coiled coils in the α-protein paramyosin. *Nature*, 218:656–659.
6. Elliott, G. F. (1960): *Electron Microscope and X-ray Diffraction Studies of Invertebrate Muscle Fibres.* PhD thesis, University of London.
7. Elliott, G. F. (1963): A two dimensional Fourier synthesis of paramyosin from low-angle data. *Acta Cryst.*, 16:A81.
8. Elliott, G. F. (1964): Electron microscope studies of the structure of the filaments in the opaque adductor muscle of the oyster. *J. Mol. Biol.*, 10:89–104.
9. Elliott, G. F. (1964): X-ray diffraction studies on striated and smooth muscles. *Proc. R. Soc. Lond. [Biol.]*, 160:467–472.
10. Elliott, G. F., and Lowy, J. (1961): Low-angle X-ray reflections from living molluscan muscles. *J. Mol. Biol.*, 3:41–46.
11. Hall, C. E., Jakus, M. A., and Schmitt, F. O. (1945): The structure of certain muscle fibrils as revealed by the use of electron stains. *J. Appl. Physiol.*, 16:459–465.
12. Lowy, J., and Vibert, P. J. (1968): The structure and organization of actin in a molluscan smooth muscle. *Nature*, 215:1254–1255.
13. Miller, A. (1968): A short periodicity in the thick filaments of ABRM. *J. Mol. Biol.*, 32:687–688.
14. Millman, B. M., and Elliott, G. F. (1973): An X-ray diffraction study of contracting molluscan smooth muscle. *Biophys. J.*, 12:1405–1414.
15. Shaw, R. K. (1977): *X-ray Diffraction Studies of the ABRM.* M.Sc thesis, University of Guelph.

Basic Biology of Muscles: A Comparative
Approach, edited by B. M. Twarog,
R. J. C. Levine, and M. M. Dewey.
Raven Press, New York © 1982.

Cross-Bridge Properties in the Rigor State

R. J. Podolsky, G. R. S. Naylor, and T. Arata

*Laboratory of Physical Biology, National Institute of Arthritis, Diabetes, and Digestive
and Kidney Diseases, National Institutes of Health, Bethesda, Maryland 20205*

Considerable progress has been made in bringing together the mechanical, biochemical, and structural approaches to the study of the contraction mechanism in
muscle in the 11 years since the last symposium on this topic was organized by
the Society of General Physiologists (20). Increased use was made of skinned muscle
fiber preparations, where the myofilament space is accessible to direct experimental
control. Several studies along these lines were made in our laboratory, and we will
describe these to you in this chapter.

The general aim of the work was to obtain information about the mechanical
properties of the actomyosin (AM) cross-bridge, the site of force generation in
muscle cells. The models that influenced our experiments are shown in Fig. 1. The
diagram on the left shows the cross-bridge model put forward some years ago by
H. E. Huxley (11). The cross-bridge in this case consists of the heavy meromyosin
moiety (HMM) of the myosin molecule; the light meromyosin moiety (LMM) is
part of the thick, myosin-containing filament. The connecting link between the
thick filament, and S1, the head of the myosin molecule, is assumed to be the S2
moiety of the molecule, which is 3 to 5 times longer than S1. Thus, the two regions
of the molecule that are sensitive to proteolytic enzymes are given hinge-like functions in this model of the organized system. In addition, it is assumed that force
is generated by a tendency of S1 to "rock" after it attaches to the thin, actin-
containing filament.

A modification of this idea, described several years later by A. F. Huxley and
Simmons (10), is shown on the right side of Fig. 1. They kept the idea that S1
generates force by "rocking" on the thin filament, and supposed that S2 contains
a compliance that is instantaneous on the submillisecond time scale. To explain the
force transients seen in length step experiments, they supposed that the different
states of S1, three of which are diagrammed here, are in thermal equilibrium, and
that the distribution among the states is affected by the force in S2.

STABILITY OF S1 ORIENTATION IN THE RIGOR STATE

These models raised the question of whether an applied force can actually change
the angle S1 makes with the thin filament. The simplest condition in which this

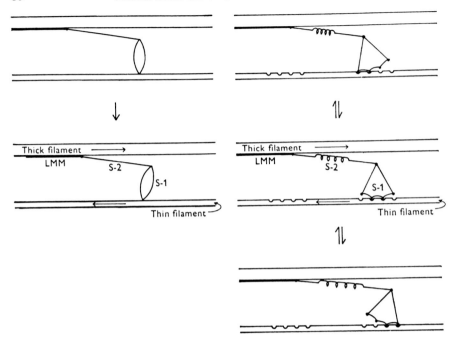

FIG. 1. Models of the actomyosin cross-bridge. **Left:** Cross-bridge model put forward by H. E. Huxley (11). S1 forms a cross-bridge by attaching to the thin, actin-containing filament, and it is linked to the surface of the thick, myosin-containing filament by S2. Force is generated by a tendency of S1 to "rock" relative to the thin filament. **Right:** Modification of **left** by A. F. Huxley and Simmons (10). S2 contains compliance and S1 rocks by moving through a series of three stable positions. (Reproduced from A. F. Huxley, ref. 9, with permission.)

can be examined is the ATP-free, or rigor, system, where the cross-bridge cycle is blocked and all the cross-bridges are in the AM state. The myofilaments in vertebrate striated muscle are arranged in an extremely regular double hexagonal lattice, and the equatorial X-ray diffraction pattern produced by the lattice is sensitive to the distribution of mass within the unit cell. The intensity of the innermost, or 10, reflection is determined mainly by the mass associated with the thick filaments, whereas that of the next reflection, the 11, is more sensitive to the mass associated with the thin filaments. The intensity ratio I_{11}/I_{10} increases when fibers are activated, presumably because of the movement of S1 toward the thin filaments when the cross-bridges are formed. It is clear from the model on the left that rotation of the S1 moiety from an acute to a less acute angle would decrease the density of the mass associated with the thin filament. The magnitude of this effect was calculated by Lymn (14). He found that a change in position of S1 from the rigor position, deduced from electron micrographs of decorated thin filaments (17), to a more perpendicular position could bring about a threefold change in the intensity ratio, I_{11}/I_{10}.

With this large potential change in mind, we put together the setup shown in Fig. 2 (18). The objective was to make simultaneous measurements of the force,

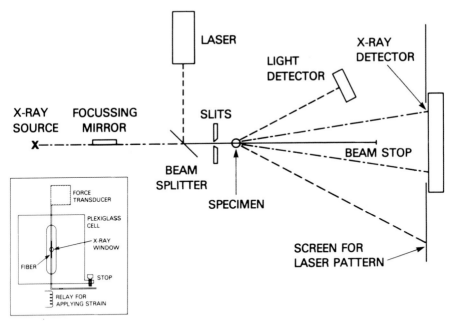

FIG. 2. Diagram of experimental setup for measuring the X-ray diffraction pattern of rigor muscle fibers as a function of strain. The X-ray source, camera, X-ray position-sensitive detector, and the laser system for measuring small changes in sarcomere length are shown. The output of the X-ray detector is computer controlled and directed to two different memories according to the strain state of the preparation. The *insert* shows the main features of the specimen holder. (From Naylor and Podolsky, ref. 18.)

the equatorial X-ray diffraction pattern, and the laser diffraction pattern before, during, and after an applied strain. The X-ray diffraction pattern was measured with a position-sensitive X-ray detector. The laser beam was passed through the same part of the preparation that was sampled by the X-ray beam, and the position of the first order laser diffraction line was measured with a position sensitive photodiode.

Figure 3 shows a typical experimental record. The preparation was a small bundle of glycerinated rabbit psoas fibers, about 250 μm in diameter. The upper trace shows the sarcomere strain, which in this case was 3 nm/half sarcomere. The lower trace shows the force. The strain was applied for about 500 msec, and the X-ray diffraction pattern was collected before, during, and after the strain. To get a reasonable diffraction pattern, the cycle was repeated several hundred times.

A typical diffraction pattern in the strained state is shown in Fig. 4. The dots are the actual counts and the line is a computer-generated best fit. The experiment was made at different sarcomere lengths, and the results are shown in Table 1. In each case the change in intensity ratio associated with the strain was calculated, and the average value for each condition was found. As shown in the last column, strain produced almost no change in intensity ratio. It would appear, then, that the angle of the cross-bridge in the rigor state is extremely stable. This is difficult to

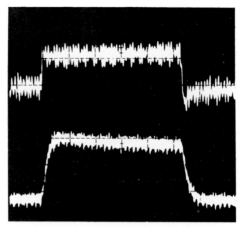

FIG. 3. Strain and force records for a fiber bundle. The *upper trace* is the output of the laser detector and shows a strain of 35 Å/half sarcomere applied for 500 msec. The *lower trace* is the output of the force transducer. The force increment was 60 mg. The bathing solution was 100 mM KCl, 2 mM MgCl$_2$, 4 mM EGTA, 7 mM KH$_2$PO$_4$, 13 mM K$_2$HPO$_4$, pH 7.0, 25°C. (From Naylor and Podolsky, ref. 18.)

reconcile with a force generating mechanism in which force is produced by the transition of cross-bridges through a series of attached states in thermal equilibrium, which rocks the S1 moiety relative to the actin filament, and where the distribution among states depends only on the force in the system (10). It is also at variance with a cross-bridge model in which the instantaneous compliance is associated with the rocking of S1 (2). In this aspect, the results are more consistent with the presence of a separate compliance that is either part of the cross-bridge (8) or a series element in the sarcomere.

This is not an entirely new finding. Haselgrove (6) stressed both frog and rabbit fibers in the rigor state and found no change in the equatorial intensity ratio. R. T. Tregear *(personal communication)* made a similar observation in insect muscle fibers. Morales and his colleagues (1) found that the angle of the intrinsic tryptophan fluorescence in rigor fibers was not affected by stress. More recently, Yoffe and Cooke [23; *see also* Cooke, R. (1981): *Nature*, 294:570–571], using the electron paramagnetic resonance spectra of spin labels attached to S1 as a probe of cross-bridge orientation, found no change in aspect when a rabbit fiber preparation was stressed. These observations, together with the present results, provide strong evidence that rocking does not occur in the rigor state.

THE LINK BETWEEN S1 AND THE MYOSIN FILAMENT

Another question that can be addressed by combining X-ray diffraction with mechanical measurements is the nature of the link between S1 and the myosin filament. The link is generally associated with the S2 moiety of the myosin molecule (12,19). This model of the cross-bridge is diagrammed on the left side of Fig. 5. The link could be either compliant (lower left) or rigid (upper left) relative to the

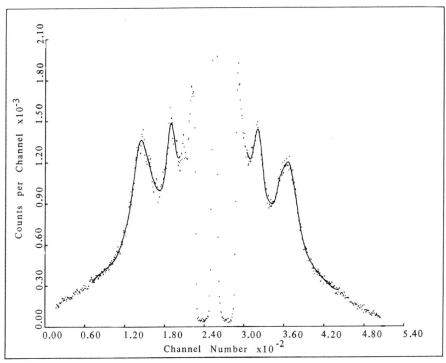

FIG. 4. A typical X-ray equatorial diffraction pattern for fiber preparation in rigor. The *dots* are the experimental data and the *smooth line* is the fitted curve assuming Gaussian peaks at the (10), Z line, (11), and (20) positions and a fifth-order polynomial as background. (From Naylor and Podolsky, ref. 18.)

TABLE 1. *The effect of applied strain on the equatorial X-ray diffraction pattern of small bundles of glycerinated rabbit psoas fibers in rigor*

Sarcomere length (μm)	Strain (nm/half sarcomere)	$\Delta I_{(11)}/I_{(10)}$[a] (mean % ± SEM)
2.0–2.2	3	5 ± 15
2.4–2.6	3	1 ± 4
2.8–3.0	3	1 ± 3
2.2–2.6	7–10	10 ± 10

From Naylor and Podolsky (18).
[a]This is the average value of the difference between the intensity ratio in the strained and unstrained states for each experiment.

other parts of the cross-bridge. However, in both cases, because S2 is long (60 nm) relative to the distance between the filament surfaces (about 15 nm at pH 8.5), there would be almost no change in fiber length if the filament lattice were compressed or expanded and the angle between S1 and the actin filament remained

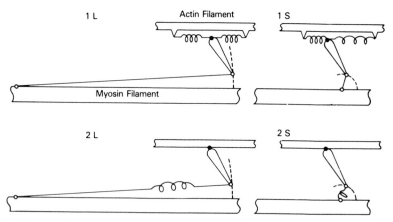

FIG. 5. Models of the actomyosin cross-bridge in the rigor state. *1L* and *2L* are similar to the models shown in Fig. 1 where the cross-bridge consists of S1 and the entire length of S2. *1S* and *2S* have a much shorter link between S1 and the myosin filament. Note that the actin filament, the myosin filament, and the cross-bridge are assumed to be in the same plane. In the *upper panels*, the cross-bridge compliance is drawn at the actin-S1 interface. In *2L*, the cross-bridge compliance is in S2 and, in *2S*, it is in the short link between S1 and the myosin filament. The *interrupted line* shows the trajectory of the S1-S2 junction when the lattice spacing is changed. Note that the curvature of this line depends on the length of the link between S1 and the myosin filament.

constant. A very different effect would be seen if the link were short, as diagrammed on the right side of Fig. 5. In this model, much larger changes in fiber length would be seen when the filament lattice is compressed or expanded.

The situation for the short link cross-bridge model is shown in more detail in Fig. 6. We consider the rigor fiber to be initially at pH 7.0 (upper right). The filament lattice spacing decreases linearly as the pH decreases (21). At low ionic strength (50 to 70 mM), we found the decrease in the distance between actin and myosin, $\Delta d_{AM} = 2/3(\Delta d_{10})$, to be about 2.0 nm when the pH was decreased from 7.0 to 5.5; this distance increased an equal amount when the pH was increased from 7.0 to 8.5. As shown in Fig. 6, these changes in filament spacing would produce changes in cross-bridge length, $\Delta\xi$. The geometry of the short link model is such that the transition from pH 7.0 to 8.5 would be expected to produce a much greater (in magnitude) change in cross-bridge length than the transition from pH 7.0 to 5.5. This change in cross-bridge length would produce a change in fiber length if the load on the fiber remained constant. The change in fiber length between pH 8.5 and 5.5 would be close to the length of the short link.

The effect of lattice spacing on fiber properties was measured in two ways. In the first method, shown in Fig. 7, the length of a glycerinated psoas fiber at a sarcomere length close to 2.3 μm was deduced from the position of its force-extension curve on the length axis. The points on a given force-extension curve represent the peak force developed by the fiber after it was stretched the indicated amount in 1 sec. The curves at pH 8.5, 7.0, and 5.5 all have the same shape but they are displaced relative to each other. The displacement between pH 8.5 and

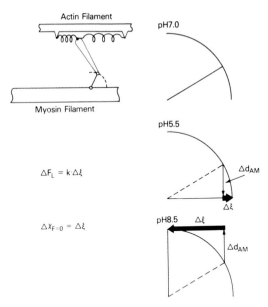

FIG. 6. Asymmetric effect of lattice spacing in the short link model. The diagrams on the *right* show the effect of changes in the distance between the actin and myosin filaments, Δd_{AM}, on changes in cross-bridge length, $\Delta \xi$. The magnitude of $\Delta \xi$ is smaller for lattice compression (pH 7.0 → pH 5.5, *middle right*) than for lattice expansion (pH 7.0 → pH 8.5, *lower right*). Depending on the mechanical conditions, $\Delta \xi$ produces a change in fiber length $\Delta \chi$ or a change in force ΔF_L; k is the apparent stiffness of a fiber kept at constant length.

7.0 is very much greater than that between pH 7.0 and 5.5. If the displacements of these curves are taken as length changes, the results appear to be consistent with the short link model diagrammed in Fig. 6. The shift of the curves between pH 8.5 and 5.5 is 0.25%, which corresponds to about 3.0 nm/half sarcomere. A similar set of force-extension curves was obtained when lattice spacing was controlled by adding dextran T-500 to the pH 8.5 solution. Thus, the shift in these curves along the length axis depends primarily on changes in lattice spacing.

The second method of measuring the effect of lattice spacing on the fiber properties is shown in Fig. 8. In this experiment, lattice spacing changes were made by both changing pH and by adding various amounts of dextran T-500 to a pH 8.5 solution. The spacing data in the lower part of the figure were obtained by X-ray diffraction. The changes in isometric force associated with these changes in lattice spacing, measured in a parallel experiment, are shown in the upper part of the figure.

At the start of the experiment, a state of stress was developed in the fiber by extending it 1% of the unstrained length. When the bathing solution was changed, the force became stable in a minute or less. Consider first the effect of dextran. The force decreased when the lattice spacing was decreased; a similar effect was reported previously by Maughan and Godt (15) for frog fibers in rigor. The change in force for a given decrease in lattice spacing from the most expanded state (pH

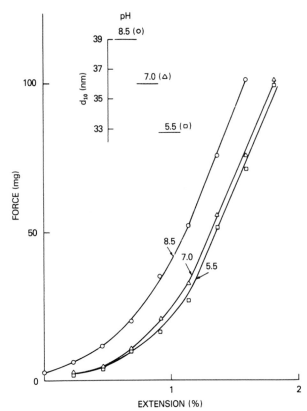

FIG. 7. Force extension curve of glycerinated psoas fibers at different lattice spacings. Lattice spacing was controlled by pH and measured by X-ray diffraction *(inset)*. Note that the force extension curves have the same shape and appear to be displaced relative to each other. Solution composition for pH 8.5: 40 mM triethanolamine HCl, 10 mM NaCl, 0.1 mM MgCl₂; pH 7.0: 40 mM imidazole, 30 mM NaCl, 0.1 mM MgCl₂; pH 5.5: 40 mM Na acetate, 30 mM NaCl, 0.1 mM MgCl₂. Temperature, 40°C. Bundle diameter, 100 μm. (T. Arata and R. J. Podolsky, *unpublished experiment.*)

8.5, no dextran) was greater than when the spacing was decreased further by the same amount (compare the change in force between 0 and 10% dextran with that between 10 and 20% dextran.) Thus, the change in force is a nonlinear function of lattice spacing. Similar changes in force were produced when the lattice spacing was changed by pH. Since it does not matter whether the lattice is controlled by dextran or by pH, the force level of the fiber in rigor appears to be a function of lattice spacing alone. According to the diagram in Fig. 6, the change in force level at a given length, ΔF_L, is proportional to $\Delta\xi$. The apparent stiffness for this effect, k, would be expected to be a function of both the cross-bridge stiffness and the compliance in series with the cross-bridges.

In the discussion so far, we have assumed that the link between S1 and the myosin filament is in the plane defined by that filament and the actin filament. A

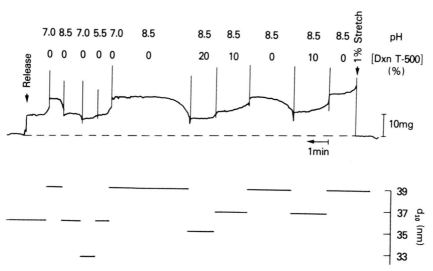

FIG. 8. Effect of lattice spacing on isometric force. **Upper panel:** Isometric force record from a glycerinated psoas fiber bundle in solutions where lattice spacing was controlled with dextran T-500 *(right side)* or change in pH *(left side)*. Bundle diameter, 100 μm. **Lower panel:** Influence of solution composition on lattice spacing. X-ray diffraction measurements were made in a separate experiment with the solutions used in the upper panel but with a 200 μm diameter bundle. Solution compositions given in legend to Fig. 7. (T. Arata and R. J. Podolsky, *unpublished experiment.*)

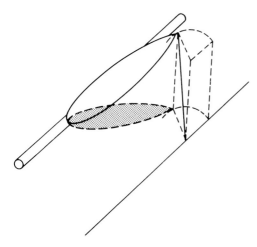

FIG. 9. Representation of the cross-bridge in three dimensions.

more realistic picture of the cross-bridge that takes account of the three-dimensional aspect of the structure is shown in Fig. 9. In this case, S1 is outside the plane of the filaments. However, the explanation we have put forward for the influence of lattice spacing on fiber length and isometric force can be applied to the projection of S1 (shaded in Fig. 9) and its connecting link in the plane of the filaments. Thus, the present data indicate that the projection of the connecting link in the plane of of the situation is such that the total length of the link cannot exceed that of S1, which is about 12 nm (16). Thus, the present data suggest that the link between S1 and the thick filament in the rigor fiber is between 3 and 12 nm in length.

We have used a simple geometric model to account for the effect of lattice spacing on the length and force of a rigor fiber. The model is based on the premise that changes in lattice spacing reorient a given set of structural elements and that these elements do not change state. In contrast, Ueno and Harrington (22) have put forward evidence, derived mainly from the reaction rates of certain bifunctional reagents with rigor fibers under various conditions, that the S2 moiety of myosin is released from the surface of the thick filament when the pH of the bathing solution is increased from 7.0 to 8.5. They argued that this displacement leads to a helix-coil transition in S2 that produces a change in fiber length (4,22). However, the extent to which the changes in structure detected by the bifunctional reagents contribute to the phenomena described here is not known, and the similarity of the force-extension curves of the fiber at pH 7.0 and 8.5 (Fig. 7) makes it unlikely that the change in fiber length is due to a phase transition in one of the structural elements.

The present results imply that in the rigor fiber the protease-sensitive region in the myosin rod at the LMM-S2 junction is considerably less flexible than that at the S2-S1 junction. Lowey et al. (13) pointed out that enzyme sensitivity is not necessarily associated with structural flexibility and fluorescence depolarization studies on the flexibility of the isolated myosin rod indicate very limited flexibility in the region between LMM and S2 at neutral pH (5). On the other hand, electro-optical measurements on the rod and its components suggest the presence of a flexible region in the rod (7) and a similar conclusion was drawn from the appearance of isolated myosin molecules in the electron microscope (3). These data can all be accommodated if it is assumed, as seems reasonable, that flexibility in a molecular structure is a graded phenomenon, and that the flexibility of myosin at the S1-rod junction is much greater than that at the LMM-S2 junction. This is consistent with the observation of Elliott and Offer (3) that the S1 moieties are free to move in all angles relative to the rod, but that only a fraction of the myosin molecules show sharp bending in the tail region.

SUMMARY

The inability of applied strain to affect the equatorial diffraction pattern of the rigor fiber indicates that the attachment angle of S1 to the actin filament is extremely stable. The effects of lattice spacing on fiber length (at constant force) and force (at constant length) indicate that the S1 moiety is connected to the surface of the myosin filament by a link that is between 3 and 12 nm in length, which is considerably less than the 60 nm length that has been assumed previously.

REFERENCES

1. Dos Remedios, C. G., Millikan, R. G. C., and Morales, M. F. (1972): Polarization of tryptophane fluorescence from single striated muscle fibers. *J. Gen. Physiol.*, 59:103–120.

2. Eisenberg, E., and Hill, T. L. (1978): A crossbridge model of muscle contraction. *Prog. Biophys. Mol. Biol.*, 33:55–82.
3. Elliott, A., and Offer, G. (1978): Shape and flexibility of the myosin molecule. *J. Mol. Biol.*, 123:505–519.
4. Harrington, W. F. (1979): On the origin of the contractile force in skeletal muscle. *Proc. Natl. Acad. Sci. USA*, 76:5066–5070.
5. Harvey, S. C., and Cheung, H. C. (1977): Fluorescence depolarization studies on the flexibility of myosin rod. *Biochemistry*, 16:5181–5187.
6. Haselgrove, J. C. (1970): *X-ray Diffraction Studies on Muscle.* Ph.D. thesis, University of Cambridge.
7. Highsmith, S., Kretzschmar, K. M., O'Konski, C. T., and Morales, M. F. (1977): Flexibility of myosin rod, light meromyosin, and myosin subfragment-2 in solution. *Proc. Natl. Acad. Sci. USA*, 74:4986–4990.
8. Huxley, A. F. (1957): Muscle structure and theories of contraction. *Prog. Biophys. Biophys. Chem.*, 7:255–318.
9. Huxley, A. F. (1974): Muscular contraction. *J. Physiol. (Lond.)*, 243:1–43.
10. Huxley, A. F., and Simmons, R. M. (1971): Proposed mechanism of force generation in striated muscle. *Nature*, 233:533–538.
11. Huxley, H. E. (1969): The mechanism of muscular contraction. *Science*, 164:1356–1366.
12. Huxley, H. E., and Brown, W. (1967): The low angle X-ray diagram of vertebrate striated muscle and its behavior during contraction and rigor. *J. Mol. Biol.*, 30:383–434.
13. Lowey, S., Slayter, H. S., Weeds, A. G., and Baker, H. (1969): Substructure of the myosin molecule. I. Subfragments of myosin by enzymatic degradation. *J. Mol. Biol.*, 42:1–29.
14. Lymn, R. W. (1978): Myosin subfragment-1 attachment to actin. Expected effect on equatorial reflections. *Biophys. J.*, 21:93–98.
15. Maughan, D. W., and Godt, R. E. (1981): Radial forces within muscle fibers in rigor. *J. Gen. Physiol.*, 77:49–64.
16. Mendelson, R., and Kretzschmar, K. M. (1980): Structure of myosin subfragment 1 from low angle X-ray scattering. *Biochemistry*, 19:4103–4108.
17. Moore, P. B., Huxley, H. E., and DeRosier, D. J. (1970): Three dimensional reconstruction of F-actin, thin filaments, and decorated thin filaments. *J. Mol. Biol.*, 50:279–295.
18. Naylor, G. R. S., and Podolsky, R. J. (1981): X-ray diffraction of strained muscle fibers in rigor. *Proc. Natl. Acad. Sci. USA*, 78:5559–5563.
19. Pepe, F. A. (1967): The myosin filament. II. Interaction between myosin and actin filaments observed using antibody-staining in fluorescent and electron microscopy. *J. Mol. Biol.*, 27:227–236.
20. Podolsky, R. J., ed. (1971): *Contractility of Muscle Cells and Related Processes.* Prentice-Hall, Englewood Cliffs, N.J.
21. Rome, E. W. (1968): X-ray diffraction studies of the filament lattice of striated muscle in various bathing media. *J. Mol. Biol.*, 37:331–344.
22. Ueno, H., and Harrington, W. J. (1981): Cross-bridge movement and the conformational state of the myosin hinge in skeletal muscle. *J. Mol. Biol.*, 149:619–640.
23. Yoffe, A., and Cooke, R. (1981): Static strain does not alter the angle of the actin-myosin bond in rigor muscle fibers. *Biophys. J.*, 33:82a.

Note added in proof: The present analysis assumes that myofibrils in the rigor state are cylindrical, and that they remain cylindrical when the d_{10} spacing of the filament lattice is varied between 33 and 39 nm. However, we have recently found evidence that the sarcomere may deviate from a cylindrical shape when d_{10} exceeds 39 nm. Preliminary data suggest that this distortion effect makes only a minor contribution to the fiber properties measured here, although further study is required to establish this point with certainty.

Basic Biology of Muscles: A Comparative
Approach, edited by B. M. Twarog,
R. J. C. Levine, and M. M. Dewey.
Raven Press, New York © 1982.

General Considerations of Cross-Bridge Models in Relation to the Dependence on MgATP Concentration of Mechanical Parameters of Skinned Fibers from Frog Muscle

†M. A. Ferenczi, ‡R. M. Simmons, and ‡J. A. Sleep

*Department of Physiology, University College London,
London WC1E 6BT, United Kingdom*

For the past several years we and our colleagues have been studying the way in which various mechanical parameters of frog muscle fibers, such as force and velocity, depend on the concentration of the substrate of actomyosin, magnesium adenosine triphosphate (MgATP). More recently we have started an investigation of the dependence of rate of hydrolysis and the distribution of bound nucleotides on MgATP concentration in both rabbit and frog muscle fibers. There is a general physiological interest in such studies, for example in relation to extreme fatigue, but we have two specific objectives that are relevant to the mechanism of contraction. First, the experiments extend observations that have already been made on intact fibers from frog muscles to a wider range of conditions, and the results should in principle help to decide between the different cross-bridge kinetic theories that these observations have engendered. Second, there is evidence from both mechanical and biochemical experiments that there are a number of states of the cross-bridge in its interaction with actin, and again in principle the experiments should help to explore the relation between the mechanically and biochemically defined states. In this chapter we shall be mainly concerned with the general explanation of our results in terms of models of the cross-bridge and the results will be presented in brief with little discussion of their limitations. Our treatment of various models here is much simplified. This is partly for ease of presentation, because complete models are not easily made intelligible, but also because, as will become apparent, our data are not yet sufficiently complete to attempt a really comprehensive model.

---

†Present address: Department of Biochemistry and Biophysics, School of Medicine G3, University of Pennsylvania, Philadelphia, Pennsylvania 19104.
‡Present address: MRC Cell Biophysics Unit, King's College, London WC2B 5RL, United Kingdom.

MECHANICAL STUDIES OF INTACT MUSCLE FIBERS

Steady State

Before describing the results of our experiments it is convenient to give in some detail the background to models of the cross-bridge cycle that has come from studies of the steady state and transient mechanical properties of intact fibers from frog muscle.

The steady state properties we consider here are the relation between force and velocity and also the rate of hydrolysis of MgATP as a function of shortening velocity. If a muscle fiber is stimulated to produce an isometric tetanus and the load on the fiber is suddenly changed, the fibre shortens at a steady velocity (after a transient period, see Fig. 2). If such isotonic releases are performed over a range of loads, the resulting relation between load and steady shortening velocity is roughly hyperbolic (Fig. 4) and can be described by Hill's equation,

$$(P + a) (V + b) = (P_0 + a)b$$

where P is the isotonic tension, P_0 the isometric tension, and V the shortening velocity. The maximum rate of shortening, V_0, can be obtained from this equation and can be used to characterise the force-velocity relation together with P_0 and a/P_0. The latter is a measure of the degree of curvature of the hyperbola; the more curved the relation, the smaller is a/P_0.

During an isometric contraction at 0°C in whole sartorius muscles from the frog, the rate of hydrolysis of ATP can be shown to be about 1.5 s^{-1} per myosin subfragment-1 head, assuming a myosin head concentration of 280 μM(4). During shortening, the turnover rate increases at first with increasing velocity to about 4.6 s^{-1} but decreases as the maximum velocity of shortening is reached (24).

The data from steady state mechanical experiments do not suggest a kinetic scheme for the cross-bridges but it is possible to construct relatively simple schemes that explain the data. In the well-known theory of Huxley (16) there are only two states of the cross-bridge—attached and detached. The model is illustrated in Fig. 1. It contains the elements of models used currently. There is a finite distance over which a cross-bridge can attach to an actin site. The cross-bridge contains an elastic element that can support both positive and negative forces. In the isometric state, the cross-bridges attach at positions producing positive force and cycle at a low rate. During shortening, attached cross-bridges shorten down their individual force-extension curves and, if they fail to detach in time, are dragged into positions in which they produce negative tension, in which positions the rate of detachment is much greater than in the isometric state.

Tension is lower during shortening in this model than in the isometric state partly because fewer cross-bridges are attached as a result of the relatively low rate of attachment. A further reduction of force arises because of cross-bridges being dragged into positions in which tension is negative (centre cross-bridge in Fig. 1C). Tension falls to zero (maximum velocity of contraction) when the positive and

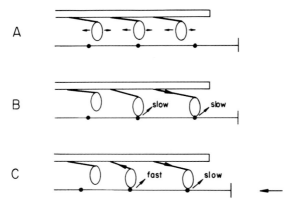

FIG. 1. Illustration of the model of Huxley (16). The cross-bridge is represented by a single subfragment-1 head with the elastic element in the subfragment-2 part of the molecule. Three cross-bridges are shown in different mean detached positions from the nearest actin site. **A:** Detached. **B:** During isometric contraction. **C:** During shortening or after a large release. It is assumed that force generation arises purely from the amount of stretch in the elastic element on attachment. Cross-bridge on right produces positive tension in isometric state, center produces no force, and left (which would produce negative tension) is assumed to be unable to attach. During shortening, cross-bridge on right produces positive force, and center, negative force (elastic element is compressed). When the elastic element is stretched, detachment is slow, and when compressed, fast (about 200 s^{-1}). Mean rate of attachment is about 25 s^{-1}.

negative forces exerted by attached cross-bridges exactly balance. It can be shown by considering the distance over which a cross-bridge remains attached during shortening that the rate constant for detachment must be about 200 s^{-1} to account for the observed maximum shortening velocity of frog muscle at 0°C; in general, the maximum shortening velocity is determined by the maximum rate of detachment and not the number of cross-bridges attached (29). Modifications have been made of the model to account for the transients that are seen before steady shortening is attained (15); one alteration is that the rate of attachment is assumed to be sufficiently high that at least as many cross-bridges are attached during shortening as in the isometric state. (In this case, tension is lower during shortening because of only the second of the two effects described above.)

Transients

It was first shown by Podolsky (27) that muscles do not immediately assume their steady state rate of shortening, if the load change is made sufficiently quickly. These "velocity" transients (Fig. 2, upper) were characterised by Civan and Podolsky (1) in intact fibers and have subsequently been examined in skinned fibres from frog muscle by Podolsky and his collaborators (see 28 for a recent summary). Corresponding "tension" transients are produced when a stimulated fiber is subjected to a rapid small length change (Fig. 2, lower); in some respects these transients are more easily interpreted than the velocity transients (although the one is the convolution of the other, albeit in a nonlinear system), and they have been studied by Huxley and his collaborators (11,12).

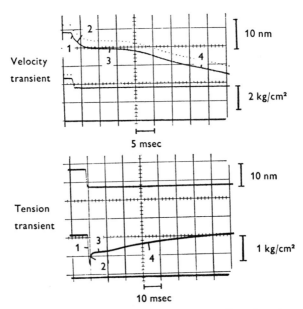

FIG. 2. Records of transients from tetanized single muscle fibers from frog muscle. **Upper:** Velocity transients when the load is suddenly changed. **Lower:** Tension transients when the length is changed. Length change was comparatively slow so that extent of immediate tension change is underestimated. (Reproduced from Huxley, ref. 17, with permission from *Journal of Physiology*.)

During and after the application of a length change, there are four phases in the tension transients. These arise wholly or nearly so from the cross-bridges. The phases are as follows:

Phase 1. A change synchronous with the length change, shown to arise from an approximately linear elastic element within each cross-bridge (12). The force-extension relation can be measured from the extent of the change of tension from a series of steps of different size or, more accurately, because of the fast rate of recovery in phase 2, from the "instantaneous" change of tension during the step itself. The slope of the force-extension relation—the stiffness—is often used as a measure of the number of attached cross-bridge, with the often tacit assumption that the stiffness is independent of the state of attachment.

Phase 2. A rapid recovery of tension, complete in a few milliseconds. The rate of phase 2, considered as a whole, depends heavily on the amplitude and direction of the step, being faster for releases than stretches. Phase 2 is not a simple exponential. Some spread of rate constants is to be anticipated because of the difference in periodicities of actin and myosin along the filaments, resulting in a distribution of stress between attached cross-bridges. Allowing for such a distribution it is, however, simpler to fit the experimental data with two sets of rate constants rather than one (R. M. Simmons, *unpublished results*).

Phase 3. A slowing or reversal of recovery, taking 10 msec or so. It is difficult to characterise in terms of a rate constant as the phase is small in amplitude and for larger releases is present only as an inflexion on the tension record.

Phase 4. The final tension redevelopment, lasting a few hundred milliseconds. The phase cannot be described by a simple exponential, but its rate constant does not depend strongly on the size or direction of the step.

Phase 2, at least for small releases and stretches in which it is clearly distinct from the later phases, apparently derives from a shift in one or more equilibrium processes. Measurements of stiffness show little change at the end of phase 2, indicating that there is little net attachment or detachment of cross-bridges, and the application of a second step before phase 2 is complete has the effect that the tension record soon follows the record that is obtained by performing a single step of the total amplitude of the two steps (10,11; L. E. Ford, J. F. Huxley, and R. M. Simmons, *unpublished results*). After the beginning of phase 3, stiffness, measured from the extent of tension change to a second step, falls and subsequently recovers in phase 4 (a result that requires confirmation from faster measurements of stiffness). The application of a second step during phase 3 produces a tension change that, the more delayed the second step, becomes progressively closer to a superimposition of the response of the second step on its own on to the tension record from the first step (L. E. Ford, J. F. Huxley, and R. M. Simmons, *unpublished results*; confirmed in skinned fibers by Iino and Simmons). Thus, during phase 3, some of the cross-bridges have undergone an irreversible step (although it cannot be ruled out the change is effected via the activation system).

Another aspect of phase 2 is the extent to which recovery occurs as a function of size and direction of step. Combined with the rate dependence, this suggested that underlying phase 2 was the step-wise movement of some part of the cross-bridge, corresponding to force-generation (19). This type of model has been generalised and extended by Hill (14). Various alternatives have been proposed to explain phase 2, some of which can probably be ruled out by the observations referred to above about the effect of a second step. Thus, theories that require a net transfer of cross-bridges between the attached and detached states are likely to be wrong, but it cannot be excluded that all or part of phase 2 is due to a rapid equilibrium between attached and detached states that takes place with no net transfer.

The scheme proposed by Huxley and Simmons (19) to account for phase 2 is shown in Fig. 3 (with the elements of Fig. 1). After attachment (Fig. 3B), the cross-bridge heads tilt, thus producing tension. Three attached states are shown, which is the minimum required to produce the observed tension and energy output. Transitions between states are fast, but the equilibrium constants for the transitions depend heavily on the size of the restraint imparted by the stretch in the elastic element (shown in the S2 part of the molecule) needed to make the transition. Thus, for a cross-bridge, which in the relaxed state lies to the left of an actin site so that

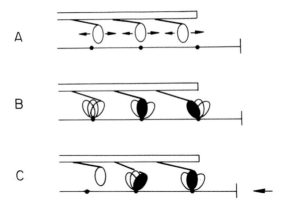

FIG. 3. Extension of model in Fig. 1 to account for tension transients (19). Cross-bridges drawn in similar positions with respect to actin sites as in Fig. 1. Three distinct states (angles) of the cross-bridge head are shown **A:** Detached. **B:** During isometric contraction. **C:** During shortening or after a release. A cross-bridge attaches to the first of the states (most counter-clockwise) and makes rapid transitions to the other states. In general, one position will be more favored than the others, according to the amount of stretch in the elastic element, and the cross-bridge will spend most of the time in that position (shown filled). Detachment occurs only from the last state, so that detachment is effectively slow unless the last state is favored (cross-bridge on left). After a release, states further clockwise are more favored (less stretch in elastic element) and head rotation occurs **(C)**, with the partial redevelopment of tension (phase 2). Detachment after a release is quicker than in isometric state because more cross-bridges are in last state, from which detachment rate is fast.

a large extension of the elastic element is needed to make transitions to higher force producing states, the equilibria favor occupancy of the first attached state. Where the relaxed cross-bridge is opposite or to the right of an actin site, the equilibria favor the second and third states. If it is assumed that cross-bridges detach with a fast rate constant from the final state, then this state will be little occupied even when the equilibria favor it (cross-bridge on left of Fig. 3). After a release (Fig. 3C), the restraints are relaxed and further positions are favored, resulting in tension redevelopment (phase 2) followed by detachment for some cross-bridges (center) that reach the final attached state (phase 3). Phase 4 is explained as the reattachment of the detached cross-bridges and the final recovery of the original cross-bridge distribution. Computations on the basis of such a model, including appropriate rate constants for attachment and detachment, have shown that the main features of the intact fiber mechanical data can be at least qualitatively reproduced (R. M. Simmons and H. D. White, *unpublished results*).

None of the individual rate constants of such a scheme need be as low as the overall turnover rate; the latter arises naturally from a combination of steps. If all the attached states equilibrate rapidly on the timescale of the detachment rate from the last state, then the turnover rate is determined approximately by the product of the equilibrium constants between attached states and the final rate constant for detachment. Turnover in the isometric state is slow but is increased during shortening when the average restraint on the equilibrium constants is lower. The fastest rate of detachment, in a large release or at the maximum velocity of shortening, is, however, likely to be determined by the detachment rate from the final state.

RESULTS OF EXPERIMENTS AT DIFFERENT CONCENTRATIONS OF MgATP

Methods

The experiments were done using mechanically skinned fibers from the semi-tendinosus muscle of the frog Rana temporaria. The experiments were at 0 to 2°C and at ionic strength 200 mM, pH 7.1. A rephosphorylating system was used to maintain the MgATP concentration in the fibers; this consisted of up to 30 mM phosphocreatine and 1 to 10 mg/ml creatine kinase. Solutions were stirred fairly vigorously to disperse boundary layers. Fibers were activated in most experiments using $10^{-4.5}$M Ca^{2+}. In most of the solutions, Mg was in excess of the ATP concentration (free Mg^{2+} = 1 mM).

Various difficulties were encountered in the course of this work. They included: force-extension relation of skinned fibers less steep and more nonlinear than intact fibers (part of which was due to extraneous compliance where the fibers were gripped); disorder of sarcomeres at full activation; decrease of velocity during shortening; and failure to redevelop full tension after a release. Where possible, the results were checked, and in some cases corrected, by recording the sarcomere length of the central part of a submaximally activated fiber, using the diffraction system described by Goldman and Simmons (13). For our present purposes we shall ignore these difficulties. Scatter of results between fibers was in most cases reduced by normalising the results from a given fiber to the values at a high MgATP concentration (5 mM).

Steady State

Experimental records for the determination of the force-velocity relation as a function of MgATP concentration are shown in Fig. 4. The results for isometric tension and for the maximum velocity of shortening (V_0) are summarized in Fig. 5. V_0 depends in a roughly hyperbolic fashion on MgATP concentration and can be described by Michaelis-Menten kinetics with a K_m of 0.44 mM and a maximum velocity of 2.37 muscle lengths/sec. At the physiological concentration of about 5 mM, the velocity is near maximal. Isometric tension shows a bell-shaped dependence on MgATP, with a peak at about 50 μM MgATP. a/P_0 (not shown in Fig. 5) shows a tendency to decrease with increasing MgATP, that is, the force-velocity relation becomes more curved. The results show a good deal of scatter, but a/P_0 may be as high as 0.3 at 10 μM, decreasing to about 0.17 at 5 mM. Qualitatively similar results have been reported for rabbit muscle fibers by Cooke and Bialek (2).

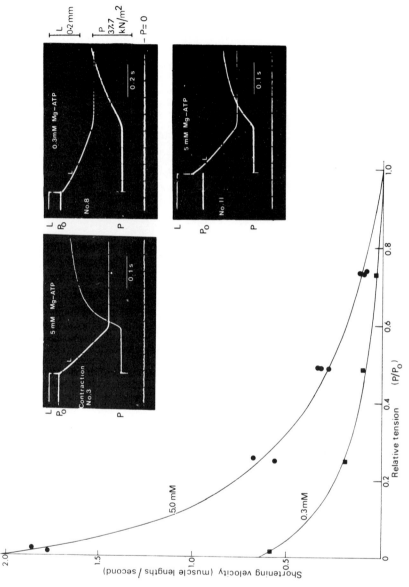

FIG. 4. Dependence of force-velocity relation on MgATP concentration. *Inset* shows experimental records for shortening steps at 5 and 0.3 mm MgATP. The graph shows experimental data and least squares fits of Hill's equation to force-velocity relation. (Reproduced from Ferenczi et al., ref. 8, with permission from the Journal of Physiology.)

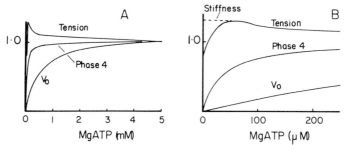

FIG. 5. Dependence of isometric tension, stiffness, V_0, and rate of recovery in phase 4 on MgATP concentration. The data are shown on two scales: **A:** Up to 5 mM; **B:** Up to 250 μM. Stiffness *(dashed line)* is omitted from **A** for clarity. Above about 50 μM MgATP, the stiffness curve follows the tension curve. Values are expressed relative to those at 5 mM MgATP.

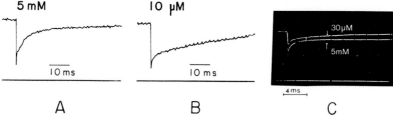

FIG. 6. Tension transients at high and low MgATP concentrations. **A, B:** Showing slow tension recovery (phase 4) after a release of about 1% of fiber length. These records were obtained with the sarcomere length of the central part of the fiber controlled using a diffraction system (13); the fiber was activated submaximally. Because of noise on the records (from the diffraction system) the early part of the tension change is not clearly seen. **A:** 5 mM; **B:** 10 μM. **C:** Superposed tension records are shown at 5 mM and 30 μM from a fiber at full activation, diffraction system not used. Faster timebase than in **A** and **B**, showing early recovery. Motor and servo-system were similar to those described in Ford et al. (11)

Transients

As shown in Fig. 5, the "instantaneous" stiffness decreases with increasing MgATP. At very low MgATP it remains high although tension is comparatively low. At higher concentrations it decreases in much the same way as tension. The clearest result for the recovery phases is for phase 4. Records of the response to releases at 5 mM and 10 μM MgATP are shown in Fig. 6 (A and B, respectively). The later phase of recovery is clearly much slower at the lower MgATP concentration, but the change occurs at comparatively low concentrations. The relation between the rate constant of phase 4 and MgATP is roughly hyperbolic with a K_m of about 30 μM and a maximum rate of about 17 s^{-1}.

The dependence of the rate constants and amplitudes of phases 2 and 3 are more complicated. Phase 3 disappears when the MgATP concentration is less than about 1 mM. We would assume from this observation that the K_m is high and this certainly seems to be the case in rabbit muscle (3,22). Earlier, phase 3 was described in terms of detachment, and, on that basis, we would equate it in MgATP-dependence wth V_0. Difficulty in dealing with phase 2 is illustrated in Fig. 6C, which shows superimposed records on a faster timebase than Fig. 6A and B. The impression at low MgATP is that there is a fast component of recovery similar in timecourse to phase 2 at high MgATP but much reduced in amplitude. This is followed by a slower process, but one that recovers during the flat part of the record at high MgATP (phase 3). This slower component is certainly sensitive to the concentration of MgATP over the range 10 to 100 μM (7), but it is difficult to subdivide the records unambiguously.

DISCUSSION

General Interpretation. Intermediates in the Cross-Bridge Cycle

Our objective in this section is to extend the type of scheme discussed earlier to account for the observed dependence on MgATP concentration of the various mechanical parameters. The interpretation of our results requires some knowledge (or guesswork) as to the predominant intermediates as a function of MgATP concentration, particularly those concerned with force production in the isometric state. The isometric turnover rate shows saturation at quite low values of MgATP and it follows that as the MgATP concentration is increased from zero, the predominant intermediate(s) must change from AM (rigor) to myosin and actomyosin states with bound ATP or products (AM.ATP, AM.Pr, M.ATP, M.Pr; we do not in general distinguish here between Pr and ADP). Unfortunately, we do not yet know what these intermediates are in frog muscle. Marston and Tregear (26) found that in activated myofibrils from rabbit muscle about two-thirds of the myosin was associated with bound products at saturating ATP concentrations, but the results were obtained at low temperature where rabbit muscle produces comparatively little tension, so it is not clear whether the products were combined with myosin or actomyosin. There are three lines of biochemical evidence that suggest that AM.Pr is the predominant attached state at high ATP concentrations. First, in the AM.ATP states, actin is not tightly bound. Second, about 60% of the energy of ATP hydrolysis is liberated between the M.Pr and AM.ADP states (31,32). Third, at least until recently, the AM.ATP states were thought not to exist in significant concentration (6,30). We shall assume that the maintenance of force is mainly associated with the AM.Pr and AM states, but will note when involvement of the AM.ATP state could help to account for our results.

It is convenient to start with the process of cross-bridge detachment and subsequently to enlarge the scheme to other parts of the cycle. This is partly because ATP is most directly involved in detachment and partly because, in isolating detachment experimentally by studying the maximum velocity of shortening, the restraints on equilibria between attached states are relaxed as far as possible, and the situation of the cross-bridges is at its closest to the interaction of myosin and actin in solution.

K_m for V_0

The K_m for the maximum velocity of shortening (V_0)—and we assume for phase 3—is about 0.4 mM. This is to be associated with a maximum rate of 200 s^{-1}. The data are most simply explained by the following scheme, where the AM and AM.ATP states correspond to the final attached state:

$$\text{AM} + \text{ATP} \underset{(1)}{\overset{0.4 \text{ mM}}{\rightleftharpoons}} \text{AM.ATP} \overset{200 \text{ s}^{-1}}{\underset{(2)}{\longrightarrow}} \text{M.ATP} + \text{A} \qquad [1]$$

Assuming $K_{-1} >> k_2$, the K_m is $1/K_1$ (0.4 mM). It is of interest to compare this with the dissociation of frog actomyosin by ATP in solution (7,9). The observed rate of dissociation showed a linear dependence on ATP up to the highest ATP concentration that it was possible to use (0.5 mM), when the rate was 375 s^{-1}. Subsequently, even faster rates have been observed for other vertebrate skeletal muscle myosins, which appear to show no saturation up to rates of over 1000 s^{-1}; however, smooth muscle myosin showed saturation with an apparent binding constant for ATP equivalent to about 0.6 mM, in close, and perhaps fortuitous, agreement with our value (25). There are a number of possible reasons for the discrepancy between the biochemical and mechanical rates at high MgATP. First, it could arise from a genuine difference between the intact cross-bridge and extracted myosin subfragment-1. Second, it could be that the restraints on cross-bridges referred to above are not fully removed during shortening. If the scheme for dissociation is expanded to AM + ATP $\gtrless$ AM.ATP $\gtrless$ AM.ATP′ $\gtrless$ M.ATP + A, where the second step is rate limiting (5) and where the AM.ATP and AM.ATP′ states are mechanically distinct, it could be that mechanical restraints slow this transition in the intact lattice. Third, it might be that the rate limiting step for detachment, or more correctly the step that limits the maximum rate of shortening at high ATP, occurs *before* the ATP binding step, for example, it might correspond to product release.

We shall assume for the present that the scheme of equation 1 correctly describes the detachment process.

K_m for Phase 4

Phase 4 was explained earlier by the reattachment of cross-bridges after detachment. Extending equation 1 to reattachment we obtain:

$$\text{AM} + \text{ATP} \underset{(1)}{\overset{0.4 \text{ mM}}{\rightleftharpoons}} \text{AM.ATP} \underset{(2)}{\overset{200 \text{ s}^{-1}}{\rightleftharpoons}} \text{M}_{\text{diss}} \overset{20 \text{ s}^{-1}}{\underset{(3)}{\longrightarrow}} \text{AM.Pr} \qquad [2]$$

The rate limiting step for reattachment is, roughly, 20 s^{-1}. M_{diss} includes all detached states. The K_m for reattachment from equation 2 can be shown to be k_3/K_1k_2, provided $K_{-2} << k_3$. This yields a value of 40 μM in close agreement with the experimental result. It should be noted that equation 2 applies to unrestrained shortening and not to the isometric state.

K_m for Isometric ATPase Rate

The rate of hydrolysis of ATP in an isometric contraction at high ATP in frog muscle is about 1 to 2 s^{-1} per myosin subfragment-1. The K_m has not yet been determined accurately, but in rabbit muscle fibers under conditions that are roughly equivalent to our experiments, the value might be as high as 40 μM (upper limit in *unpublished experiments* by J. A. Sleep). At first sight this would seem to be inconsistent with the argument in the previous section, in which the addition of a slower step after detachment produced a much lower K_m; it might be expected if a further slow process of 2 s^{-1}, corresponding to the turnover rate, is added to the end of equation 2 so that the K_m for turnover should be about 4 μM. If, however, we assume that in the isometric state the equilibrium constant for ATP binding is restricted by mechanical constraints, then K_1 in equation 2 will be decreased (to become K_1'). The K_m is then given by $v_{max}/K_2'k_2$, where v_{max} is the maximum turnover rate. If the K_m is indeed as high as 40 μM, $1/K_1'$ would have to be 4 mM, a factor of 10 greater than that during unrestrained shortening. Just as for the argument above in relation to restraints on the maximum velocity of shortening, it is more likely that the present restriction should be applied to an isomerisation of AM.ATP states after ATP binding. (It should be noted here that we have implicitly assumed that all cross-bridges are able to interact with actin sites. If this is not the case, then the turnover rate is increased appropriately.)

It was shown earlier that the slow turnover rate in the isometric state does not necessarily imply a correspondingly slow rate limiting step. Rate limitation can occur by the combination of an unfavorable fast equilibrium preceding an irreversible step. For example, in the scheme

$$\text{AM.Pr}_1 \underset{1{,}000 \text{ s}^{-1}}{\overset{100 \text{ s}^{-1}}{\rightleftharpoons}} \text{AM.Pr}_2 \underset{200 \text{ s}^{-1}}{\overset{20 \text{ s}^{-1}}{\rightleftharpoons}} \text{AM.Pr}_3 \overset{\text{fast}}{\longrightarrow} \text{AM}$$

the flux through the second step is slowed by a factor of 10 by the preceding step, so the net effect is a rate limitation of 2 s^{-1} (a limitation that might be removed during shortening). The steps can, nevertheless account for phase 2 because the rates of the transitions are the sum of the forward and backward rate constants, i.e., 1,100 and 220 s^{-1}. We would stress that this example serves only to illustrate a general point.

Phase 2

We have so far been unable to determine a K_m for phase 2 because of the difficulties referred to in the results section. Cox and Kawai (3) have shown in rabbit muscle fibers that the process corresponding to phase 2 has a K_m intermediate between that for phase 4 and phase 3. The very fact of a dependence of the rate of phase 2 on MgATP suggested to Kawai (21,22) that the underlying process most likely corresponded to detachment and not to transitions between attached states.

It is, in general, not true that a change in the rate of phase 2 rules out an origin in transitions between attached states because it is possible that the transition arises from the scheme AM + ATP $\gtrless$ AM.ATP $\gtrless$ AM.ATP'. However, since phase 2 is associated with a higher rate constant than phase 3, it would be expected to have a higher K_m and not lower as observed. Our provisional interpretation is that phase 2 derives from transitions both between AM.Pr states and AM.ATP states. At high ATP, most of the cross-bridges would be in the AM.Pr states, but at low ATP the AM state becomes more important and recovery after a release might be accounted for by transitions between AM.ATP states. As noted in the results section, at low ATP there appear to be two distinct processes in phase 2.

Several lines of argument suggest that one or more AM.ATP states might play an important part in explaining the steady state and transient mechanical events observed in this study. Whether the state plays a role in the production of isometric tension at high MgATP is not clear, but it is a distinct possibility, and an accurate determination of the occupancy of bound nucleotide states would help to clarify the issue.

Fall of Tension and Stiffness with Increasing MgATP

Figure 5 shows that, although stiffness falls as MgATP is raised, tension at first increases and reaches a peak before falling. We shall deal with the initial rise of tension below. It is also noticeable in Fig. 5 that, although tension and stiffness fall most rapidly between about 50 μM and 0.5 mM MgATP, there is a continued downward "creep" at higher MgATP concentrations. We shall ignore this creep in what follows; it is perhaps due to a nonspecific action of high MgATP or to the action of a binding site with very low affinity. The range over which the main fall in tension and stiffness occur makes it likely that the fall is associated with a shift in occupancy of intermediates. The simplest explanation would be that there is partial rate limitation of the isometric turnover rate by a comparatively slow step between detached intermediates or from detachment to attachment. To explain the results in this way, the rate of such a step has to be 5 s^{-1} or less, which is clearly in conflict with the requirement of a maximum rate of about 20 s^{-1} for phase 4, unless a fair proportion of the cross-bridges do not contribute to the total isometric turnover rate, or there is a pool of detached cross-bridges that can attach at a fast rate after a release. An alternative explanation is that the fall represents a shift from the AM to the AM.Pr state and that the latter contributes less to stiffness and tension than the former.

A further difficulty in explaining the fall may arise because our results seem to require a K_m of the underlying process of at least 100 μM. Cooke and Bialek (2) came to similar conclusions for rabbit muscle fibers, in which the K_m for the isometric ATPase rate is almost certainly 40 μM or less (26; J. A. Sleep, *unpublished results*). The expected correlation between fall of tension and ATPase rate may not exist.

a/P_0

With increasing MgATP, a/P_0 becomes smaller. The general trend is as anticipated from the cross-bridge model of Huxley (16), but the observed extent of change is less than expected if the effect of lowered MgATP is simulated by changing the detachment rate to match the observed maximum velocity of shortening (2,7). The reason for this is that at zero tension (V_0) it is only the rate of detachment that determines the value of V_0, but at moderate loads the rate of attachment is also important. Therefore, velocities at loads greater than zero are less affected than the maximum velocity by a change of the detachment rate, resulting in a considerable change of curvature. Models in which the attachment rate is very high give little change of curvature with a change of detachment rate, since this dominates the force produced at all velocities. It is not clear to us if this is the correct explanation for the result, because the curvature is also affected by the mean distance over which cross-bridges remain attached. It is conceivable that this alters with MgATP concentration (see below).

Rise of Tension Between 0 and 50 μM MgATP

The rise of tension at low MgATP as the concentration is increased is accompanied by a relatively small change in stiffness, so it is likely that the number of cross-bridges that are attached remains roughly constant or even falls. Thus, each attached cross-bridge must be exerting less tension, on average, at very low MgATP. An explanation for the low tension can be advanced in terms of slippage of cross-bridges in the AM state (regarded as contributing to tension) to another state, AM', which contributes to stiffness but not to tension. There is some evidence that such a slippage does indeed take place, since frog muscle fibers in rigor maintain a tension that is only about 30% of that in a contraction at high MgATP, and a stretch produces a rise of tension that is followed by a decline to the original level (Y. E. Goldman and R. M. Simmons, *unpublished results*). One possibility is that cross-bridges slip on stretching from one actin site to the next (23). If an alternate pathway for detachment is introduced as in the following scheme

$$AM.Pr \rightarrow AM \rightarrow AM'$$
$$\nwarrow \quad \downarrow \quad \swarrow$$
$$M.pr$$

where, for convenience, several states have been omitted, an approximate fit to the experimental data can be obtained (for a full version of this model, see 7). At zero MgATP, all the cross-bridges will be in the AM' state, but as the concentration is increased, the AM.Pr and AM states become populated, producing tension. The contribution of the route via AM' to the total flux becomes negligible when the route from AM to M.Pr becomes much faster.

A reasonable assumption in this scheme is that the second order rate constant from AM' to M.Pr exceeds that from AM to M.Pr, reflecting the lower restraints on the "slipped" cross-bridges. It is, however, unnecessary to make an arbitrary assumption that this occurs in an alternate pathway; a similar result is obtained more naturally by considering the distribution of attachment of cross-bridges (we thank our colleague Dr. J. F. Rondinone for reminding us of this argument). (The scheme above also suffers from the defect that slippage might be expected to occur to a greater extent in the AM.Pr state than the more tightly bound AM state.) In rigor, or more strictly at equilibrium, a muscle fiber cannot in the long term produce tension if the filament lattice is such that there is no coincidence between the myosin and actin periodicities (14). That is not to say that all cross-bridges produce zero tension, rather that some will produce positive and others negative tension, in a similar way to the situation at the maximum velocity of shortening. For example, referring to Fig. 3, cross-bridges that lie to the right of the nearest actin site do not attach or do not remain attached to such sites at high ATP, whereas they are favored at very low ATP, and some of these will produce negative tension. As the ATP concentration is raised from near zero, these cross-bridges will detach (and given other actin sites within range, may reattach to a site that will result in positive tension). The loss of negative tension results in a rise of tension.

CONCLUSIONS

In the discussion section we have tried to explain our results in terms of a single cycle for the hydrolysis of ATP for actomyosin, combined with reasonable mechanical considerations of the cross-bridge. In terms of the model for intact fiber mechanics of Fig. 3, as a result of these experiments, we would extend the number of attached states from three to at least four. The first two would be AM.Pr states. The third would correspond to both AM and to an AM.ATP state, with a binding step interpolated that does not depend on the force exerted by the cross-bridge. The fourth would be a further AM.ATP state. Whether or not a further detached state needs to be added is not yet clear. The resulting scheme is similar to that described by Huxley (18). The interpretation of the results given here is heavily dependent on the identification of the processes underlying the transient phases in intact fibers (and at high MgATP concentration in skinned fibers), particularly of phase 2. If the identification is correct, then the outlines of the scheme follow with a certain inevitability, although it is probably an oversimplification to consider a simple cycle with detachment from only one state. However, it does seem likely that the type of scheme considered here can successfully explain the rates and K_m' s for V_0 and phases 3 and 4, the rise of tension at low MgATP and the rate (and probably K_m) of the isometric ATPase, although there are some doubts about explaining the corresponding data for the fall of tension and stiffness at higher ATP concentrations and for phase 2. (Also, a detailed computation of a cross-bridge model is required to see if the empirical approach adopted here can be sustained.)

A more accurate and complete set of mechanical data would no doubt help to clarify some of the uncertainties. A complete set of data involves a study of the effect of a second step at various times during transients at a range of MgATP concentrations. It is, for example, particularly important to discover whether the slower process in phase 2 at low MgATP is reversible. It would also be helpful to have data for the occupancy of intermediates as a function of MgATP concentration, not only during the steady state but also on a comparable timescale to the transients.

Lastly, it is worth considering whether, if all the experiments listed above were to be completed, an unambiguous explanation of the results would emerge. We think this is unlikely, partly because of the difficulty of interpreting results from perturbations of the steady state when the steady state itself is changed by altering the substrate concentration, and partly because the choice of restraint to apply to attached cross-bridges and to rate constants in general is somewhat arbitrary. It would be far preferable to be able to isolate the attached states or, at the very least, to start our experiments at some defined part of the cycle in the equivalent of a biochemical stopped flow experiment. The advent of "caged-ATP" (20), an inert compound that can be photolysed to produce ATP, makes some of these experiments possible, which previously were impossible because of diffusional limitations.

ACKNOWLEDGMENTS

Many of the results referred to above were obtained in collaboration with Drs. Y. E. Goldman, B. B. Hamrell, and J. F. Rondinone. We should like to thank them and also Professor Sir Andrew Huxley and Dr. L. L. Ford for their permission to quote unpublished results. This work was supported by the Medical Research Council and by the Muscular Dystrophy Association.

REFERENCES

1. Civan, M. M., and Podolsky, R. J. (1966): Contraction kinetics of striated muscle fibers following quick changes in load. *J. Physiol. (Lond.)*, 184:511–534.
2. Cooke, R., and Bialek, W. (1979): Contraction of glycerinated muscle fibers as a function of the ATP concentration. *Biophys. J.*, 28:241–258.
3. Cox, R. N., and Kawai, M. (1981): Alternate energy transduction routes in chemically skinned rabbit psoas muscle fibers: A further study of the effect of MgATP over a wide concentration range. *J. Mus. Res. Cell Motil.*, 2:203–214.
4. Curtin, N. A., Gilbert, C., Kretzschmar, K. M., and Wilkie, D. R. (1974): The effect of performance of work on the total energy output and metabolism during muscle contraction. *J. Physiol. (Lond.)*, 238:455–472.
5. Eccleston, J. F., Geeves, M. A., Trentham, D. R., Bagshaw, C. R., and Mrwa, U. (1975): The binding and cleavage of ATP in the myosin and actomyosin ATPase mechanisms. In: *Molecular Basis of Motility (Mosbach Colloq. 26th)*, edited by L. M. G. Heilmeyer, J. C., Rüegg, and Th. Wieland, pp. 42–52. Springer-Verlag, Berlin.
6. Eisenberg, E., and Greene, L. E. (1980): The relation of muscle biochemistry to muscle physiology. *Annu. Rev. Physiol.*, 42:293–309.
7. Ferenczi, M. A. (1978): Kinetics of contraction in frog muscle. Ph.D. thesis, University of London.
8. Ferenczi, M. A., Goldman, Y. E., and Simmons, R. M. (1979): The relation between shortening velocity and the magnesium adenosine triphosphate concentration in frog skinned muscle fibers. *J. Physiol. (Lond.)*, 292:71–72P.
9. Ferenczi, M. A., Homsher, E., Trentham, D. R., and Simmons, R. M. (1978): The reaction mechanism of the Mg^{2+} dependent ATPase of frog myosin and subfragment 1. *Biochem. J.*, 171:165–175.
10. Ford, L. E., Huxley, A. F., and Simmons, R. M. (1974): Mechanism of early tension recovery after a quick release in tetanized muscle fibers. *J. Physiol. (Lond.)*, 40:42–43P.

11. Ford, L. E., Huxley, A. F., and Simmons, R. M. (1977): Tension responses to sudden length change in stimulated frog muscle fibers near slack length. *J. Physiol. (Lond.)*, 269:441–515.
12. Ford, L. E., Huxley, A. F., and Simmons, R. M. (1981): The relation between stiffness and filament overlap in stimulated frog muscle fibers. *J. Physiol. (Lond.)*, 311:219–249.
13. Goldman, Y. E., and Simmons, R. M. (1979): A diffraction system for measuring muscle sarcomere length. *J. Physiol. (Lond.)*, 292:5–6P.
14. Hill, T. L. (1974): Theoretical formalism for the sliding filament model of contraction of striated muscle. Part I. *Prog. Biophys. Mol. Biol.*, 28:267–340.
15. Hill, T. L., Eisenberg, E., Chen, Y., and Podolsky, R. J. (1975): Some self-consistent two-state sliding filament models of muscle contraction. *Biophys. J.*, 15:335–372.
16. Huxley, A. F. (1957): Muscle structure and theories of contraction. *Prog. Biophys. Mol. Biol.*, 7:255–318.
17. Huxley, A. F. (1974): Review lecture. Muscular contraction. *J. Physiol. (Lond.)*, 243:1–43.
18. Huxley, A. F. (1980): *Reflections on Muscle*. Liverpool University Press, Liverpool.
19. Huxley, A. F., and Simmons, R. M. (1971): Proposed mechanism of force generation in striated muscle. *Nature*, 233:533–538.
20. Kaplan, J. H., Forbush, III, B., and Hoffman, J. F. (1978): Rapid photolytic release of adenosine 5'-triphosphate from a protected analogue: Utilization by the Na : K pump of human red blood cell ghosts. *Biochemistry*, 17:1929–1935.
21. Kawai, M. (1978): Head rotation or dissociation? A study of exponential rate processes in chemically skinned rabbit muscle fibers when MgATP concentration is changed. *Biophys. J.*, 22:99–103.
22. Kawai, M. (1979): Effect of MgATP on cross-bridge kinetics in chemically skinned rabbit psoas muscle fibers as measured by sinusoidal analysis technique. In: *Cross-bridge Mechanism in Muscle Contraction*, edited by H. Sugi and G. H. Pollack, pp. 149–169. University of Tokyo Press, Tokyo.
23. Kuhn, H. J. (1978): Cross-bridge slippage induced by the ATP analogue AMP-PNP and stretch in glycerol-extracted fibrillar muscle fibers. *Biophys. Struct. Mech.*, 4:169–168.
24. Kushmerick, M. M., and Davies, R. E. (1969): The chemical energetics of muscle contraction. II. The chemistry, efficiency and power of maximally working sartorius muscles. *Proc. R. Soc. Lond. [Biol.]*, 174:315–353.
25. Marston, S. B., and Taylor, E. W. (1980): Comparison of the myosin and actomyosin ATPase mechanisms of the four types of vertebrate muscles. *J. Mol. Biol.*, 139:573–600.
26. Marston, S. B., and Tregear, R. T. (1974): Nucleotide binding to myosin in calcium activated muscle. *Biochim. Biophys. Acta*, 333:581–584.
27. Podolsky, R. J. (1960): Kinetics of muscular contraction. The approach to the steady state. *Nature*, 188:666–668.
28. Podolsky, R. J., and Tawada, K. (1980): A rate limiting step in muscle contraction. In: *Muscle Contraction: Its Regulatory Mechanisms*, edited by S. Ebashi, K. Maruyama, and M. Endo, pp. 65–78. Japan Scientific Societies Press, Tokyo.
29. Simmons, R. M., and Jewell, B. R. (1974): Mechanics and models of muscular contraction. In: *Recent Advances in Physiology, Vol. 31*, edited by R. J. Linden, pp. 87–147. Churchill-Livingstone, London.
30. Stein, L. A., Chock, P. B., and Eisenberg, E. (1981): Mechanism of the actomyosin ATPase: Effect of actin on the ATP hydrolysis step. *Proc. Natl. Acad. Sci. USA*, 78:1346–1350.
31. White, H. D. (1977): Magnesium ADP binding to actomyosin-S1 and acto-heavymeromyosin. *Biophys. J.*, 17:40a.
32. White, H. D., and Taylor, E. W. (1976): Energetics and mechanism of actomyosin adenosine triphosphatase. *Biochemistry*, 15:5818–5826.

Basic Biology of Muscles: A Comparative Approach, edited by B. M. Twarog, R. J. C. Levine, and M. M. Dewey. Raven Press, New York © 1982.

Correlation Between Exponential Processes and Cross-Bridge Kinetics

Masataka Kawai

H. Houston Merritt Clinical Research Center for Muscular Dystrophy and Related Diseases, College of Physicians and Surgeons of Columbia University, New York, New York 10032

There is an important difference between a reconstituted protein system and a structured muscle system. In a reconstituted system, there is no obvious way to modify chemical reactions by imposed mechanical change, whereas in the structured physiological system, a change in the length of the muscle perturbs chemical reactions. This coupling between mechanics and chemical reactions enables the muscle to transduce energy stored in MgATP into work. An influence of such coupling was first observed by Fenn (11) who reported that isotonically contracting muscle liberated more energy than one held isometric during a tetanus. An early mathematical treatment of such a system was made by Huxley (16) who assumed that the attachment and detachment rate constants of cross-bridges are functions of the longitudinal distance between actin and myosin sites. Later, this idea was developed to signify that reaction rate constants are sensitive to the strain on the cross-bridges (18,21,42).

To detect rate constants, the length of muscle is changed and the concomitant changes in tension time course are followed. Although a step length change is most commonly used for this purpose, the length change imposed to influence cross-bridge kinetics is not limited to this waveform. It can also be sinusoidal, ramp, or white noise. The choice of the waveform depends on the purpose of the experiment. Our emphasis is to find the mapping relationship between experimentally observed rate constants and the rate constants of underlying cross-bridge reactions in skinned fibers. For this purpose, it is most appropriate to use sinusoidal length changes because of the high resolving power of the method, and because we do not need at this time the nonlinear information more readily obtained by the step length change methods. A single fiber of rabbit psoas has the diameter of 60 to 80 μm and generates only 40 to 80 dyn tension. With a perturbation method, we must detect small variations in this tension, hence we need a high signal-to-noise ratio. A skinned psoas fiber can be kept active for many tens of seconds, which is also suitable for the sinusoidal analysis. This technique detects strain-sensitive rate constants in the vicinity of zero length change when P-P (peak-to-peak) amplitude is kept small ($\leq$0.3%).

Our method is in contrast to that of Huxley and Simmons (18), who used the step length change technique because their emphasis was to correlate the nonlinear profile of the "phase 2" rate constant to that of their model, and because intact fibers cannot be kept active for long times. There is, however, a tight correlation between the sinusoidal and step methods: one has to be very secure in the theory behind the techniques before criticizing the applicabilities of either method, since many assets are common to both techniques. The correlation is graphically represented in Fig. 3.

There is another important method for perturbing the contractile system besides length change, namely chemical modification. Owing to the invention of skinned fibers (31,36), we can change the chemical environment in which contractile proteins are bathed and interactions between actin, myosin, and other ions take place. We can change an ionic species whose site of action within the cross-bridge cycle is known. We can then follow concomitant change in the rate constants. In this way we can find a mapping relationship between intrinsic rate constants and the experimentally observed rate constants. In actual muscle there are only a few known specific reactions that we can manipulate by modification of the chemical environment: the MgATP binding and concomitant detachment steps, and the work-producing (often referred to "head rotation") step, which is now known to be a Ca-sensitive step (7). We also have some idea about the mechanism behind the effect of ionic strength change (6). Thus, we have examined the effects of MgATP, Ca, and ionic strength, and have attempted to correlate the results with the underlying mechanochemistry.

METHODS

Preparation

Adult rabbit psoas is used to facilitate comparison of our results with those in the biochemical literature. Strips of psoas about 1 to 2 mm in diameter and 30 mm in length are lifted from the central portion of the muscle and tied at two places to small wooden sticks. They are chemically skinned in a saline containing (mM): 5 Na_2H_2EGTA (ethyleneglycol-bis-(β-aminoethyl ether) N,N'-tetraacetic acid), 2 Na_2MgATP (adenosine 5'-triphosphate), 180 KProp (propionate), 5 Imid (imidazole), pH 7.0, for 48 hr at 0 to 4°C. A detailed description of the chemical skinning procedure and resultant ultrastructure were published (9). The preparation remains viable in skinning solution for up to 2 weeks, or it can be kept at -20°C in an iso-ionic glycerol saline (skinning saline + 6 M glycerol) for several months. At the time of the experiment, a segment of small bundles (1 to 3 fibers) $\sim$10 mm long is dissected from a stock bundle and transferred to the experimental chamber (Fig. 1A). The ends of the preparation are wrapped around ss (stainless steel) tubes and wedged under small clips (5). In our experience, this way of mounting the fiber is the best for maintaining the viability of the preparation for repeated activations. The ss tubes are respectively connected to the length driver and the force transducer (Fig. 1A). The sarcomere length is adjusted to 2.5 to 2.6 μm (as determined by laser diffraction at 632.8 nm). Muscle at this length is defined as L_o, which is approximately 10% above the slack length and corresponds to the peak of the active length-tension diagram in rabbit psoas.

Solutions

Compositions of solutions are as follows. The "relaxing" solution contains (mM): 5 Na_2H_2EGTA, 2 Na_2MgATP, 8 NaPi (phosphate), 41–43 NaProp, 39 Na_2SO_4. EGTA is washed out by applying a solution of 2 Na_2MgATP, 8 NaPi, 41 NaProp, 44 Na_2SO_4 ("wash" solution). The standard "activating" solutions contain: 3 mM Na_2CaATP, 5 Na_2MgATP, 5 Na_4ATP, 16 Na_2CP (creatine phosphate), 74 unit/ml CPK (kinase), 8 NaPi, 43 NaProp, and 7 Na_2SO_4. Details of solution composition are given under appropriate figure legends. The concentrations of multivalent ionic species are calculated after solving the multiple equilibria of two metals (Ca, Mg) and two ligands (EGTA, ATP) by using apparent association constants as follows (log values at pH 7.00): CaEGTA 6.28, MgEGTA 1.60, CaATP 3.70, MgATP 4.00. In order to maintain ionic strength constant (standard: 200 mM) and to minimize change in monovalent cation (Na^+) concentration, change in monovalent anion ($H_2PO_4^-$, $MOPS^-$, $imidazole^-$) is replaced with propionate, and change in polyvalent anion ($HPO_4^=$, $MgATP^=$, $CaATP^=$, $H_2EGTA^=$, $CaEGTA^=$, $HATP^{3-}$, ATP^{4-}) with proper equivalence of sulfate. All experimental solutions are buffered with 10 mM MOPS (morpholinopropane sulfonic acid) or 6 mM Imid together with 7.5 to 8 mM phosphate to pH 7.00 $\pm$ 0.01, temperature is controlled to 20.0 $\pm$ 0.1°C, and the solution is constantly stirred to minimize local heterogeneities in ionic concentrations.

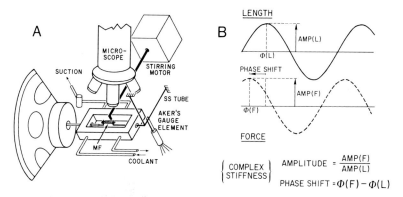

FIG. 1. **A:** Sketch of the experimental apparatus. Shown in this diagram are the muscle fiber (MF), chamber, stirrer, strain gauge assembly, a part of the length driver, and other accessories. (Reproduced from Kawai and Brandt, ref. 25.) **B:** Graphic representation of the definition of two components of complex stiffness.

Length Driver

The length driver consists of two loudspeakers and a position sensor. One speaker is used to drive a ss rod that interconnects all the moving elements including the muscle clamp. The other speaker is used to sense velocity. These signals are hybridized and fed back to the driving speaker to improve its stability and frequency response. Bandwidth is DC-300 Hz, and a step length change is complete in less than 1 msec. By use of an amplitude control circuit we can easily obtain a flat frequency response up to 1 kHz. The whole unit is practically free of drift, total excursion is ± 1.0 mm, and a length change as small as 0.1 μm is possible. With feedback, the system has a low compliance (<0.5 nm/dyn). A photoelectric position detector is used for calibration.

Force Transducer

The force transducer is a homemade unit (5); it is characterized by low compliance (2 nm/dyn), high sensitivity (0.1 dyn change can be detected), high stability, and little drift (0.005%/°C). The resonant frequency is 2kHz with a muscle clamp attached. The transducer is comprised of a base, two 20 mm pieces of #22 ss tube, and two AE801 gauge elements (Aker Microelectronics, Horten, Norway). Its framework is depicted in Fig. 1A. A #26 ss tube is bent to a right angle and firmly clamped to the base at the free ends. The cantilever thus formed can only move in one radial direction. The transducer element is adjusted to touch one leg just before the right angle junction. At this point, a small length of the tube is brazed normal to the cantilever plane and allowed to enter the chamber. Tension generated in the muscle deforms the cantilever tube and the transducer element to produce a proportionate change in resistance.

Activation Procedure

A single fiber soaked in relaxing saline was first washed to remove EGTA, then twice replaced with experimental saline containing all ingredients necessary for activation, except Ca. The volume of experimental saline was 500 to 1000 μl (this depends on the experiment). After collecting a control base-line record, a concentrated stock (50 μl) of CaProp$_2$ (MgATP and ionic strength experiments) or Ca-EGTA/EGTA (Ca experiment) was introduced. pH of injection solution was adjusted so that pH 7.00 is maintained after mixing. As soon as steady tension developed, the minicomputer was triggered to collect time course data of length and tension (see below). In rigor experiments, the preparation was then washed with the rigor saline for several (~3) times to produce the "high rigor state" (24). It was subsequently relaxed and this sequence was repeated for each experimental condition.

Sinusoidal Analysis Technique

In this technique, the length of the muscle is changed according to sinusoidal waveforms of varying frequencies, and the concomitant tension time course is analyzed in terms of amplitude and phase to yield "complex stiffness" (Fig. 1B). Peak-to-peak length oscillation is limited to a small amplitude such as 0.25% L_o. At the cross-bridge level, this corresponds to ± 1.6 nm length change per half sarcomere. Historically, sinusoidal analysis has been employed mostly on insect muscles to observe the phenomenon of "oscillatory work" (for review, see 33,34,46). We have significantly modified the technique by introduction of computer control. The waveform and magnitude of the length change are created in a minicomputer (NOVA 4S, Data General Corp.) and stored in memory, and directly applied to the length driver via D/A converters. Both tension and length time course signals are digitized and recorded in memory, and amplitude and phase elements are resolved in the programmed digital logic. We minimized the time needed for a measurement, and made the nonlinear analysis possible.

Calculation of Complex Stiffness

As soon as the periodic force time course data $F(t,f)$ are collected at each frequency f, they are expanded into a Fourier series:

$$F(t,f) = \sum_{k = -\infty}^{\infty} P_k(f) \cdot \exp(2\pi kfti) \qquad [1]$$

or in *real* nomenclature,

$$F(t,f) = P_o + 2 \sum_{k = 1}^{\infty} [Re(P_k)\cos(2\pi kft) - Im(P_k)\sin(2\pi kft)] \qquad [2]$$

where *Re* and *Im* represent the real and imaginary parts, respectively. The coefficients of the series are calculated by:

$$P_k(f) = \frac{1}{T} \int_0^T F(t,f) \cdot \exp(-2\pi kfti) \, dt \qquad [3]$$

where T is the duration of data collection $(T = n/f)$, and n (integer) is the number of cycles during this period. The term corresponding to $k = 0$ represents the average force (P_o), $k = \pm 1$ represents the linear component, and $k = \pm 2, \pm 3, \pm 4, \ldots$ represent nonlinear components. Another Fourier expansion is repeated for the length time course $L(t,f)$ to obtain $L_1(f)$, i.e., the amplitude and phase information on the length driver. This step is important because an active muscle modifies the performance of the length driver particularly at the oscillatory work frequencies.

Complex stiffness, a frequency response function relating force to length change, is defined by:

$$Y(f) = P_1(f)/L_1(f) \tag{4}$$

Complex stiffness contains both amplitude (through $|Y(f)|$) and phase information (through $\arg[Y(f)]$). These quantities are the same as the amplitude and the phase shift of complex stiffness as graphically shown in Fig. 1B, except that the above method (equations 3 and 4) is an analytical way to obtain these parameters by an automatic machine. This method also rejects noise, because equation 3 intrinsically involves a digital-filtering technique as well as the signal averaging procedure for n cycles. The complex stiffness data are corrected against the system response. Data are standardized to the size (length L_o, cross-sectional area A_o) of the preparation: $Y_M(f) = Y(f) \cdot L_o/A_0$. This quantity is called the "complex modulus." Its real part is called the "elastic modulus" and its imaginary part the "viscous modulus." Total elapsed time to collect data at 16 frequencies (0.25 to 167 Hz) is 22 sec, including waiting periods for steady state (0.25 sec × 2 at each frequency), data collection (≥ 0.4 sec), complex modulus calculation, and storage of the results in a disk file. Technical details of the sinusoidal analysis method have been published (25).

Analysis for Nonlinearity

It is our common experience that the sinusoidal waveform of the force time courses is minimally distorted in our experimental conditions. In order to validate qualitatively the application of sinusoidal analysis, we calculated the power of the nonlinear terms in the following way. From equation 3, the relative power of k-th harmonic is calculated by:

$$I_k(f) = |P_k(f)|^2/I(f) \tag{5}$$

where $k = 1,2,3\ldots$; $I_1(f)$ is the relative linear power, and $I(f)$ is half the oscillation power:

$$I(f) = \sum_{k=1}^{\infty} |P_k(f)|^2 = \frac{1}{2T} \int_0^T [F(t,f) - P_o]^2 dt$$

P_o is the average force. The relative nonlinear power is calculated by:

$$N(f) = \sum_{k=2}^{\infty} I_k(f) = 1 - I_1(f)$$

RESULTS

Complex Modulus of Relaxed, Active, and Rigor Muscles

Figure 2A–C shows complex modulus data from three conditions (rest, active, and rigor). Figure 2A is a plot of dynamic modulus (= $|Y_M(f)|$) versus frequency; Fig. 2B shows phase shift (= $\arg[Y_M(f)]$) versus frequency. In Fig. 2C ("Nyquist plot"), the data are plotted as elastic modulus (abscissa) versus viscous modulus (ordinate). Different points correspond to different frequencies. The Nyquist plot is useful in judging both the quality and the character of the modulus data.

The complex modulus data are relatively simple and constant for relaxed or rigor muscles. At L_0, relaxed muscle is least stiff (Fig. 2A), and points scatter around

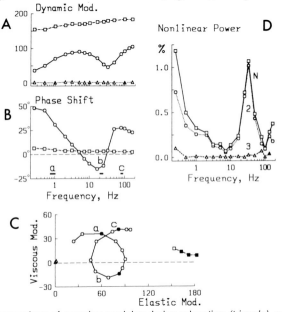

FIG. 2. **A-C,** Comparison of complex modulus during relaxation *(triangle),* activation *(circle),* and high-rigor *(square).* Data from 5/23/78, experiment #2, record #46–49; a bundle of two fibers; P_0 = 1.35 Mdyn/cm². (Data are reproduced from Kawai and Brandt, ref. 25.) Unit of moduli is Mdyn/cm². Frequencies used are 0.25, 0.5, 1, 2, 3.13, 5, 7.14, 10, 16.7, 33, 50, 80, 100, 133, and 167 Hz (some of the frequency points are omitted from the plots for simplicity). P-P length change: 0.23% L_0. In **B,** *horizontal bars along abcissa* indicate characteristic frequencies *a, b, c* and their 95% confidence ranges obtained after Fisher's F-statistics. Phase shift of relaxed muscle is indeterminant and not entered here. In **C,** approximate locations of characteristic frequencies *a, b, c* are marked in Nyquist plot; *filled symbols* correspond to 1, 10, and 100 Hz frequency points. **D:** Plot of relative nonlinear powers for the active fiber. N = Sum of all nonlinear terms; 2 = second order term; 3 = third order term. Higher order terms (not plotted) are less than that plotted for the third order term. The relative linear power $l_1(f)$ is just the mirror image of the curve N (equation 6). In **A–D,** activating solution contained (mм): 3 Na₂CaATP, 5 Na₂MgATP, 5 Na₄ATP, 8 NaPi, 16 Na₂CP, 7 Na₂SO₄, 43 NaProp, 6 Imid, 74 unit/ml CPK, pCa 4.00. Rigor saline contained: 1 CaProp₂, 8 NaPi, 45 Na₂SO₄, 43 NaProp, 6 Imid. (Solution composition for the active and rigor data of Figs. 2, 3, and 6a of ref. 25 is in error.)

the origin of the Nyquist plot (Fig. 2C). Rigor muscle is most stiff, and the modulus has a weak frequency dependence (Fig. 2A and B). This pattern has been observed in rabbit psoas (25) and crayfish walking leg (26) muscles. The fact that the complex modulus of rigor muscle is approximately constant over the frequency range examined establishes two points: there are no artifactual time constants in our measuring system, and in rigor, there is no chemical transition between cross-bridge states in our frequency range.

The Nyquist plot of activated muscle consists of three hemicircles (Fig. 2C). This is compatible with the existence of at least three "exponential processes" (26). These are called exponential processes because the same underlying mechanism shows exponential time courses on step input (29). We name the respective processes (A), (B), (C) in the order of slow to fast (marked a, b, c in Fig. 2B and C). Process (A) is a low frequency exponential advance, and centers around 1 Hz ($\sim a$) where the muscle absorbs net work from the length driver. Process (B) is a medium frequency exponential delay centering around 17 Hz ($\sim b$); here the muscle generates oscillatory work. Process (C) is a high frequency exponential advance centered at around 100 Hz ($\sim c$), and again the muscle absorbs work. We also identified the existence of three processes in fully activated frog semitendinosus muscle and crayfish walking leg muscle (25,26). It seems that presence of these processes is a universal property of fast skeletal muscles. A transfer function that involves three exponential processes is:

$$\overset{\text{Process(A)}}{}\quad\overset{\text{Process(B)}}{}\quad\overset{\text{Process(C)}}{}$$
$$Y(f) = Y_A + Y_B + Y_C = H + Aif/(a + if) - Bif/(b + if) + Cif/(c + if)\,[7]$$

where A, B, C are magnitude parameters that are population factors; a, b, c are characteristic frequencies of respective processes; and H is the DC stiffness. Note that process (B) has a negative polarity. As shown in equation 8 below, 2π times a, b, c represent apparent rate constants of the respective processes. Data from activated muscle in Fig. 2 are fitted to equation 7, and their theoretical curve is plotted in Fig. 3A in the form of a Nyquist plot. An equivalent mechanical model for the response predicted by equations 7 and 8 is shown in Fig. 3C.

Instantaneous Elasticity

In equation 7, instantaneous elasticity can be obtained by extrapolating f to infinity: $Y_\infty = H + A - B + C$. Actually Y_∞ thus obtained is an underestimate of the true value, because the complex modulus data are better described by adding more exponential advance (positive) terms to equation 7 at the higher frequency end (cf., 2,12). Instantaneous elasticity is generally assumed to be proportionate to the number of cross-bridges present at the time of the length change. In this case, we can calculate the number of cross-bridges made during activation compared to the number during rigor by taking the ratio of Y_∞ from the two conditions. We use the 100 Hz-stiffness of the "high-rigor" state as the reference point because this rigor is the stiffest of all rigor conditions (24) (stiffness and tension of rigor depend on the pathway to induce it). We obtained this ratio to be 0.57 ± 0.10 (1 SD, $N = 7$) for the activating condition given in Fig. 2. Thus we can conclude as much as 57% of cross-bridges can be attached during maximal activation if compared to the rigor case. Again, this value is an underestimate (above), and it depends on activating conditions: it becomes higher in a saline that has less MgATP and/or Pi than our standard condition.

The ratio P_0/Y_∞ indicates the amount of length release required to abolish full tension, and generally assumed to be a measure of the working distance of cross-bridges. We find P_0/Y_∞ to be $0.92 \pm 0.09\% \, L_0$ ($N = 12$) during our standard activation, and $0.56 \pm 0.11\% \, L_0$ ($N = 12$) during rigor. The ratio during activation is larger than that obtained from intact preparations, because (i) the ratio approximately doubles on skinning (M. Kawai, *unpublished observation* on crayfish fibers), (ii) Y_∞ is an underestimate, and (iii) the activating condition probably does not match to that in intact preparation.

Nonlinearity

The results of calculations of the relative nonlinear power in data from activated muscle preparations are summarized in Fig. 2D. Only terms corresponding to $I_2(f)$, $I_3(f)$, and $N(f)$ are shown since higher order terms vanish rapidly. As shown in Fig. 2D, most of the nonlinearity is explained by the second order term. In rabbit psoas, when P-P length oscillation is $0.23\% \, L_0$, the nonlinear power does not exceed 1.2%, so that the linear power is $\geq 98.8\%$. The maximum of nonlinearity takes place at frequencies around process (B), and peaks approximately at the frequency where the dynamic modulus becomes minimum (cf., Fig. 2A and D). Nonlinear power associated with processes (A) and (C) are smaller. In Fig. 2D, this amounts to $<0.4\%$. Nonlinear power of the rigor muscle is likewise small and we obtained $<0.2\%$ for the experiment in Fig. 2.

Calculation of $P_1(f)$ by means of the Fourier expansion (equations 1–3) corresponds to fitting a periodic force time course $F(t,f)$ to sine and cosine functions of the same periodicity. This Fourier transform is a linear regression method involving three fitting parameters (equation 2 with a truncation at $k = 1$), and, therefore, regression coefficients can be used to examine goodness of fit. The regression coefficient is the square root of the relative linear power. We obtained that the regression coefficient is ≥ 0.994 for the active muscle, and 0.999 for rigor muscle. From this observation it is apparent that, under our experimental conditions, the force response is so linear that its time course is described adequately by sine and cosine functions of the same periodicity.

Correlation Between Sinusoidal and Step Analyses

It is useful to correlate the processes we observe in sinusoidal analysis to the "phases" or "stages" (17,18) of step analysis. The correlation uses equation 7 and inverse Laplace transforms. This procedure was published (23,25) and is presented

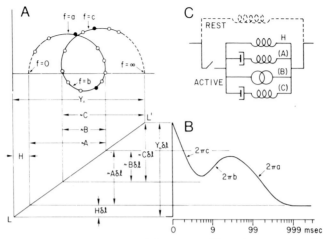

FIG. 3. Correlation between sinusoidal **(A)** and step **(B)** analyses, and equivalent diagram **(C)**. (Reproduced from Kawai and Brandt, ref. 25.) **A:** Theoretical curve to describe complex modulus data of an active muscle is represented in a Nyquist plot. The curve and points (corresponding to our experimental frequencies) are calculated by using equation 7 with best fit parameters to the active curve of Fig. 2A–C: a = 0.87 Hz, b = 21.9 Hz, c = 83.8 Hz; H = 15.1, A = 80.8, B = 86.3, C = 118 (unit in Mdyn/cm²). Approximate magnitude or position of these parameters are indicated. **B:** Simulated force time course following a step length increase. The data are calculated with equation 8 with same parameters used to construct **A**. Rate constants of each phase are indicated by *arrows*. Note logarithmic time scale. *Diagonal line LL'* symbolizes the inverse Laplace transformation. **C:** Equivalent mechanical setup of three rate processes in equations 7 and 8. Note that **(A)** and **(C)** have different time constants.

graphically in Fig. 3. The force time course for a given step length increase δl is given by:

$$\text{(Phase 2)} \qquad \text{(Phase 3)} \qquad \text{(Phase 4)}$$
$$\Delta F(t) = [C\exp(-2\pi ct) - B\exp(-2\pi bt) + A\exp(-2\pi at) + H] \cdot \delta l \quad [8]$$

for $t \geq 0$. The equation is rearranged in the order fast to slow, and is plotted in Fig. 3B. The unique (negative) polarity of process (B) (phase 3) allows it to be identified unambiguously. We define process (C) (phase 2) as the one faster than (B), and process (A) (phase 4) as that slower than (B). Instantaneous elasticity Y_∞ ($= H + A - B + C$) corresponds to phase 1, and DC stiffness H to the "residual tension" studied by Sugi (40) and by many others. This method of identification and correlation of each process or phase is useful when different muscle preparations are compared under varying experimental conditions. The correlation is limited to the vicinity of small length change.

MgATP Effect

MgATP is known to be the substrate of actomyosin ATPase (4,36,44). It binds to the myosin moiety of the rigor cross-bridges and promotes dissociation of the myosin head from actin (28). Therefore, we can follow MgATP binding and dissociation reactions by changing the substrate concentrations and observing the changes in the exponential processes. The experiment was carried out with MgATP concentration in the range of 0.25 to 10 mM. Free ATP was kept either at 1 or 5 mM in order to determine whether free ATP competes with MgATP for the substrate site, and whether or not the ATP is adequately supplied. Fig. 4A and B represents plots of complex modulus versus frequency at varying substrate concentrations and fixed free ATP (1 mM); Fig. 4C represents Nyquist plots of the same data. As seen in Fig. 4B, the characteristic frequencies *b* and *c*, represented by peaks of the plot,

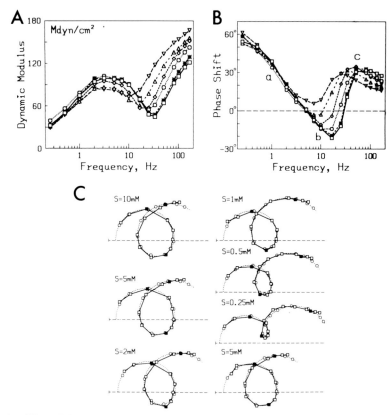

FIG. 4. Effect of MgATP on complex modulus. Experiment on a bundle of two fibers of 5/25/78, record #15–29. Activating solution contained (mM): 3 Na$_2$CaATP, 1 Na$_4$ATP, 10 Na$_2$MgATP/Na$_2$SO$_4$ to achieve a desired substrate concentration, 14 Na$_2$SO$_4$, 8 NaPi, 16 Na$_2$CP, 43 NaProp, 6 Imid, 74 unit/ml CPK, pCa 3.42–3.53. P-P length change: 0.22% L_o. Active tension ranged 1.34–1.47 Mdyn/cm^2, and peaked at 1 mM MgATP. **A:** Plot of dynamic modulus versus frequency. **B:** Phase shift versus frequency. In both **A** and **B**, MgATP concentrations are: 10 *(square)*, 5 *(star)*, 2 *(circle)*, 1 *(diamond)*, 0.5 *(triangle)*, and 0.25 mM *(inverted triangle)*. **C:** Nyquist plot of the same data. The number on the *upper left* corner of each panel indicates MgATP concentrations. Experiments were done in the decrement order of MgATP concentration, and the condition at 5 mM was repeated at the end. *Thick full lines* and *squares* are from experimental results; *thin dotted lines* and *circles* are results calculated from equation 7 based on best fit parameters. Filled squares signify 1, 10, and 100 Hz frequency points. Axes are not entered for visual simplicity (*abscissa* is elastic modulus, and *ordinate* is viscous modulus).

change continuously in this substrate range. It is seen in Nyquist plots that the effect of MgATP is largest on the rate constant and magnitude of process (B), whereas substantial effect is noticed on the rate constant of process (C). Process (A) seems to be little affected in this substrate concentration range.

The data are fitted to equation 7 to obtain apparent rate constants. The theoretical values of complex stiffness are entered in thin dotted lines in Fig. 4C. The fit is relatively good in the middle frequency range, but it is approximate in the high frequency range. The apparent rate constants $2\pi b$ and $2\pi c$ are plotted against MgATP concentrations in Fig. 5A and C for two different free ATP conditions. To see how well the rate constant fits to Michaelis-Menten enzyme kinetics (30) (equation 10 below), data are also plotted in double reciprocal scales (Fig. 5B and D). As seen in these plots, rate constant $2\pi b$ fits very well to the Michaelis-Menten kinetics with a regression coefficient of 0.99 (Table 1), whereas $2\pi c$ has an approximate fit with a regression coefficient ranging from 0.94 to 0.98. K_m, k_1, k_2 are extracted from such data, and their average values are listed in Table 1. From this table it is clear that K_m for b and c is different (0.7 to 0.8 and 0.2 mM, respectively), whereas the effect of free ATP on this parameter is minimal.

We found that the kinetics are very different in the μM substrate concentration range (8). The complex modulus data continuously change and approach those of rigor as the substrate concentration is lowered; K_m for the rate constant of process

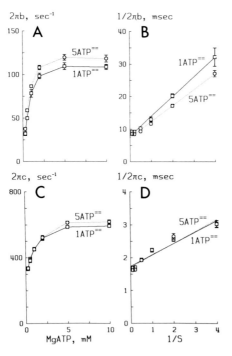

FIG. 5. Plots of MgATP concentration vs rate constants $2\pi b$ (in **A**, **B**) and $2\pi c$ (in **C**, **D**) at two different free ATP concentrations (indicated). Averaged data are plotted in linear scales (**A**, **C**), and in double reciprocal scales (**B**, **D**). $N = 10$ for 5 mM free ATP, $N = 5$ for 1 mM free ATP experiments. SEM bars are indicated when they are larger than the symbol size.

TABLE 1. K_m and intrinsic rate constants[a]

Rate constants	Free ATP (mM)	N	K_m (mM) $\pm$ 1 SD)	k_2 (s^{-1} $\pm$ 1 SD)	k_1 (10^6 M^{-1} s^{-1})	Regression coefficient
2πc (phase 2)	1	5	0.20 ± 0.03	576 ± 25	2.9	0.95–0.98
	5	10	0.21 ± 0.04	590 ± 35	2.8	0.94–0.98
2πb (phase 3)	1	5	0.79 ± 0.19	130 ± 7	0.16	0.99
	5	10	0.68 ± 0.11	138 ± 11	0.20	0.99
2πa (phase 4)	1		0.01 ± 0.005[b]	~6	0.4–1.2	

[a]k_1 is calculated by k_2/K_m. Regression was carried out for each experiment.
[b]Data from Cox and Kawai, ref. 8.

(A) is about 10 μM (Table 1). The mapping relationship is necessarily different in this extremely low substrate range. Since our current interest is to find out mapping relationship at the physiological condition, we shall not further consider this low substrate experiment in this report.

Ca Effect

Ca experiments were carried out in the presence of EGTA (6 mM total) and Ca to achieve the desired pCa ($-\log[\text{Ca}^{2+}]$). The purpose of these experiments is to find out a process that is affected by Ca change, and to determine whether or not apparent rate constants change in a graded manner with Ca concentration. There has been some controversy over the effect of Ca on cross-bridge kinetics (14,19,20,32,41,48). We hope our sinusoidal analysis methodology provides an independent insight to this problem. Ca experiments are also interesting because of the new evidence of Chalovich et al. (7), who showed that the Ca sensitive reaction is not the attachment step, but the Pi-release step that presumably accompanies head rotation and work production. Therefore, by changing Ca concentration we must be able to map the kinetics of the Pi-release/work-production reaction to exponential processes.

Results of Ca experiments are reported (27) and reproduced in Fig. 6. It is seen from Fig. 6 (left) that tension threshold occurs at pCa 6.0 and maximum tension is observed by pCa 5 to 5.5 for the standard activating conditions. The pCa-tension relation is very steep with a Hill coefficient of 4 or larger (5).

The phase-frequency plot from fully activated preparations is V-shaped (Fig. 6 right), i.e., there is only one region of negative phase (a phase-delay). It is in this frequency range that the muscle generates oscillatory work. As Ca concentration is lowered, the position of this minimum does not shift, indicating that there is little change in the rate constant 2πb. In the plots from partially activated fibers, however, there is an additional minimum in the lower frequency range (arrow in Fig. 6 right). The main minimum centers at around 14 Hz, the extra minimum at 3 Hz. Hence, the phase-frequency plot has a W-shaped appearance. We infer from this observation that process (B) splits into two processes (B) and (B′). Thus, our

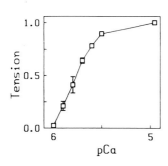

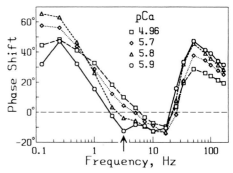

FIG. 6. Results of Ca experiment obtained on a bundle of three fibers. Data from 4/6/79, experiment #1, record #3–13. (Data reproduced from Kawai et al., ref. 27.) Activating solution contained (mM): 6 $Na_2H_2EGTA/Na_2CaEGTA$ to achieve a desired pCa, 5 Na_2MgATP, 5 Na_4ATP, 7.5 NaPi, 4 Na_2SO_4, 42 NaProp, 10 MOPS, 16 Na_2CP, 80 unit/ml CPK. P-P length change: 0.17% L_o. P_o: 1.91 Mdyn/cm^2. Frequency range: 0.125–167 Hz. **Left:** pCa-tension plot. **Right:** Phase-frequency plot (the *arrow* indicates an extra valley evident at partial activation).

(B) term in equation 7 should be rewritten to include this additional term for partial activation:

$$Y_B(f) = -B'fi/(b' + fi) - Bfi/(b + fi)$$

where $b' = 3$ Hz, $b = 14$ Hz approximately, and these values appear to remain constant as Ca concentration changes. The effect of Ca is, therefore, very different from that of MgATP, which in low mM concentrations causes a graded shift in both b and c frequencies (Figs. 4B and 5). We also noticed that change in rate constants of processes (A) and (C) is minimal with Ca concentration (Fig. 6 right). We obtained very similar results with Ca experiments at two additional ionic strengths (128 and 265 mM).

The effect of Ca ion on the optimum frequency of oscillatory work—equivalent to process (B)—was also observed by Abbott (1) in glycerinated insect flight muscle, and on the time course of "delayed tension"—equivalent to phase 3—in glycerinated caridac preparations by Herzig and Herzig (15). Qualitative aspect of their results is the same as the present report, although there is some difference in details.

Ionic Strength Effect

The ionic strength experiment was carried out at maximum Ca activation (pCa 4.96). Ionic strength was changed between 128 and 265 mM. At 128 mM, the saline did not contain sulfate nor propionate. Figure 7 is the result of such an experiment. As is seen in Fig. 7C, tension and instantaneous elasticity ($Y_∞$) decrease monotonically at the higher ionic strength. Magnitude parameters *(A, B, C)*, represented by the arks of Nyquist plots, likewise decrease to the higher ionic strength, except that C decreases faster than A or B (Fig. 7A and C). It appears that there is no big effect on the position of the peaks in phase-frequency plot (Fig. 7B). When data are fitted to Eqn. 7, we find almost identical rate constants in all ionic strength range examined (Fig. 7D). This rate constant result is generally consistent with the observations of Gulati and Podolsky (14) on "null time," or Julian and Moss (20) on V_{max} mea-

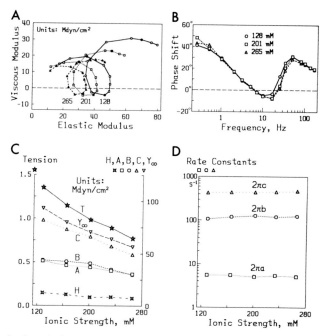

FIG. 7. Results from ionic strength experiments at (mM): 6 Na$_2$CaEGTA, 5 Na$_2$MgATP, 5 Na$_4$ATP, 7.5 NaPi, 9.4 Na$_2$CP, 80 unit/ml CPK, 0–40 NaProp, 0–33 Na$_2$SO$_4$, 10 MOPS, pCa 4.96. Data of 5/12/80, experiment #2, record #22–31; a bundle of three fibers; P-P length change: 0.20% L_o. **A:** Nyquist plot from three different ionic strengths (indicated in mM). Results from 16 frequencies between 0.25 and 167 Hz (data from 1, 10, 100 Hz are indicated by *filled symbols*). **B:** The same data shown in phase shift versus frequency plot. **C:** Tension *(left ordinate)* and magnitude parameters *(right ordinate)* plotted against the ionic strength. **D:** Three rate constants versus ionic strength. Magnitude parameters and rate constants are obtained after fitting the complex modulus data to equation 7.

surement. The fact that relative magnitude C is smaller at higher ionic strength is consistent with the observation by Gulati and Podolsky (14) who reported that the magnitude of "length transient" is larger at higher ionic strength (since the tension and length transients are reciprocally correlated).

DISCUSSION

It is now known that labile myosin cross-bridges cyclically interact with actin to convert chemical energy into work. To achieve this transduction, the actin-myosin-substrate complex goes through several internal changes. The cross-bridge states of physiological interest are (see Fig. 8): (i) rigor-like state AM, (ii) substrate-bound state to AM, (iii) dissociated state A + MS, (iv) non-refractory state, (v) weakly attached state, and (vi) tension state. Cross-bridges are "made" in states (i), (ii), (v), and (vi); they are "open" in states (iii) and (iv). This scheme is based on biochemical work (7,28,39,43,47). From one state to an adjacent state there is a finite transition probability, and this can be represented by an intrinsic rate constant. The apparent rate constants we observe are complicated functions of the intrinsic rate constants. It is the purpose of the present report to establish their functional relationship.

Mechanisms of MgATP Action

The most significant observation in our recent work is that the rate constants $2\pi b$, $2\pi c$ of processes (B) and (C) are affected by a low mM change of MgATP (substrate) concentration (22,23) and that this effect is graded. This observation is confirmed again by the experiments shown in this chapter (Fig. 4). It is further shown that rate constant versus substrate relationship follows Michaelis-Menten kinetics (Fig. 5) with a K_m of 0.7 to 0.8 mM for $2\pi b$, and 0.2 mM for $2\pi c$ (Table 1). This kinetics (30) is schematically represented as follows:

$$AM + S \underset{k_{-1}}{\overset{k_1}{\rightleftharpoons}} AMS \overset{k_2}{\rightarrow} \text{Product}$$

$$r = \frac{1}{AM_T} \cdot \frac{d}{dt} \text{(product)} = \frac{k_2}{1 + K_m/S} \; ; \; K_m = \frac{k_{-1} + k_2}{k_1} \qquad [10]$$

where AM is actomyosin, S is substrate, $AM_T = (AM) + (AMS)$, and r is a composite rate constant we assume is measured by $2\pi b$ or $2\pi c$. We show only one scheme here, but it applies to both $2\pi b$ and $2\pi c$ with different sets of intrinsic rate constants, k_1, k_{-1}, and k_2. The substrate binding step (k_1) is a second order reaction and its rate is proportionate to the substrate concentration. To calculate k_1 we assumed the back reaction (k_{-1}) to be negligible based on biochemical observations (47; $k_{-1} \ll k_2$). An approximate fit (Fig. 5, Table 1) of the data to equation 10 suggests that the above scheme is a gross representation of the reality, although the mechanisms that produce processes (B) and (C) may be complicated ones.

In the case of process (C), we have additional evidence that the step involving k_1 is substrate binding to AM, and k_2 is the subsequent dissociation step. This is based on our result (Table 1) that $k_1 \sim 3 \times 10^6 \text{ M}^{-1}\text{s}^{-1}$, and $k_2 \sim 600\text{s}^{-1}$. The k_1 is within the range of biochemically observed value ($1 - 4 \times 10^6 \text{ M}^{-1}\text{s}^{-1}$) by Lymn and Taylor (28), and White and Taylor (47); k_2 is also within the range of that found by the above authors ($\geqslant 500\text{s}^{-1}$). Abbott and Steiger (2) argued that phase 2 involves dissociation based on its temperature sensitivity.

Process (B) is more complicated than (C) because of its involvement in oscillatory work and energy transduction (see below); $k_2 = 130 - 138\text{s}^{-1}$ suggests that this rate limiting step is somewhere in the work-producing step. Process (B) has a smaller k_2 and a larger K_m (Table 1). This results in a 15 times difference in k_1 values between $2\pi b$ and $2\pi c$. This difference may indicate that the mechanism of substrate binding reaction is different at two different MgATP concentrations, such that MgATP binds to rigor-like state AM (step 1 in Fig. 8) at lower concentrations, whereas MgATP molecule displaces ADP molecule in order to bind to the site at higher concentrations (steps 1 and 6 are combined). Alternatively, it is possible

Process (B)

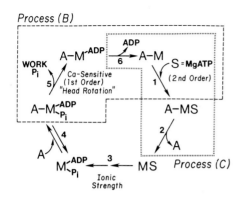

FIG. 8. Cross-bridge states and reaction steps. The scheme is based on Tonomura et al. (43), Lymn and Taylor (28), White and Taylor (47), Stein et al. (39), and Chalovich et al. (7). A, actin; M, myosin; S, MgATP; Pi, phosphate. Only states and reaction steps presumably involved in the actomyosin cross-bridge cycle are shown, and some states are lumped together. *Arrows* are entered (except for step 4) to indicate direction of ATP hydrolysis and work production, but most of the reaction steps are reversible. Reactions included in processes (B) and (C) are surrounded by *broken* and *dotted lines*, respectively.

that the AMS state in the scheme leading to equation 10 includes all nucleotide-bound states with a larger K_m.

Adequacy of ATP Buffering

Based on the observation that rate constants are invariant with Ca concentration (Fig. 6C), we can convince ourselves that ATP is adequately supplied in the core of the fiber. If not, then the rate constants $2\pi b$ and $2\pi c$ should decrease with increasing Ca concentration, because of an increased ATP demand at higher tension in which more oscillatory work is output. This inference is based on the observation that both rate constants are functions of physiological MgATP concentrations (Fig. 5), and that oscillatory power output and ATP hydrolysis rate are linearly correlated (35,37,38). The present observation that almost no effect of free ATP (Fig. 5, Table 1) on rate constants and their K_m values is consistent with our conclusion that ATP is adequately supplied in the fiber. The small effect of free ATP may correlate with a specific effect of free ATP, or with that of Mg^{2+}, which changes together with the free ATP concentration change.

Free ATP Effect as Evidence Against Competitive Inhibition

There is a possibility that free ATP competes with MgATP for the substrate site on the myosin head. If this is the case, then apparent K_m should *increase* with increase in free ATP concentration (pp. 295–296, ref. 30). This clearly is not the case (Table 1). Thus, we can conclude that free ATP is not a competitive inhibitor.

Ca Regulatory Mechanism and Work-Producing Step

In vertebrate skeletal muscles, actomyosin ATPase is regulated via a control mechanism on the thin filaments in which Ca^{2+} plays a central role (3,10,45). Ca binding to a TnC (troponin C) molecule turns on a segment of thin filament, consisting of seven monomeric actins, via the troponin-tropomyosin mediated control system; increasing Ca^{2+} concentration presumably results in an increased number of active actin units.

A new approach to the Ca regulatory mechanism can be constructed from the recent report of Stein et al. (39). They concluded that the "dissociated" actin and myosin species are, in fact, in rapid equilibrium with a "weakly associated" species (step 4 in Fig. 8). Furthermore, Chalovich et al. (7) concluded that Ca binding to the regulatory proteins affects a reaction subsequent to cross-bridge attachment, rather than the attachment itself (see also 13). In other words, the attachment transition (step 4) is fully reversible and does not require Ca, whereas the subsequent reaction (step 5), which releases Pi, requires Ca-activated actin. After this second reaction, myosin heads are strongly bound to actin and the back dissociation becomes a very low probability event. This reaction is a first-order reaction, and it is generally assumed to be the work-producing step. A large free energy drop occurs somewhere in steps 5 and 6 (47) at physiological substrate/product concentrations.

This new discovery by Chalovich and others suggests a way to identify a process that involves the work production step. A change in Ca concentration should affect only the process associated with head rotation. Our experimental results (Fig. 6B) shows that the phase-frequency plot appears V-shaped at full activation; it appears W-shaped at partial activation. From this observation we inferred that the thin filament may exist in three forms: relaxed, partially activated, and fully activated (27). This Ca effect provides evidence that the work production step is involved in process (B) (phase 3), rather than in process (C) (phase 2) as originally proposed by Huxley and Simmons (18). This conclusion may be more apparent when we realize that net work can be extracted from a muscle at frequencies around process (B).

Our observation that none of the rate constants we measure are altered by Ca concentration (Fig. 6B) requires further consideration. We interpret this result as follows: since head rotation (step 5) is a first-order reaction (cross-bridges are already made), its rate constant could not be modified by increased availability of actin caused by Ca binding to TnC. The rate constant of the transduction step may be governed by the condition of actin, however. Since activated actin can be in one of two states (either partially or fully active) (27), there must be two classes of rate constants. When the myoplasmic Ca ion concentration increases in the low concentration range, the population of partially active actin increases, and so does the number of cross-bridge heads rotating with a lower rate constant. When the Ca ion concentration increases further, the population of fully active actin increases in part at the expense of partially active ones, and this is accompanied by an increased number of cross-bridges rotating with a higher rate constant. This number reaches a maximum as all TnC sites are occupied by Ca ions. In this mechanism, the kinetics would not be modified by Ca ion concentration if the partially active form of actin were not present; the appearance of an extra process (B') helps to localize the transduction step among the experimentally observed processes (A), (B), and (C).

A Comment on the Maximum Velocity of Shortening

Our current observation that there is an extra slow process (B') at partial Ca activation predicts that V_{max} would be slower if the force-velocity experiments were carried out under these conditions. This is because V_{max} depends on both slowly, and quickly cycling cross-bridges, whereas our measurement separately detects the two bridges acting in parallel. The above prediction is consistent with observations by Julian (19) on skinned frog fibers, and by Wise et al. (48) on glycerinated rabbit psoas fibers, which showed that V_{max} increases with tension in the full Ca range. Our finding, however, that none of the rate constants change with Ca concentration, is consistent with a switch mechanism similar to the one proposed by Podolsky and Teichholz (32). A new addition in our case is that there have to be two qualitatively different switches to account for the data. Our observation of an existence of slowly cycling cross-bridges at partial activation may bridge the gap between observation of graded and all-or-none behavior.

There is a possibility that V_{max} and rate constants do not correlate. This is an argument that the strain on a cross-bridge is very different in two experimental conditions, hence the kinetics can be different. The significance of this argument depends on the degree of strain sensitivity of the reaction rate constants, and thus the relationship between the two measurements depends on assumptions of a model. Therefore, it may be unrealistic to expect too tight a correlation between the results of the two techniques.

Refractory States and the Ionic Strength Effect

Brandt et al. (6) argue that the major effect of the ionic strength change is to modify refractory to nonrefractory transition (step 3 in Fig. 8; see ref. 39 for the terminology) based on their pCa-tension data and kinetic scheme, which involves Ca binding and unbinding to TnC. Under the present observation, it appears that, for the first approximation, lowering ionic strength recruits more cross-bridges into the active pool without affecting kinetics of individual cross-bridges. This inference is based on the result that all population parameters increase toward the lower ionic strength (Fig. 7C) without changing the respective rate constants (Fig. 7D). The recruited cross-bridges must have come from detached states, otherwise the instantaneous elasticity Y_∞ would not increase toward the lower ionic strength (Fig. 7C). The refractory state is the most likely source for the extra cross-bridges. The reaction, involving refractory to nonrefractory transition, is unlikely to be detected readily by our rate constant measurements, because the refractory transition is a slow reaction, and because many cross-bridges are detached during this transition (cf. 39). Thus, the present observation is consistent with the hypothesis of Brandt et al. (6) in terms of the site of action of the ionic strength change within the cross-bridge cycle.

CONCLUSIONS

We are able to establish a functional relationship between two important cross-bridge reactions and observed processes. Process (B) (phase 3) involves the work-producing step (head rotation) and substrate binding step; process (C) (phase 2) involves substrate binding and subsequent dissociation steps (Fig. 8). Since the rate constant of process (C) is larger than that of (B), it follows that the rate constant of the binding/dissociation reaction must be larger than that of head rotation. This functional relationship will play an important role in our future conceptualization and modeling of actomyosin interaction and muscle contraction. The sinusoidal analysis technique has proven to be a powerful tool in the study of cross-bridge kinetics.

ACKNOWLEDGMENTS

The author is grateful to Drs. H. Grundfest and P. W. Brandt for their useful comments and discussions during the preparation of the manuscript. The present work is supported in part by grants from NIH (AM21530, NS11766), NSF (PCM80–14527), and MDA.

REFERENCES

1. Abbott, R. H. (1973): The effects of fibre length and calcium ion concentration on the dynamic response of glycerol extracted insect fibrillar muscle. *J. Physiol. (Lond.)*, 231:195–208.
2. Abbott, R. H., and Steiger, G. J. (1977): Temperature and amplitude dependence of tension transients in glycerinated skeletal and insect fibrillar muscle. *J. Physiol. (Lond.)*, 266:13–42.
3. Adelstein, R. S., and Eisenberg, E. (1980): Regulation and kinetics of the actin-myosin-ATP interaction. *Annu. Rev. Biochem.*, 49:921–956.
4. Bozler, E. (1956): The effect of polyphosphates and magnesium on the mechanical properties of extracted muscle fibers. *J. Gen. Physiol.*, 39:789–800.
5. Brandt, P. W., Cox, R. N., and Kawai, M. (1980): Can the binding of Ca^{++} to two regulatory sites on troponin-C determine the steep pCa/tension relation of skeletal muscle? *Proc. Natl. Acad. Sci. USA*, 77:4717–4720.
6. Brandt, P. W., Cox, R. N., Kawai, M., and Robinson, T. (1982): Regulation of tension in skinned muscle fibers: Effect of cross-bridge kinetics on apparent Ca^{2+} sensitivity. *J. Gen. Physiol.*, *(in press)*.
7. Chalovich, J. M., Chock, P. B., and Eisenberg, E. (1981): Mechanism of action of troponin · tropomyosin. *J. Biol. Chem.*, 256:575–578.
8. Cox, R. N., and Kawai, M. (1981): Alternate energy transduction routes in chemically skinned rabbit psoas muscle fibers: A further study of the effect of MgATP over a wide concentration range. *J. Mus. Res. Cell Motil.*, 2:203–214.
9. Eastwood, A. B., Wood, D. S., Bock, K. L., and Sorenson, M. M. (1979): Chemically skinned mammalian skeletal muscle. I. The structure of skinned rabbit psoas. *Tissue Cell*, 11:553–566.
10. Ebashi, S., and Endo, M. (1968): Calcium ion and muscle contraction. *Prog. Biophys. Mol. Biol.*, 18:123–183.
11. Fenn, F. O. (1923): A quantitative comparison between the energy liberated and the work performed by the isolated sartorius muscle of the frog. *J. Physiol. (Lond.)*, 58:175–203.
12. Ford, L. E., Huxley, A. F., and Simmons, R. M. (1977): Tension responses to sudden length change in stimulated frog muscle fibers near slack length. *J. Physiol. (Lond.)*, 269:441–515.
13. Greene, L. E., and Eisenberg, E. (1980): Cooperative binding of myosin subfragment-1 to the actin-troponin-tropomyosin complex. *Proc. Natl. Acad. Sci. USA*, 77:2616–2620.
14. Gulati, J., and Podolsky, R. J. (1978): Contraction transients of skinned muscle fibers: Effects of calcium and ionic strength. *J. Gen. Physiol.*, 72:701–715.
15. Herzig, J. W., and Herzig, U. B. (1974): Effect of Ca-ions on contraction speed and force generation in glycerinated heart muscle. *Symp. Biol. Hung.*, 17:85–88.
16. Huxley, A. F. (1957): Muscle structure and theories of contraction. *Prog. Biophys.*, 7:255–318.
17. Huxley, A. F. (1974): Muscular contraction. *J. Physiol. (Lond.)*, 243:1–43.
18. Huxley, A. F., and Simmons, R. M. (1971): Proposed mechanism of force generation in striated muscle. *Nature*, 233:533–538.
19. Julian, F. J. (1971): The effect of calcium on the force-velocity relation of briefly glycerinated frog muscle fibres. *J. Physiol. (Lond.)*, 218:117–145.

20. Julian, F. J., and Moss, R. L. (1981): Effects of calcium and ionic strength on shortening velocity and tension development in frog skinned muscle fibres. *J. Physiol. (Lond.)*, 311:179–199.
21. Julian, F. J., Sollins, K. R., and Sollins, M. R. (1974): A model for the transient and steady-state mechanical behavior of contracting muscle. *Biophys. J.*, 14:546–562.
22. Kawai, M. (1978): Head rotation or dissociation? A study of exponential rate processes in chemically skinned rabbit muscle fibers when MgATP concentration is changed. *Biophys. J.*, 22:97–103.
23. Kawai, M. (1979): Effect of MgATP on cross-bridge kinetics in chemically skinned rabbit psoas fibers as measured by sinusoidal analysis technique. In: *Cross-bridge Mechanism in Muscle Contraction*, edited by H. Sugi and G. H. Pollack, pp. 149–169. University of Tokyo Press, Tokyo.
24. Kawai, M., and Brandt, P. W. (1976): Two rigor states in skinned crayfish single muscle fibers. *J. Gen. Physiol.*, 68:267–280.
25. Kawai, M., and Brandt, P. W. (1980): Sinusoidal analysis: a high resolution method for correlating biochemical reactions with physiological processes in activated skeletal muscles of rabbit, frog and crayfish. *J. Muscle Res. Cell Motil.*, 1:279–303.
26. Kawai, M., Brandt, P. W., and Orentlicher, M. (1977): Dependence of energy transduction in intact skeletal muscles on the time in tension. *Biophys. J.*, 18:161–172.
27. Kawai, M., Cox, R. N., and Brandt, P. W. (1981): Effect of Ca ion concentration on cross-bridge kinetics in rabbit psoas fibers: Evidence for the presence of two Ca-activated states of thin filament. *Biophys. J.*, 35:375–384.
28. Lymn, R. W., and Taylor, E. W. (1971): The mechanism of adenosine triphosphate hydrolysis by actomyosin. *Biochemistry*, 10:4617–4624.
29. Machine, K. E. (1964): Feedback theory and its application to biological systems. *Symp. Soc. Exp. Biol.*, 18:421–445.
30. Mahler, H. R., and Cordes, E. H. (1971): *Biological Chemistry*. Harper & Row, New York.
31. Natori, R. (1954): The property and contraction process of isolated myofibrils. *Jikeikai Med. J.*, 1:119–126.
32. Podolsky, R. J., and Teichholz, L. E. (1970): The relation between calcium and contraction kinetics in skinned muscle fibres. *J. Physiol. (Lond.)*, 211:19–35.
33. Pringle, J. W. S. (1967): The contractile mechanism of insect fibrillar muscle. *Prog. Biophys. Mol. Biol.*, 17:1–60.
34. Pringle, J. W. S. (1978): Stretch activation of muscle: Function and mechanism. *Proc. R. Soc. Lond. [Biol.]*, 201:107–130.
35. Pybus, J., and Tregear, R. T. (1975): The relationship of adenosine triphosphatase activity to tension and power output of insect flight muscle. *J. Physiol. (Lond.)*, 247:71–89.
36. Reuben, J. P., Brandt, P. W., Berman, M., and Grundfest, H. (1971): Regulation of tension in the skinned crayfish muscle fiber. *J. Gen. Physiol.*, 57:385–407.
37. Rüegg, J. C., and Tregear, R. T. (1966): Mechanical factors affecting the ATPase activity of glycerol-extracted fibers of insect fibrillar flight muscle. *Proc. R. Soc. Lond. [Biol.]*, 165:497–512.
38. Steiger, G. J., and Rüegg, J. C. (1969): Energetics and "efficiency" in the isolated contractile machinery of an insect fibrillar muscle at various frequencies of oscillation. *Pflugers Arch.*, 307:1–21.
39. Stein, L. A., Schwarz, R. P., Chock, P. B., and Eisenberg, E. (1979): Mechanism of actomyosin adenosine triphosphatase. Evidence that adenosine 5'-triphosphate hydrolysis can occur without dissociation of the actomyosin complex. *Biochemistry*, 18:3895–3909.
40. Sugi, H. (1972): Tension changes during and after stretch in frog muscle fibers. *J. Physiol. (Lond.)*, 225:237–253.
41. Thames, M. D., Teichholz, L. E., and Podolsky, R. J. (1974): Ionic strength and the contraction kinetics of skinned muscle fibers. *J. Gen. Physiol.*, 63:509–530.
42. Thorson, J. W., and White, D. C. S. (1969): Distributed representations for actomyosin interaction in the oscillatory contraction of muscle. *Biophys. J.*, 9:360–390.
43. Tonomura, Y., Nakamura, H., Kinoshita, N., Onish, H., and Shigekawa, M. (1969): The presteady state of the myosin-adenosine triphosphate system. X. The reaction mechanism of the myosin-ATP system and a molecular mechanism of muscle contraction. *J. Biochem.*, 66:599–618.
44. Weber, A., Herz, R., and Reiss, I. (1969): The role of magnesium in the relaxation of myofibrils. *Biochemistry*, 8:2266–2270.
45. Weber, A., and Murray, J. M. (1973): Molecular control mechanisms in muscle contraction. *Physiol. Rev.*, 53:612–673.

46. White, D. C. S., and Thorson, J. (1973): The kinetics of muscle contraction. *Prog. Biophys. Mol. Biol.*, 27:173–255.
47. White, H. D., and Taylor, E. W. (1976): Energetics and mechanism of actomyosin adenosine triphosphatase. *Biochemistry*, 15:5818–5826.
48. Wise, R. M., Rondinone, J. F., and Briggs, F. N. (1971): Effect of calcium on force-velocity characteristics of glycerinated skeletal muscle. *Am. J. Physiol.*, 221:973–979.

Basic Biology of Muscles: A Comparative
Approach, edited by B. M. Twarog,
R. J. C. Levine, and M. M. Dewey.
Raven Press, New York © 1982.

A Study of Demembranated Muscle Fibers under Equilibrium Conditions

*R. T. Tregear, *M. L. Clarke, **S. B. Marston, †C. D. Rodger, ‡J. Bordas, and ‡M. Koch

*Institute of Animal Physiology, Babraham, Cambridge CB2 4AT, United Kingdom; **Cardiothoracic Institute, London W1N 2DX, United Kingdom; †Roehampton Institute of Higher Education, Roehampton Lane, London SW15, United Kingdom; and ‡European Molecular Biology Laboratory Outstation, DESY, Hamburg, West Germany

There are essentially two ways of looking for the intermediates in the tension producing cycle of actomyosin. The first and most direct is to add the substrate Mg.ATP and see what happens. This approach entails the biochemist performing transient kinetic experiments (25), whereas the physiologist looks at the transient mechanics (11) and the molecular biologist must employ high speed data collection methods.

The second approach is to apply a competitive inhibitor of the reaction, the most potent of which is Mg.adenylyl imidodiphosphate (Mg.AMPPNP) (35), and to observe the subsequent effect on the system. To the molecular biologist, this means observing the resultant structure by X-ray diffraction (3) or by electron microscopy (22). The physical chemist may also adopt this approach to study the thermodynamics of the coupling between the nucleotide binding and the mechanical potential energy (13).

We have chosen the second approach because its relative simplicity gave hope of positive results in the structural field and we had been disappointed by the small return gained thus far from the work on the X-ray diffraction of active insect flight muscle (1). The combined efforts of several groups have now demonstrated the elements of the response of insect flight muscle to the addition of AMPPNP. The nucleotide binds specifically to the enzymatic sites of myosin in the muscle fiber. This causes the muscle's isometric tension to fall without affecting its isotonic stiffness, as if the actomyosin interactions had changed in form but not in total number (16,17). Reciprocally, the application of tension to the muscle promotes nucleotide binding as if there were a close coupling between the sites of myosin involved in nucleotide binding and the mechanical interaction with actin (12,17). The mechanical effect of Mg.AMPPNP also occurs in glycerinated vertebrate skeletal muscle. In early work on rabbit psoas fibers we saw a relatively small reduction in the isometric tension compared to insect muscle (16), but more recent experiments

(M. L. Clarke, *unpublished obervations*) using single fiber preparations have now demonstrated a loss of tension with Mg.AMPPNP, some three-quarters of that seen with insect fibers. Dr. R. Padron *(personal communication)* has also found such a reduction of tension in skinned (as distinct from glycerinated) frog striated muscle in the present of Mg.AMPPNP.

The X-ray diffraction pattern from insect flight muscle in the presence of AMPPNP shows elements of both rigor and relaxed patterns (2,3,8,16,29). However, the electron microscopic appearance is different from either of the other two states (22) as if a different cross-bridge conformation had been produced.

Similar, but smaller effects can also be observed when other nucleotides bind to myosin in demembranated fibers. Both H.AMPPNP and Mg.ADP lower the iso-metric tension of insect flight muscle by approximately one-third as much as Mg.AMPPNP (17). The X-ray diffraction pattern of insect flight muscle is hardly affected by the addition of Mg.ADP (8,23) but the inner equatorial reflections are shifted toward the relaxed values by the addition of H.AMPPNP, although again by a lower amount than observed with Mg.AMPPNP (17). Small changes were seen in the meridional pattern from insect muscle in the presence of H.AMPPNP (29). No change in tension was observed on addition of AMP or H.ADP. for single insect fibers (M. L. Clarke, *unpublished observations*).

Thus, the binding of an unhydrolysed nucleotide that contains at least two phosphate moieties appears to distort the myosin molecule in such a way as to reduce the mechanical energy retained by the rigor muscle. Our working hypothesis for the mechanism of this process is that the nucleotide binding promotes the formation of a distinct conformation. Therefore, the quantitative differences between the various nucleotide species we have employed would represent different positions of the equilibrium between the two types of cross-bridge (28).

There is some evidence from experiments with pure actin and myosin in solution that the structure of the complexes AM and AM-ligand are different. Marston (15) observed a difference in light scatter between the two complexes after modifying the actin, and Trybus and Taylor (30) found the ternary complex had greater tryptophan fluorescence than AM; their experiments also implied the presence of two ternary complexes with differing fluorescence. Shriver and Sykes (24) and Béchet et al. (4) have observed temperature-dependent changes in n.m.r. spectrum and tryptophan fluorescence of myosin-ligand complexes. These changes have been interpreted as being due to two different myosin-ligand complexes whose equilibrium constant is close to unity at room temperature and is highly temperature dependent.

Similar "nucleotide-bound" cross-bridges may occur as intermediates in the active cross-bridge cycle hydrolysing Mg.ATP. At normal temperatures (0 to 25°C), however, they would be present only transiently. In an attempt to isolate and identify them, we have begun measurements on insect flight muscle fibers cooled to well below 0° with some encouraging preliminary results. The muscle fibers were cooled in a medium employing 50% ethylene glycol as an anti-freeze solvent and then Mg.ATP. was added. The muscle fibers retained high stiffness despite apparent binding of the nucleotide to the enzymatic site, and the low angle X-ray diffraction pattern was little altered by the addition of the nucleotide (5,27).

We found that the addition of ethylene glycol per se produced mechanical and structural effects on the muscle at $+10°$, and, therefore, the effects of glycol were studied prior to continuing the low-temperature phenomena. We summarise the result of this work below.

MECHANICS

Ethylene glycol causes the isometric tension in a muscle fiber to fall without loss of isotonic stiffness in a similar manner to Mg.AMPPNP. This effect is graded with glycol concentration and appears to saturate at a high glycol:water proportion (Fig. 1). The isometric tension drop in insect flight muscle is equivalent to an increase in muscle rest length by 0.2 to 0.3%, whereas the effect of Mg.AMPPNP is to increase rest length by 0.16% (17). The effect on rabbit psoas muscle is to increase rest length by approximately the same amount.

Addition of ethylene glycol to a rabbit muscle fiber whose tension has already been reduced by a saturating concentration of Mg.AMPPNP still further reduces that tension, until at 45% glycol:55% water concentration it reaches zero. At lesser concentrations, the isotonic stiffness of the muscle remains unimpaired, but at 45% glycol concentration the muscle does not hold isometric tension, and when the muscle fiber is extended the isotonic stiffness measured before the rapid fall of tension is found to be reduced (Fig. 2). Insect flight muscle shows the same

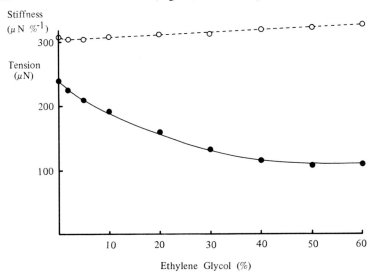

FIG. 1. The effect of replacing water with ethylene glycol on the mechanical properties of insect flight muscle. Single glycol-extracted *Lethocerus* doral longitudinal flight muscle fibers immersed in variable glycol concentration without passage through a meniscus. Isometic tension *(closed circles)* and isotonic stiffness *(open circles:* response to a 105 Hz sinusoidal length vibration of 0.03% amptitude); $\mu = 0.1$, pH 7.0, 10°.

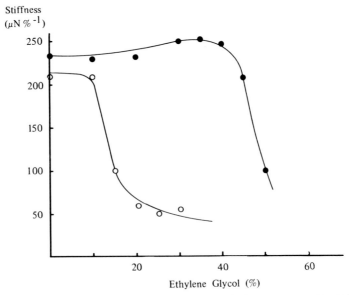

FIG. 2. The effect of replacing water with ethylene glycol in the presence of Mg.AMPPNP on the isotonic stiffness of insect flight *(open circles)* and rabbit psoas *(closed circles)* muscle. Single fibers, protocol as in Fig. 1.

phenomena at a lower glycol concentration. Therefore, it appears that the muscle is relaxed by the combined addition of Mg.AMPPNP and ethylene glycol. We have termed the state "equilibrium relaxation" to distinguish it from physiological relaxation.

Removal of the glycol completely abolishes equilibrium relaxation. The tension in the muscle quickly rises and the isotonic stiffness reverts to its original value. Removal of AMPPNP while retaining the glycol does not have these effects. An insect flight muscle fiber remains relaxed in 50% ethylene glycol at 10° for 15 hr after the removal of the Mg.AMPPNP but reverts at once to rigor on removal of the glycol. Thus, the effects of glycol and AMPPNP on their own are similar, and together they summate to relax the muscle. Relaxation can only be reversed by removal of the glycol.

X-RAY DIFFRACTION

The actin-based X-ray diffraction pattern of insect flight muscle in rigor is well known (Fig. 3A), and has been interpreted as a measure of attached and angled rigor type cross-bridges, arranged in group at each turn of the actin helix (10). In rigor the only indication of thick filament regularity is a low-intensity 14 nm meridionally confined reflection. In physiological relaxation, when Mg.ATP is present, the situation is reversed; the actin-based layer lines are weak and the 14 nm meridional and its associated layer lines are strong (Fig. 3C). This relaxed system has been interpreted in terms of the helical regularity of detached cross-bridges around the thick filament (20).

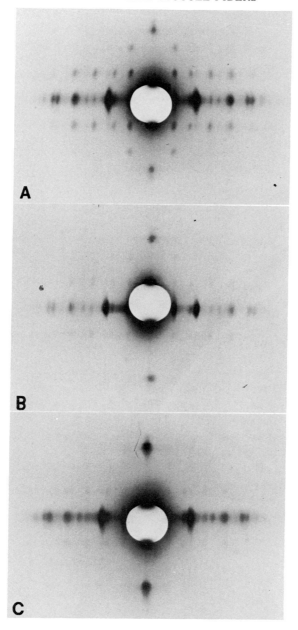

FIG. 3. Low-angle X-ray diffraction patterns from insect flight muscle. Bundle of glycerol-extracted *Lethocerus* dorsal longitudinal flight muscle fiber in rigor **(A)**, in 50% ethylene glycol + 2 mM Mg.AMPPNP **(B)**, and in 5 mM aqueous Mg.ATP **(C)**. Conditions as for previous figures.

As noted earlier, Mg.AMPPNP induces an intermediate situation. The actin-based layer lines are weaker than in rigor and the myosin reflections are weaker than in relaxed muscle (2,29). A similar situation is seen in diffraction patterns from insect flight muscle in 50% ethylene glycol (Fig. 4). The pattern as a whole is significantly weaker than that from rigor muscle in an aqueous medium. This is to be expected as a result of the increased scattering density of the glycol solution. We also note a 5% decrease in the lattice spacing when the muscle fibers are immersed in 50% glycol. We have observed no other differences between the diffraction pattern in glycol and that obtained using Mg.AMPPNP in aqueous solution.

When Mg.AMPNP is used in addition to the glycol medium, thus promoting equilibrium relaxation, the diffraction patterns become similar to that of physiologically relaxed muscle (Fig. 3B). The intensity of the 14 nm meridional reflection is high, whereas that of the actin-based layer line system is low (Fig. 4). This experiment has been repeated several times and given consistent results, indicating that equilibrium relaxation structurally resembles physiological relaxation.

On removal of the AMPPNP, the intensities of the reflection do not revert quickly to their original values. When the glycol is removed, then they change back immediately to their original values in aqueous rigor solution (Fig. 4). These experiments have to be performed on a high-intensity X-ray beam at a synchrotron source, in order to collect the information in a reasonable time.

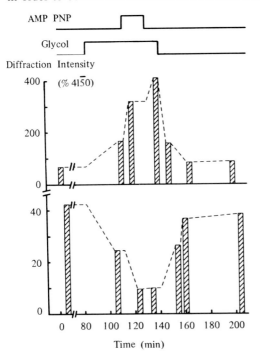

FIG. 4. The intensity of X-ray diffraction from insect flight muscle when ethylene glycol and AMPPNP are added and removed; 50 to 100 fiber bundle of *Lethocerus* dorsal longitudinal flight muscle fibers immersed sequentially in rigor solution, 50% ethylene glycol, 50% ethylene glycol +Mg.AMPPNP, glycol and rigor solution (see upper markers). Reflection intensities recorded from the 31 peak of the 38 nm layer line *(lower graph)* and the 00 peak of the 14 nm layer line *(upper graph)*, 2°; otherwise conditions as for previous figures. Each record took 2 to 3 min. Breaks in connecting lines shown at solution changes.

BIOCHEMISTRY

Actin Binding

Mg.AMPPNP reduces the affinity of rabbit subfragment-1 for actin in solution by two orders of magnitude (Table 1). HAMPPNP and Mg.ADP have a lesser effect (17). Ethylene glycol also reduces the interprotein affinity. The effect increases with concentration and shows no tendency to saturate. At 50% it is similar in size to that of saturating Mg.AMPPNP (Table 1).

Addition of glycol to acto-subfragment-1 saturated with Mg.AMPPNP still further reduces the interprotein affinity. Again, the effect is graded with concentration. At 50% glycol interprotein affinity is similar to that produced by Mg.ATP in aqueous solution (31) (Table 1).

The reduction in interprotein affinity is correlated with the mechanical effect of the same treatment. At an affinity of 10^5 M^{-1} muscle retains its isotonic stiffness; by 10^3 M^{-1} rabbit muscle has relaxed (compare Table 1 and Fig. 4). From the greater mechanical response of insect fibers to a given relaxing condition, it follows that either insect myosin's actin affinity is more sensitive to glycol and/or AMPPNP, or the structure of the protein lattice reduces the effective actin concentration within it relative to that of the rabbit.

Nucleotide Kinetics

Mg.AMPPNP binds to insect muscle fibers with an affinity of some 3×10^4 M^{-1}, and maximal binding is some 200 μM (17). As this is close to the amount expected from the structure of the insect flight muscle thick filaments (21,34) and is approximately the same as that found for Mg.ATP (14), we believe that it is binding principally at the enzymatic site of myosin. Similar meaurements in the presence of 50% ethylene glycol showed a similar maximum binding, with an affinity increased by one to two orders of magnitude.

TABLE 1. *Binding constants for the interaction of rabbit skeletal muscle actin and subfragment-1[a]*

Conditions	K (M^{-1})
Aqueous rigor buffer	50×10^6
50% glycol rigor buffer	0.5×10^6
Aqueous rigor buffer + 1 mM Mg AMPPNP	0.19×10^6
50% glycol rigor buffer + 1 mM Mg AMPPNP	0.0016×10^6

[a]Rigor buffer: 60 mM KCl, 5 mM Mg Cl$_2$, 5 mM EGTA, 5 mM PIPES, 0.2 mM DTT, pH 7.1, 10°C. (^{14}C) iodoacetamide-labeled S-1 was used, acto-S-1 sedimented, and the quantity of free (^{14}C) S-1 remaining in the supernatant measured (18). For the very weak binding in 50% glycol + 1 mM Mg AMPPNP, the quantity of (^{14}C) S-1 pelleted with actin was measured and bound (^{14}C) S-1 determined by using a (^{3}H) glucose volume marker to correct for unbound (^{14}C) S-1.

AMPPNP bound to insect flight muscle fibers in aqueous solution is washed out rapidly as the fast regain of tension on removing the nucleotide shows (17). In 50% ethylene glycol, however, the washout of bound AMPPNP is low (Fig. 5, closed circles). This may explain the retention of relaxation, although the critical experiments remain to be done.

Mg.AMPPNP binds to myosin in 50% glycol with an affinity of around 10^6 M^{-1}, which is similar to the affinity in aqueous solution. However, the rate of disassociation of the nucleotide from myosin is much slower in glycol. We have obtained half-lives of minutes at 25°C and over an hour at 14°C in 50% glycol (Fig. 5, open circles). The association rate in 50% glycol is correspondingly slow (around 10^3 M^{-1} sec^{-1} at 14°C). These rates with pure myosin roughly correpond to the very slow rate of Mg.AMPPNP disassociation from myosin in intact fibers.

DISCUSSION

These three lines of evidence based on tension measurements, X-ray diffraction, and binding studies have shown that ethylene glycol displaces the equilibrium between cross-bridge configurations in the same direction as Mg.AMPPNP.

It is clear that an organic solvent such as glycol is likely to have an effect on protein conformation at many sites in the myofilament lattice. However, what has

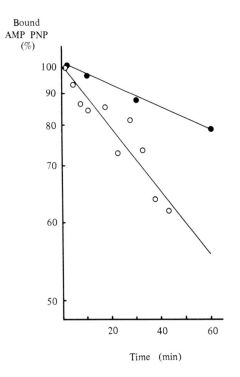

FIG. 5. Dissociation of Mg.AMPPNP from myosin in muscle fibers and in solution in 50% glycol. Conditions as given in Table 1, except temperature 14°C. Closed circles: Single Lethocerus dorsal longitudinal flight muscle fiber pre-incubated in 12 M (H³) AMPPNP for 3 min. The washout of radioactivity was measured by serial incubation in 50% glycol. Open circles: Rabbit myosin (3 mg/ml) pre-incubated in 50 μM Mg.AMPPNP for 10 min at 25°C. Dissociation measured at 14°C. The occupancy of the myosin active site by Mg.AMPPNP was measured by the degree of inhibition of myosin Mg.ATPase; 10 mM Mg.ATP was added at zero time and the rate of ATP hydrolysis was measured by taking aliquots at a series of time intervals.

aroused our interest in the case of glycol is the clear indication of specific effects that are correlated with the active sites on the myosin molecule. For example, glycol can achieve the same alteration in actomyosin geometry as seen when Mg.AMPPNP binds to the enzymatic site on myosin. Furthermore, glycol can increase the effectiveness of Mg.AMPPNP by promoting detachment of the cross-bridges, resulting in equilibrium relaxation. This is apparently a direct effect on the nucleotide binding site itself, since the rate of nucleotide release is greatly reduced.

Organic solvents differ from water both in terms of their polarity (6) and their hydrophobicity. Maurel (19) has shown that the principal effect of these solvents on enzymes is via their hydrophobic properties. A hydrophobic solvent will enhance the formation of α helices and β sheets and thus stabilise a globular protein's secondary structure. It is suggested, on the basis of spectroscopic studies, that the effect of ethylene glycol on myosin is purely local and does not result in any large scale structural change (4,26). However, even subtle structural changes may well affect the catalytic activity of an enzyme. In the case of myosin, it is found that 40% ethylene glycol has hardly any effect on the Mg.ATPase activity of myosin S1 (26) or heavy meromyosin (4). However, glycol does have profound effects involving the active site because it causes a significant reduction in Ca ATPase activity (4,26). This might relate to the slow release of AMPPNP that we have observed in the presence of glycol.

Organic solvents are not the only way in which nucleotide "trapping" may be achieved. Goodno (7) showed that vanadate will irreversibly inhibit rabbit myosin ATPase by forming a stable complex, MADP.V_i. This was later shown to produce "structural" relaxation in insect flight muscle. In this case, the effect of vanadate is mediated directly via the nucleotide binding site of myosin (9). Synthetic cross-linking of the SH1 and SH2 thiol groups of rabbit myosin will also "trap" bound nucleotide at the enzymatic site (33) and the cross-linking is itself enhanced by such binding (32).

CONCLUSION

There is considerable evidence that several disparate intermolecular events may have a common effect on myosin by triggering off a series of important changes. The nucleotide binding site becomes tighter, the actin binding site becomes weaker, and mechanical energy is lost from the system. We believe that the solution of the structure of the equilibrium state we have observed is crucial to understanding how muscle contracts because the transition between states is known to produce muscle shortening at the expense of nucleotide binding energy. These transitions are thus likely to be close analogues of the active contractile event and, since they take place at equilibrium, it should be possible to study them in detail.

REFERENCES

1. Armitage, P. M., Tregear, R. T., and Miller, A. (1975): Effect of activation by calcium on the X-ray diffraction pattern from insect flight muscle. *J. Mol. Biol.*, 92:39–53.

2. Barrington-Leigh, J., Goody, R. S., Hofmann, W., Holmes, K. C., Mannherz, H-G., Rosenbaum, G., and Tregear, R. T. (1977): The interpretation of X-ray diffraction from glycerinated flight muscle fiber bundles: New theoretical and experimental approaches. In: *Insect Flight Muscle*, edited by R. T. Tregear, pp. 137–146. North-Holland, Amsterdam.
3. Barrington-Leigh, J., Holmes, K. C., Mannherz, H. G., Rosenbaum, G., Eckstein, F., and Goody, R. S. (1972): Effects of ATP analogues on the low-angle X-ray diffraction pattern of insect flight muscle. *Cold Spring Harbor Symp. Quant. Biol.*, 37:443–448.
4. Béchet, J. J., Breda, C., Guinand, S., Hill, M., and d'Albis, A. (1979): Magnesium-ion dependent ATPase activity of heavy meromyosin as a function of temperature between +2 and −15°C. *Biochemistry*, 18:4080–4089.
5. Clarke, M. L., Rodger, C. D., Tregear, R. T., Bordas, J., and Koch, M. (1980): The effect of ethylene glycol and low temperature on the structure and function of insect flight muscle. *J. Muscle Res. Cell Motil.*, 1:195–196.
6. Douzou, P. (1977): *Cryobiochemistry*. Academic, London.
7. Goodno, C. C. (1979): Inhibition of myosin ATPase by vanadate ion. *Proc. Natl. Acad. Sci. USA*, 76:2620–2624.
8. Goody, R. S., Barrington-Leigh, J., Mannherz, H-G., Tregear, R. T., and Rosenbaum, G. (1976): X-ray titration of binding of β, γ, imido ATP to myosin in insect flight muscle. *Nature*, 262:613–615.
9. Goody, R. S., Hofmann, W., Reedy, M. K., Magid, A., and Goodno, C. C. (1980): Relaxation of glycerinated insect flight muscle by vanadate. *J. Muscle Res. Cell Motil.*, 1:198–199.
10. Holmes, K. C., Tregear, R. T., and Barrington-Leigh, J. (1980): Interpretation of the low-angle X-ray diffraction from insect flight muscle in rigor. *Proc. R. Soc. Lond. [Biol.]*, 207:13–33.
11. Huxley, A. F. (1974): Muscular contraction. *J. Phyiol. (Lond.)*, 243:1–43.
12. Kuhn, H. J. (1977): Reversible transformation of mechanical work into chemical free energy by stretch-dependent binding of AMP.PNP in glycerinated fibrillar muscle fibers. In: *Insect Flight Muscle*, edited by R. T. Tregear, pp. 307–316. North Holland, Amsterdam.
13. Kuhn, H. J. (1981): The mechanochemistry of force production in muscle. *J. Muscle Res. Cell Motil.*, 2:7–44.
14. Marston, S. B. (1973): The nucleotide complexes of myosin in glycerinated muscle fibers. *Biochem. Biophys. Acta*, 305:397–412.
15. Marston, S. B. (1980): Evidence for the altered structure of actin-subfragment-1 complexes when magnesium-adenylyl imidodiphosphate binds. *J. Muscle Res. Cell. Motil.*, 1:305–320.
16. Marston, S. B., Rodger, C. D., and Tregear, R. T. (1976): Changes in muscle crossbridges when β,γ-imido-ATP bind to myosin. *J. Mol. Biol.*, 104:263–276.
17. Marston, S. B., Tregear, R. T., Rodger, C. D., and Clarke, M. L. (1979): Coupling between the enzymatic site of myosin and the mechanical output of muscle. *J. Mol. Biol.*, 128:111–126.
18. Marston, S. B., and Weber, A-M. (1975): The dissociation constant of the actin-heavy meromyosin subfragment-1 complex. *Biochemistry*, 14:3868–3873.
19. Maurel, P. J. (1978): Relevance of dielectric constant and solvent hydrophobicity to the organic solvent effect in enzymology. *J. Biol. Chem.*, 253:1677–1683.
20. Miller, A., and Tregear, R. T. (1972): Structure of insect fibrillar flight muscle in the presence and absence of ATP. *J. Mol. Biol.*, 70:85–104.
21. Reedy, M. K., Leonard, K. R., Freeman, R., and Arad, T. (1981): Thick myofilament mass determination by electron scattering measurement with the scanning transmission electron microscope. *J. Muscle Res. Cell Motil.*, 2:45–64.
22. Reedy, M. K., Reedy, M. K., and Goody, R. S. (1981): Crossbridge structure in rigor and AMP.PNP states of insect flight muscle. *Biophys. J.*, 33:22a.
23. Rodger, C. D., and Tregear, R. T. (1974): Crossbridge angle when ADP is bound to myosin. *J. Mol. Biol.*, 86:495–497.
24. Shriver, J. W., and Sykes, B. D. (1981): Phosphorus-31 nuclear magnetic resonance evidence for two conformations of myosin subfragment-1. nucleotide complexes. *Biochemistry*, 20:2004–2012.
25. Taylor, E. W. (1979): Mechanism of actomyosin ATPase and the problem of muscle contraction. *CRC Crit. Rev. Biochem.*, 6:103–164.
26. Traver, F., and Hillaire, D. (1979): Cryoenzymological studies on myosin subfragment-1. *Eur. J. Biochem.*, 89:293–299.
27. Tregear, R. T. (1981): Activation and action in insect flight muscle. In: *Development and Specialisation of Skeletal Muscle*, edited by D. F. Goldspink, pp. 107–122. Cambridge University Press, Cambridge, England.

28. Tregear, R. T., and Marston, S. B. (1979): The crossbridge theory. *Annu. Rev. Physiol.*, 41:723–736.
29. Tregear, R. T., Milch, J. R., Goody, R. S., Holmes, K. C., and Rodger, C. D. (1979): The use of some novel X-ray diffraction techniques to study the effect of nucleotides on crossbridge in insect flight muscle. In: *Crossbridge Mechanism in Muscle Contraction*, edited by H. Sugi and G. Pollack, pp. 407–423. University Park Press, Baltimore.
30. Trybus, K. M., and Taylor, E. W. (1979): Kinetic of ADP and AMP.PNP binding to subfragment-1. *Biophys. J.*, 25:21a.
31. Wagner, P. D., and Weeds, A. (1979): Determination of the association of myosin subfragment-1 with actin in the presence of ATP. *Biochemistry*, 18:2260–2266.
32. Wells, J. A., Knoeber, C., Sheldon, M., Werber, M. M., and Yount, R. G. (1980): Cross-linking of myosin subfragment-1. Nucleotide-enhanced modification by a variety of bifunctional reagents. *J. Biol. Chem.*, 255:11135–11140.
33. Wells, J. A., Sheldon, M., and Yount, R. G. (1980): Magnesium nucleotide is stoichiometrically trapped at the active site of myosin and its active proteolytic fragments by thiol cross-linking reagents. *J. Biol. Chem.*, 255:1598–1602.
34. Wray, J. S. (1979): Filament geometry and the activation of insect flight muscles. *Nature*, 280:325–326.
35. Yount, R. G., Ojala, D., and Babcock, D. (1971): Interaction of PNP and PCP analogues of ATP with heavy meromyosin, myosin and actin. *Biochemistry*, 10:2490–2496.

*Basic Biology of Muscles: A Comparative
Approach*, edited by B. M. Twarog,
R. J. C. Levine, and M. M. Dewey.
Raven Press, New York © 1982.

Contractile Mechanism of Single Isolated Smooth Muscle Cells

*Fredric S. Fay, *Kevin Fogarty, **Keigi Fujiwara, and
†Richard Tuft

*University of Massachusetts Medical School, Department of Physiology, Worcester,
Massachusetts 01605; **Harvard Medical School, Department of Anatomy, Boston,
Massachusetts 02115; and †Worcester Polytechnic Institute, Department of Physics,
Worcester, Massachusetts 01609

The function of smooth muscle is vital to the normal operation of all organ systems and, thus, an understanding of smooth muscle contraction is of great importance from a biomedical point of view. Smooth muscle contraction is also of interest to students of cell motility since this muscle may provide a better model of this basic cellular process than striated muscles, which are highly ordered and evolved. Despite the importance of this problem, the events responsible for contraction of smooth muscle are quite poorly understood (12). We do know that actin jnd myosin, with generally similar properties to that found in skeletal muscle, are found within the smooth muscle cell (16), and it has thus been assumed that the interaction of these two proteins via a cross-bridge mechanism is responsible for the generation of force. The contraction of smooth muscle, although similar in some regards to that of skeletal muscle, is much slower in its time course and is also distinguished by a much lower ATP cost for force maintenance (22). In addition, although smooth and striated muscles generate equivalent forces/unit cross-sectional area, smooth muscle is remarkable in doing this with ⅕ to ⅒ the myosin/g muscle (18). Surely, an understanding of these functional characteristics must depend on a firm understanding of the organization of the contractile machinery and the details of the cross-bridge cycle. Insights into these aspects of smooth muscle structure and function have been limited to date. Certainly, the fact that smooth muscle cells exhibit no readily discernible order has slowed development of insights into the organization of the contractile machinery. Both structural and functional analyses of the contractile process in smooth muscle have also been greatly hampered by the complexities of the intact tissues containing smooth muscle cells. Such tissues are characterized by a complex mechanical and electrical interconnection of cells, a dense connective tissue matrix, and the existence of numerous neurons and other nonmuscle cell types. We have developed techniques for studying the contraction of single smooth muscle cells (7,11,23) in the belief that significant insights into

143

the process of smooth muscle contraction might be obtained if this process could be directly studied at the level of a single cell. The studies we have performed have all been carried out on single smooth muscle cells isolated by enzymatic disaggregation of the stomach muscularis of the toad *Bufo marinus*; the method of isolation (11) is a modification of the procedure originally described by Bagby et al. (4). The smooth muscle cells within the stomach of this amphibian bear great structural (9) and contractile similarities (3,6,30) to those from mammalian vascular and visceral smooth muscle, and we believe that results obtained in these amphibian smooth muscle cells may be of significance to our understanding of mammal smooth muscle cells as well. An extensive series of studies on the isolated smooth muscle cells after enzymatic disaggregation from the tissue have been performed to assess their integrity. Their membrane structure (8,9), ion permeability (19), and electrophysiological (24,27–29) properties are remarkably similar to those found in the intact tissue. In addition, we find that the electrical (28,31), biochemical (17,20), and contractile (13,17) responses to neurotransmitters are basically unaltered following cell isolation, further attesting to the integrity of the cell following the isolation procedure. We believe that this isolated single cell is a valid physiological and structural model of the smooth muscle cell within the intact tissue and appropriate for studying the mechanisms underlying the generation and regulation of force in smooth muscle.

Our analysis of the events responsible for contraction of smooth muscle began with and has been guided by an initial series of observations with the light microscope in which the contraction/relaxation cycles of single smooth muscle cells were analyzed (9). Figure 1 contains typical results from such studies, and shows that contraction of the isolated smooth muscle cell is associated with evagination of a portion of the cell membrane. The formation of evaginations is a reversible process as these membrane blebs are lost when the cell reextends following cessation of stimulation. Analysis of movie records of the contraction/relaxation cycle in 19 cells revealed that the cells were completely covered by evaginations after they had actively shortened to 57% of their initial length on the average; on the other hand, following cessation of stimulation as the cells relaxed, they were devoid of evaginations after reextending back to 44% of their initial length on the average. The appearance of evaginations during the early stages of contraction and their disappearance during the earliest stages of relaxation was also strikingly confirmed when smooth muscle cells were fixed at a defined contractile state and examined with the scanning electron microscope, as can be seen in Fig. 2. Although the relaxed cell (Fig. 2a) was devoid of evaginations, they are clearly evident over the entire surface of the cell (Fig. 2b) fixed after shortening to 72% of its initial length— approximately the half-maximal extent of shortening seen in response to electrical stimulation. These evaginations appear to increase in size as cells continue to shorten, yielding the large bulbous evaginations characteristic of cells (Fig. 2c) that are actively shortening and have reached lengths 45 to 30% of that initially. As typified by the scanning electron micrograph in Fig. 2d, following slight reextension of the cell after cessation of electrical stimulation, the characteristic evaginations seen in actively contracting cells were lost. The surface of this cell is not entirely

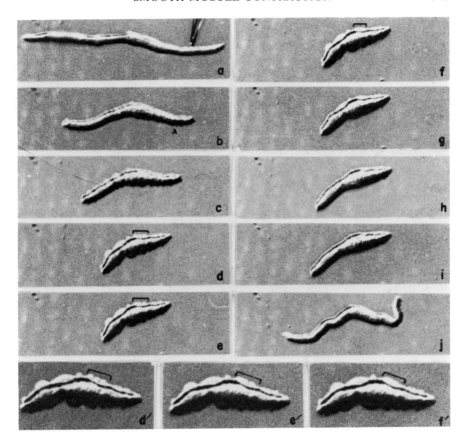

FIG. 1. Ten frames from a film record showing the contractile response of an isolated smooth muscle cell to brief electrical stimulation. Time intervals: **a**, just prior to stimulation; **b**, 1 sec after initiation of contraction, cell at 66% of its initial length L_i; **c**, 2 sec after initiation of contraction, 51% L_i; **d,d'**, 9 sec after initiation of contraction, 38% L_i; **e,e'**, 10 sec after initiation of contraction, 38% L_i; **f,f'**, 11 sec after initiation of contraction, 38% L_i-cell maximally contracted; **g**, 20 sec after initiation of contraction, 40% L_i; **h**, 26 sec after initiation of contraction, 44% L_i; **i**, 31 sec after initiation of contraction, 45% L_i; **j**, 10 min after initiation of contraction, 74% L_i. Note the appearance of evaginations after shortening to 66% L_i (**b**, *dart*), whereas the loss of such evaginations during relaxation is almost complete at 45% L_i. In frames **d**, **d'**, **e**, **e'**, **f**, and **f'**, three small evaginations *(brackets)* are seen to fuse in a stepwise manner. **a–i**, ×200; **d'**, **e'**, **f'**, ×295. (From Fay et al., ref. 8, with permission.)

smooth but this probably reflects the need to accomodate excess surface membrane in this cell, which is more spherical than in its initial relaxed state and thus requires less surface area. These studies of changes in the surface of single isolated smooth muscle cells during contraction revealed that the striking and characteristic changes in surface features are not correlated with relative cell length per se. Rather, they are associated with the active process of contraction.

We then undertook studies to determine what structural changes take place within the cell that might be responsible for the changes in the surface appearance during contraction/relaxation cycles. Cells were again fixed at defined stages of their

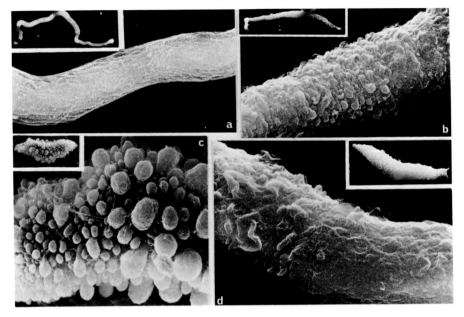

FIG. 2. Scanning electron micrographs of four smooth muscle cells fixed at different stages of their contraction/relaxation cycle. **a:** Cell fixed without prior stimulation; note the generally smooth surface of this cell. ×3,635. *Inset* shows entire cell. ×307. **b:** Cell fixed after approximately half-maximal shortening in response to brief electrical stimulation. Cell had contracted to 72% of its initial length (L_i). Note that the cell surface is completely covered by small evaginations. ×3,180. *Inset* shows the entire cell. ×265. **c:** Cell fixed after maximally shortening to 35% L_i in response to brief electrical stimulation. Note that most of the evaginations are connected to the main body of the cell by a slender neck; occasionally evaginations appear connected by multiple stalks. ×2,925. *Inset* shows entire cell. ×490. **d:** Cell fixed following slight relaxation after contracting in response to brief electrical stimulation. Cell length at time of fixation, 47% L_i. Cell length at point of maximum shortening, 40% L_i. Note that the bulbous and moundlike evaginations characteristic during contraction are largely absent at this early stage of relaxation. Occasionally, however, folding of the cell surface is evident. ×3,180. *Inset* shows entire cell. ×590. (From Fay et al., ref. 8, with permission.)

contraction/relaxation cycle, and prepared for viewing with the transmission electron microscope. Figure 3 shows electron micrographs of two smooth muscle cells cut in longitudinal section. The cell shown in Fig. 3a was fixed at the point of maximum contraction. An evagination is seen in this thin section to be generally devoid of myofilaments that predominate in the nonevaginated portion of the cell. In contrast to the membrane surrounding this evagination, those membrane areas not forming these evaginations are associated with an underlying amorphous density, the plasma membrane dense body. Although thick and thin filaments as well as cytoplasmic dense bodies are seen to be rather obliquely oriented with respect to the long axis of the contracted cell (Fig. 3c), in relaxed cells both of these filaments types as well as the cytoplasmic dense bodies run generally parallel to the long axis (Fig. 3b): the angle that these filaments and cytoplasmic dense bodies make with the long axis of the cell in the relaxed state ranges from 1 to 15°. In contracted cells,

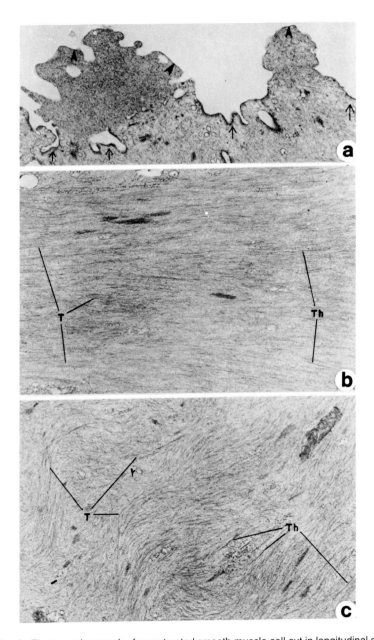

FIG. 3. **A:** Electron micrograph of a contracted smooth muscle cell cut in longitudinal section. The plasma membrane of this contracted cell is subtended by amorphous dense material along the bases and "neck regions" of the surface evaginations *(arrows)*. The membrane covering these evaginations *(darts)* is not subtended by these dense areas. × 17,280. **B:** Electron micrograph of a contracted smooth muscle cell cut in longitudinal section. The cytoplasm of this cell is characterized by a nonparallel orientation of thick (Th) and thin (T) myofilaments at various angles to the long axis of the cell. × 18,675. **C:** Electron micrograph of a relaxed smooth muscle cell cut in longitudinal section. Note the generally uniform orientation of thick (Th) and thin (T) myofilaments parallel to the long axis of the cell. × 18,675. (From Fay et al., ref. 8, with permission.)

many of the filaments and cytoplasmic dense bodies are inclined at much steeper angles.

Further evidence supporting the view that the contractile filaments become obliquely oriented during active shortening is obtained from observations of the isolated cells with polarized light as they contract and relax. As seen in Fig. 4, taken from the work of Fisher and Bagby (14), a relaxed smooth muscle cell is strongly birefringent as evidenced by the intensity of the image of this cell when viewed between crossed polarizer and analyzer with the cell at 45° to the axis of polarization of the incident light. The birefringence of the cell diminishes as the cell begins to shorten actively and, at the point of maximum shortening, the birefringence associated with the long axis of the cell is virtually absent. Although the birefringence associated with the long axis of the cell is lost in the highly shortened cell, birefringence associated with an axis oblique to that axis is now detectable. These changes in birefringence are expected to follow reordering of filaments of the sort shown in fixed cells.

These structural observations may be accomodated by the model for the arrangement of the contractile elements within the smooth muscle cell shown in Fig. 5. The essential features of this model are that contractile elements attach to points on the cell membrane, the plasma membrane dense bodies, and that these elements run for short distances relative to the length of the cell. As a consequence of this arrangement, as the cell actively shortens, the contractile elements become more obliquely oriented. This model also suggests that evaginations form during active cell shortening as a consequence of two opposing tendencies. As the cell shortens, there is a need to increase cell diameter to accommodate the volume displaced by cell shortening, but this is opposed by a vector component of force of the contractile

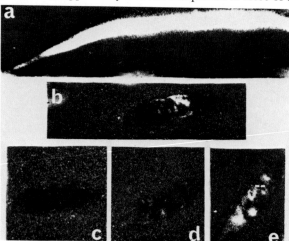

FIG. 4. Changes in birefringence in a single muscle cell actively shortening in response to electrical stimulation. Cell was oriented in **a** so that maximum birefringence was seen prior to contraction. In **b** the cell is shown 12.7 sec after being stimulated. After contraction was completed and entire cell exhibited minimum birefringence (**c**), stage was rotated 20° (**d**) and 43° (**e**). ×315. (Reproduced from Fisher and Bagby, ref. 14, with permission.)

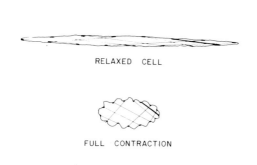

RELAXED CELL

FULL CONTRACTION

FIG. 5. Schematic representation of a smooth muscle cell showing how contractile units are attached to the cell surface. The densities along the cell membrane represent plasma membrane dense bodies and the lines connecting them represent the contractile units. One of these lines has been widened to facilitate identification in the two contractile states. Note that contractile units run for only relatively short lengths within the cell and that the angle between contractile units and the long axis of the cell increases during contraction. (From Fay and Delise, ref. 9, with permission.)

elements pulling inward at the sites of attachment to the plasma membrane, the dense body. As a consequence, only those areas not subject to these inwardly directed forces are able to move outward and, thus, evaginations form along those portions of the membrane devoid of plasma membrane dense bodies.

Although this model fits the data available at the time it was proposed (9), we had not been able to visualize directly contractile elements of the sort depicted in Fig. 5. Instead, this model was deduced from data regarding structural changes associated with contraction of smooth muscle cells. We have recently undertaken a series of studies (10) utilizing immunocytochemical techniques to probe for the existence of contractile elements of the sort depicted in our model. Our initial studies have focused on ascertaining the distribution of α-actinin. We reasoned that if contractile elements of the sort shown in our model existed, they must be at least 35 μm long given the distribution of filament angles observed in relaxed cells. Thus, it was unlikely that these elements were composed of only one sarcomere-like unit given reasonable estimates for thick and thin filament length in smooth muscle (1,12). Instead, it seemed more likely that the contractile elements were composed of several sarcomere-like units in series. One would expect that dense bodies having been shown to collimate bundles of actin filaments (1) might mark the boundaries of adjacent sarcomere-like units. Since dense bodies are known to contain relatively large amounts of α-actinin (21), we have used fluorescently labelled antibodies against α-actinin to probe further the organization of the contractile machinery in smooth muscle.

As can be seen in Fig. 6a, single smooth muscle cells stained with anti-α-actinin exhibit two different patterns of staining. Within the inner regions of the cell, numerous fusiform bodies stained with anti-α-actinin are evident. These structures have a mean axial ratio (length:width) of 4.4 and a mean length of 1.2 μm in relaxed cells. Along the outer margins of the cells somewhat larger and more irregularly shaped patches are stained. A similar staining pattern has recently been reported by Bagby (2) for glycerinated chicken gizzard smooth muscle cells utilizing anti-α-actinin. In our studies (10), the irregular stained patches found along the outer margins of the cell are found at the bases of evaginations in contracted cells and are not associated with those portions of the cell surface forming blebs. We believe that the antibody is staining proteins associated with the dense bodies. The specificity of the antibodies against α-actinin is evidenced by the complete absence of staining in smooth muscle cells exposed to antisera preabsorbed by α-actinin. In the cell stained with anti-α-actinin in Fig. 6a, regions of the cell may be observed where the stained fusiform bodies within the cell are observed to run in a string-like manner with a fairly regular longitudinal repeat (darts). Occasionally, several of these strings appeared to run in parallel for some distance with the stained fusiform bodies in register laterally. Also in some images strings of stained fusiform bodies were observed to head toward the large irregular stained patches on the cell margins.

The pattern of fluorescence in a single image is unfortunately subject to considerable uncertainty because the effective depth of field even using our highest power objectives (100 times, N. A. = 1.30) is approximately 2.0 μm and, thus, apparent order may merely be the consequence of the projection onto a single image plane of elements at quite different depths within the specimen. To define more precisely the positions of each stained structure within the cell, pictures were taken at 0.5 μm intervals along the thickness of the cell and the position of each stained structure within each image plane traced onto a transparency. Each stained structure could then be tracked as it moved into and out of focus as the objective position was varied, and it was then assigned to the central plane in the series of images that contained it. The position and size of all such structures was then entered into our computer and the resulting three-dimensional matrix displayed as illustrated in Fig. 6b. As can be seen, strings of the fusiform stained structures are now quite clearly evident. Some run into large stained patches at the cell periphery where they apparently terminate, and in some regions the stained structures within the cell interior are quite clearly in lateral register. The mean center to center distance between the stained bodies organized into these string-like arrays was 2.2 ± 0.1 μm (SEM; n = 35 stained pairs). These strings of anti-α-actinin stained bodies are very much like the contractile elements that were originally postulated, and we decided to test the possibility that these strings are indeed contractile by performing a similar "deblurring" and cell reconstruction process on sequential images obtained from a cell fixed while contracting in response to acetylcholine. In the one such

FIG. 6. a: Fluorescence photomicrograph of the distribution of anti-α-actinin in a single isolated smooth muscle cell fixed while relaxed with 3.7% formaldehyde, and treated for 10 min at $-20°$C with acetone to enhance antibody penetration as described by Fujiwara et al. (15). Note that the stained areas within the interior of this cell are fusiform and often appear to be arranged in strings *(darts)* characterized by a fairly regular repeat. The stained elements in such strings have similar alignments. Along the outer margins of the cell larger more irregular stained patches are evident. $\times$2,800. b: Computer reconstruction of the distribution of anti-α-actinin in a segment of a single isolated smooth muscle cell. The data for this reconstruction were obtained by "deblurring" images of the distribution of fluorescence at 0.5 μm intervals along the thickness of the cell. Each stained fusiform body within the cell is represented as a vector whose length and orientation corresponds to the major axis of the stained body. The larger stained bodies along the periphery of the cell are represented by crossed vectors ending at one of four corners describing that patch. Six vectors have been highlighted to demonstrate that strings of stained fusiform bodies are a prominent feature in such reconstructions. $\times$3,600.

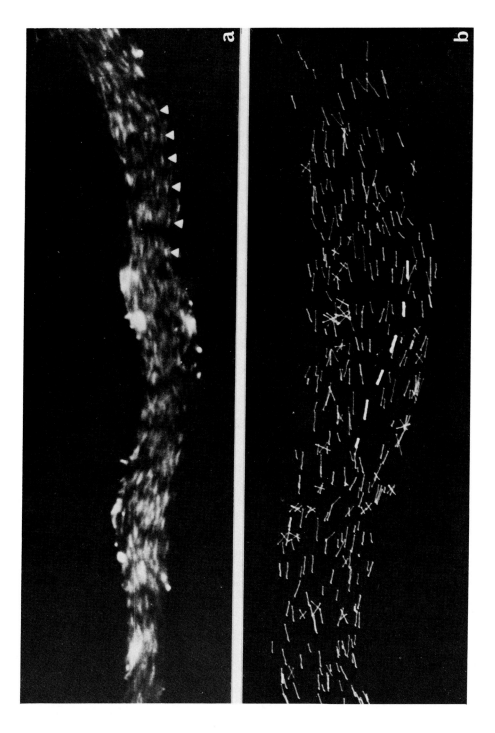

cell studied to date, the string-like arrays of stained fusiform bodies were still evident, although these strings were often inclined more steeply relative to the cell's long axis. Registration of the fusiform stained structures into laterally ordered groups was still apparent. The mean distance between the fusiform stained structures in these string-like elements was 1.4 ± 0.1 μm (SEM; n = 22 stained pairs) in this contracted cell. These data are clearly consistent with the view that the material between dense bodies represents a contractile unit, several of which are strung together to form a myofibril-like contractile element anchored at either end to a point on the plasma membrane, presumably the plasma membrane dense body. Although additional experiments will be needed to substantiate this hypothesis, the present results suggest that the contractile proteins within the smooth muscle cell may be more highly ordered than originally suspected. Furthermore, the immunocytochemical approaches demonstrated here promise to shed much needed light on the details of this organization.

The organization of the contractile proteins into elements inserting on the membrane, and as a consequence changing their orientation with active shortening, would be expected to have profound consequences for the smooth muscle cell's mechanical properties. This reorientation of contractile elements with decreased length would be expected to result in a decrease in the force or velocity of shortening of the cell as the vector component of force or velocity associated with its long axis diminishes. To evaluate to what extent such reorientation contributes to the fall in force at short lengths, we began an investigation of the force developed by these single smooth muscle cells.

To investigate the isometric contractile properties of a single smooth muscle cell, we developed methods for attaching a smooth muscle cell at one end to an ultrasensitive force transducer (5,7) and at the other end to a system for regulation of cell length. This connection is achieved by wrapping a cell in a knot-like manner around a glass probe at either end as seen in Fig. 7a. In response to a single brief (1 msec) electrical stimulus, isometric force in such cells increases, typically reaching its peak in about 4 sec and then declining back toward its prestimulus level with a t ½ of 3 to 9 sec. The magnitude of the contractile response in a single cell to a single brief electrical stimulus increased up to a maximum as the stimulus strength was raised. The peak force measured in a single smooth muscle cell (7) when normalized for its cross-sectional area was 2.6 kg/cm^2, quite similar to that reported both in intact tissues and in isolated frog skeletal muscle fibers (12).

The resting force in these single smooth muscle cells prior to stimulation is extremely low (typically 1 to 5 μg), and as shown in Fig. 7b, exhibited at most a small dependency on cell length. Although resting force in a single smooth muscle cell exhibited little dependency on cell length, active force due to sustained submaximal activation exhibited a pronounced dependency on cell length. Each smooth muscle cell exhibited a distinct length where peak force was developed: at lengths greater or less than this active force production was reduced. As fibers were allowed to shorten below the length for optimum active force production (L_{opt}), force declined

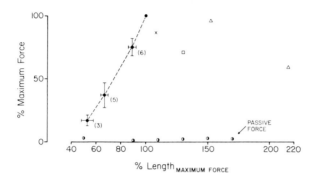

FIG. 7. Top: Photomicrograph of a single smooth muscle cell mounted for isometric measurement of force. The smooth muscle cell was tied in a knot-like manner around a probe at either end as described by Fay (6). One probe (fp) was attached to the force transducer and the other to a system for regulation of cell length. ×250. **Bottom:** The relation between length and force in three isolated smooth muscle cells. For each cell all forces have been expressed relative to the maximum obtained. Small anionic exchange resin particles were used to mark a portion of the cell (approximately half its length) and the distance between these markers when peak force was achieved denoted as Lopt and marker distances for other forces expressed relative to this marker interdistance. The length dependence of active force was assessed by stretching the cell beyond its rest length and then maintaining the cell in an activated state by continuous submaximal stimulation. When a stable force was achieved, the cell was allowed to shorten by a fixed amount, time allowed for the cell to redevelop a new stable force, and then this procedure repeated until active force had dropped below 20% of the maximum. A single cell held at constant length was capable of maintaining stable force for up to 60 sec with this method of activation. In this time, force development at least six different lengths could be assessed. The data have been grouped into length intervals of 20%; the *points* shown are means; the *bars* indicate the SEMs. Beyond Lopt the individual data points for each cell have been plotted. (From Fay, ref. 6, with permission.)

reaching its half-maximal value when the cell had shortened to 70 of L_{opt}. Although active force also diminished at lengths beyond L_{opt}, the fall off with length is far less precipitous. Rather remarkably we have found that cells stretched up to 10 times their original relaxed length are still capable of generating active force. We were particularly interested in determining if the fall in force at short lengths might to some degree reflect contractile element reorientation of the sort predicted by our model.

To test for this, we have utilized the birefringence of the cell as a first indicator of changes in filament orientation with length. As noted above, the birefringence associated with the long axis of a smooth muscle cell diminishes as it actively shortens and this has been correlated with both thick and thin filaments becoming

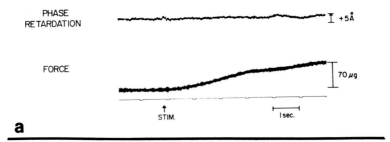

PHASE
RETARDATION   ⊥ +5Å

FORCE   } 70 μg

                    ↑   ├─┤
                  STIM.   I sec.

a

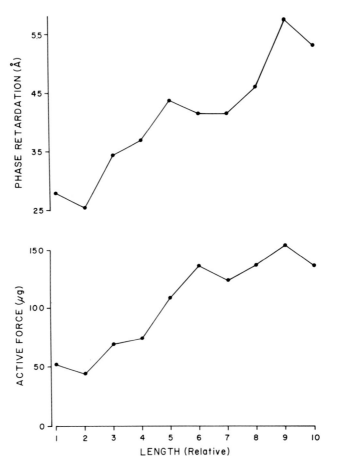

PHASE RETARDATION (Å)

5,5

45

35

25

ACTIVE FORCE (μg)

150

100

50

0

1   2   3   4   5   6   7   8   9   10
LENGTH (Relative)

b

more obliquely oriented with respect to the long axis of the cell (14). Although other structural changes may also contribute to their loss of birefringence, the reorientation of contractile filaments must contribute significantly to these optical changes. We thus set out to measure simultaneously both birefringence and isometric force as a function of length in a single smooth muscle cell. To assess birefringence quantitatively, we utilized the "Polar Eye" device described by Taylor and Zeh (26), which allows one to measure phase retardation (birefringence × specimen thickness) with a high degree of accuracy (resolution = 1 Å) and good temporal responsiveness ($\tau = 1$ msec). For these studies, a smooth muscle cell was mounted in the usual manner but the cell was illuminated with a narrow slit of monochromatic plane polarized light obtained from a 5 mW He-Ne laser. As illustrated in Fig. 8a, there is virtually no change in the phase retardation measured in a single smooth muscle cell during the development of force following electrical stimulation if the cell is maintained at constant length. This indicates that the decrease in birefringence associated with active shortening of isolated smooth muscle cells after electrical stimulation is not related to activation processes per se, rather it must reflect length-dependent changes in structures within active smooth muscle cells. To determine if these optical or structural changes might be related to length-dependent changes in mechanical properties, experiments were performed in which we simultaneously monitored isometric force and phase retardation in a single activated smooth muscle cell at various lengths. The results of a typical experiment of this sort are shown in Fig. 8b. As can be seen, both phase retardation and force fell as length was decreased. Although the optical changes require further analysis of the contribution of form and intrinsic birefringence (25) before they can be fully interpreted, these results point out that changes in contractile element orientation may well play a significant role in the length-dependence of force in smooth muscle, especially at short lengths. Further studies of the structural basis for the length-dependence of force in smooth muscle should benefit from the use of these methods for measuring birefringence. In addition, the introduction into living cells of immunocytochemical probes described earlier coupled with simultaneous mechanical measurements may

---

FIG. 8. a: Phase retardation and isometric force in a single smooth muscle cell following activation of contraction by electrical stimulation. Phase retardation was measured utilizing the "Polar Eye" device of Taylor and Zeh (26). The axis of polarization of the incident light was at 45° to the long axis of the cell. The light was recollected by the microscope objective and passed through a ¼ wave plate oriented with its slow axis parallel to the plane of polarization of the incident light. The linearly polarized light emerging from the ¼ wave plate is rotated due to the birefringence of the cell and the degree of this rotation is measured by the Polar Eye. Note that there is no change in phase retardation associated with force development following stimulation at constant length. **b:** The variation in phase retardation and active force with length in a single smooth muscle cell. The length dependence of these parameters were simultaneously assessed in a cell subject to the protocol described in Fig. 7. No markers were placed on the cell during this experiment and thus the lengths at which stable force and phase retardation were measured are denoted on an arbitrary scale of 1 to 10. The length of the cell between the probes was 188 μm at length 10; phase retardation and force were assessed as the cell was allowed to shorten successively in 8 μm intervals.

also provide a critical means for investigating the relation between structure and mechanical function in smooth muscle.

ACKNOWLEDGMENTS

Supported in part by grants from the NIH (HL14523) and the Muscular Dystrophy Association.

REFERENCES

1. Ashton, F. T., Somylo, A. V., and Somlyo, A. P. (1975): The contractile apparatus of vascular smooth muscle: Intermediate high voltage stero electron microscopy. *J. Mol. Biol.*, 98:17–29.
2. Bagby, R. M. (1980): Double-immunofluorescent staining of isolated smooth muscle cells: Preparation of anti-chicken gizzard α-actinin and its use with anti-chicken gizzard myosin for co-localization of α-actinin and myosin in chicken gizzard cells. *Histochemistry*, 69:113–130.
3. Bagby, R. M., and Fisher, B. A. (1973): Graded contraction in muscle strips and single cells from *Bufo marinus* stomach. *Am. J. Physiol.*, 225:105–109.
4. Bagby, R. M., Young, A. M., Dotson, R. S., Fisher, B. A., and McKinon, K. (1971): Contraction of single smooth muscle cells for *Bufo marinus* stomach. *Nature*, 234:351–352.
5. Canaday, P. G., and Fay, F. S. (1976): An ultrasensitive isometric force transducer for single smooth muscle cell mechanics. *J. Appl. Physiol.*, 40:243–246.
6. Fay, F. S. (1975): Mechanical properties of single isolated smooth muscle cells. In: *Smooth Muscle Physiology and Pharmacology, Vol. 50*, pp. 327–342. INSERM Colloguia, Paris.
7. Fay, F. S. (1977): Isometric contractile properties of single isolated smooth muscle cells. *Nature*, 265:553–556.
8. Fay, F. S., Cooke, P. M., and Canaday, P. G. (1976): Contractile properties of isolated smooth muscle cells. In: *Physiology of Smooth Muscle*, edited by E. Bulbring and M. F. Shuba, pp. 249–264. Raven Press, New York.
9. Fay, F. S., and Delise, C. M. (1973): Contraction of isolated smooth muscle cells—structural changes. *Proc. Natl Acad. Sci. USA*, 70:641–645.
10. Fay, F. S., and Fujiwara, K. (1981): Organization of α-actinin in relaxed and contracted single isolated smooth muscle cells. *J. Cell. Biol.*, 99:358a.
11. Fay, F. S., Hoffman, R., Leclair, S., and Merriman, P. (1981): The preparation of individual smooth muscle cells from the stomach of *Bufo marinus*. *Methods Enzymol. (in press)*.
12. Fay, F. S., Rees, D. D., and Warshaw, D. M. (1981): The contractile mechanism of smooth muscle. In: *Membrane Structure and Function, Vol. 4*, edited by E. E. Bittar, pp. 79–130. Wiley, New York.
13. Fay, F. S., and Singer, J. J. (1977): Characteristics of response of isolated smooth muscle cells to cholinergic drugs. *Am. J. Physiol.*, 232:C144–C154.
14. Fisher, B. A., and Bagby, R. M. (1977): Reorientation of myofilaments during contraction of a vertebrate smooth muscle. *Am. J. Physiol.*, 232:C5–C14.
15. Fujiwara, K., Porter, M. E., and Pollard, T. D. (1978): Alpha-actinin localization in the cleavage furrow during cytokinesis. *J. Cell. Biol.*, 79:268–275.
16. Hartshorne, D. J., and Gorecka, A. (1980): Biochemistry of the contractile proteins of smooth muscle. In: *Handbook of Physiology. Section 2: The Cardiovascular System, Vol. 2*, edited by D. F. Bohr, A. P. Somlyo, and H. V. Sparks, Jr., pp. 93–120. Williams & Wilkins, Baltimore.
17. Honeyman, T. W., Merriam, P., and Fay, F. S. (1978): Effect of isoproterenol on cyclic AMP levels and contractility of isolated smooth muscle cells. *J. Mol. Pharmacol.*, 14:86–98.
18. Murphy, R. A. (1976): Contractile system function in mammalian smooth muscle. *Blood Vessels*, 13:1–23.
19. Scheid, C. R., and Fay, F. S. (1980): Control of ion distribution in isolated smooth muscle cells. I. Potassium. *J. Gen. Physiol.*, 75:163–182.
20. Scheid, C. R., Honeyman, T. W., and Fay, F. S. (1979): Mechanism of β-adrenergic induced relaxation of smooth muscle. *Nature*, 277:32–36.
21. Schollmeyer, J. E., Furcht, L. T., Goll, D. E., Robson, R. N., and Stromer, M. H. (1976): Localization of contractile proteins in smooth muscle cells and in normal and transformed fibroblasts.

In: *Cell Motility, Vol. A*, edited by R. Goldman, T. Pollard, and J. Rosenbaum, pp. 361–388. Cold Spring Harbor Press, Cold Spring Harbor.

22. Siegman, M. J , Bulter, T. M., Mooers, S. U., and Davies, R. E. (1980): Chemical energetics of force development, force maintenance, and relaxation in mammalian smooth muscle. *J. Gen. Physiol.*, 76:609–629.
23. Singer, J. J., and Fay, F. S. (1977): Detection of contraction of isolated smooth muscle cells. *Am. J. Physiol.*, 232:C138–C143.
24. Singer, J. J., and Walsh, J. V. (1980): Passive properties of the membrane of single freshly isolated smooth muscle cells. *Am. J. Physiol.*, 239:C153–C161.
25. Slayter, E. M. (1970): *Optical Methods in Biology.* Wiley-Intersciences, New York.
26. Taylor, D. L., and Zeh, R. (1976): Methods for the measurement of polarization optical properties. I. Birefringence. *J. Microsc.*, 108:251–259.
27. Walsh, J. V., and Singer, J. J. (1980): Calcium action potentials in single freshly isolated smooth muscle cells. *Am. J. Physiol.*, 239:C162–C174.
28. Walsh, J. V., and Singer, J. J. (1980): Slow cholinergic (muscarinic) depolarization associated with a conductance decrease in freshly isolated smooth muscle cells. *Physiologist*, 23:111.
29. Walsh, J. V., and Singer, J. J. (1981): Voltage clamp of single freshly dissociated smooth muscle cells: Current-voltage relationships for three currents. *Pfluegers Arch.*, 390:207–210.
30. Warshaw, D. M., and Fay, F. S. (1982): Tension transients in single isolated smooth muscle cells: Insight into the cross-bridge mechanism. In: *Smooth Muscle Contraction*, edited by N. L. Stephens. Marcel Dekker, New York *(in press)*.
31. Yamaguchi, H., Honeyman, T. W., and Fay, F. S. (1981): Mechanisms of beta-adrenergic membrane hyperpolarization in single isolated smooth muscle cells. *J. Gen. Physiol.*, 78:31a.

Basic Biology of Muscles: A Comparative Approach, edited by B. M. Twarog, R. J. C. Levine, and M. M. Dewey. Raven Press, New York © 1982.

Chemical Energetics, Mechanics, and Phosphorylation of Regulatory Light Chains in Mammalian Fast- and Slow-Twitch Muscles

M. J. Kushmerick and *M. T. Crow

Department of Physiology, Harvard Medical School, Boston, Massachusetts 02115

Myosin contains two classes of light chains, essential and regulatory (16). A functional role for the regulatory light chains has been shown clearly only in systems where the primary control of contraction is myosin-linked, such as in scallop myofibrils (17,18,29) or smooth muscle (1,6). In smooth muscle the regulatory nature of the light chains is expressed only on phosphorylation (6,19,27). By way of contrast, in skeletal muscle the primary control is exerted through the thin filament (11). Here, however, a regulatory role for the light chains has not yet been established. Removal of light chains from whole myosin has little effect on its enzymatic activity (30). Yet, vertebrate skeletal muscles possess enzymes (23,26) that are capable of reversibly phosphorylating regulatory light chains, which thereby may exert some regulatory effects. No unified regulatory role is evident in the several reports that phosphorylation of light chains of rabbit skeletal myosin decreases the K_{app} of actin for myosin (25), is correlated with the phenomenon of posttetanic potentiation (22), decreases ATPase activity in glycerinated rabbit psoas fibers on thiophosphorylation (7), but does not alter PCr splitting in rat EDL during isometric tetani (4). We hypothesize that expression of their regulatory role depends on their phosphorylation. Physiological and biochemical measurements on isolated mouse muscles suggest a regulatory function for phosphorylation of regulatory light chains associated with a reduction in apparent actomyosin ATPase.

METHODS

Male CD-1 mice (Charles River Breeding Laboratories, Wilmington, Mass.) between the ages of 21 to 28 days were used. The muscles used were the isolated

---

*Dr. Crow's present address is Department of Medicine, Stanford University, Stanford, California 94305.

and intact slow-twitch soleus (75% SO fibers and 25% FOG fibers) and fast-twitch extensor digitorum longus (EDL) (36% FG fibers, 63% FOG fibers, and 1% SO fibers) muscles of the mouse. All experiments were performed at 20°C in a Ringers solution of the following composition: NaCl 116 mM; KCl 4.6 mM; KH_2PO_4 1.16 mM; $CaCl_2$ 2.5 mM; $MgSO_4$ 1.16 mM, $NaHCO_3$ 25.3 mM; gentamicin sulphate 10 μg/ml. The solution was gassed with a mixture of 95% O_2/5% CO_2 (v/v) to obtain a pH of 7.4. The relationship between fiber length and total muscle length and the histochemical characterization of the fiber types in these muscles are described in detail elsewhere (8).

Phosphorylation of the Myosin Light Chains

The extent of myosin light chain phosphorylation was monitored in extracts of the muscles by two-dimensional gel electrophoresis. Muscles were rapidly frozen (within 50 msec) under either resting conditions or immediately following an isometric tetanus with a device similar in operation and design to that described by Kretzschmar and Wilkie (20). The muscles were then pulverized with an aliquot of frozen urea lysis buffer (24) at −196°C and then warmed to 40°C for ½ hr to denature enzymes and to solubilize the proteins. Under these conditions, the myosin light chain kinase and phosphatase activities present in the muscle were effectively inhibited. Two-dimensional gel electrophoresis was performed following the procedures of O'Farrell (24). The positions of proteins in the two-dimensional matrix were determined by co-electrophoresis of purified rabbit muscle standards. Phosphorylation of the regulatory light chains of both fast- and slow-twitch myosin (designated LC2f and LC2s, respectively) resulted in a shift in the isoelectric focusing point of the light chains toward the acidic region of the gel as expected from the added negative charge. The extent of light chain phosphorylation is expressed as the ratio of the phosphorylated light chain content to total regulatory light chain content and was computed from densitometric tracings of the light chains in the Coomassie blue-stained (R250) gels.

The identity of these spots as the phosphorylated derivatives of the regulatory light chains was checked by making autoradiograms of two-dimensional gels of muscles incubated for 8 hr in ^{32}P-orthophosphate to label the γ-phosphate of the ATP to a constant specific activity. Autoradiography (data not shown) showed in control muscle phosphorylation of tropomyosin, LC2f, and LC2s (the last in soleus muscles only) and of two other unidentified proteins to a small extent. Only the intensity of LC2f increased with contraction.

Energetics of Contraction

The energy utilization associated with contraction was estimated by the extent of high-energy phosphate splitting occurring during contraction ($\Delta \sim P_{init}$) and by the extent of recovery resynthesis ($\Delta \sim P_{rec}$) (8). Muscles were stimulated at the length for optimal tension production at a frequency of 68 Hz. From changes in metabolite contents of stimulated muscles that were rapidly frozen at the end of isometric tetani (1 to 15 sec), $\Delta \sim P_{init}$ was calculated; $\Delta \sim P_{rec}$ was estimated from the extent of oxygen consumption for aerobic oxidations and of glycolytic lactate production, both of which occurred during the recovery period following each tetanus. Details of the methods and procedures are given elsewhere (8,21). In each case, the rate of energy utilization (in μmoles $\cdot$ g^{-1} $\cdot$ sec^{-1}) was adjusted for differences in the force generated per cross-sectional area in these muscles as well as the fatigue of force during prolonged stimulation. This was done by expressing the rate with respect to the time integral of tension (Newtons $\cdot$ m $\cdot$ sec $\cdot$ g^{-1}) (8,14,21).

Mechanical Characterization

The maximum velocity of shortening (units:fiber lengths $\cdot$ sec^{-1}) was estimated from the velocity of unloaded shortening (V_{us}) by the quick release method described by Edman (12). This velocity (V_{us}) was determined from the time course of events following a rapid release of the muscle to zero force at various times during an isometric tetanus. Control and measurement of the position of the muscle was achieved with a "moving iron" type galvonometer equipped with a differential-capacitor position detector (Model 305, Cambridge Technology, Cambridge, Mass.). The total system compliance (motor, lever, and attachments) was 0.5 μm $\cdot$ g^{-1}. Rapid release of the muscle was accomplished by applying an analog signal to the position input of the servo-control loop. This signal consisted of a voltage ramp corresponding to a velocity 5 times greater than the maximum velocity of shortening (V_{max}) of the faster muscle (EDL). In an independent series of experiments, it was verified that the velocity of unloaded shortening was equivalent to the V_{max} estimated from force-velocity data obtained independently on these muscles by isotonic releases.

RESULTS

Myosin Light Chain Phosphorylation in Isometric Tetanic Contraction

The phosphorylated derivatives of the regulatory myosin light chains present in slow-twitch soleus and fast-twitch EDL muscles are marked by the arrows in Fig. 1. The regulatory light chains (LC2f and LC2s) in both the soleus and EDL were phosphorylated only to a small extent (Fig. 1A and C) in unstimulated muscles. This basal level of phosphorylation, expressed as the ratio of phosphorylated to total regulatory light chain content of its type, amounted to 0.11 ± 0.04 (x $\pm$ SD; n = 12) in both muscles. On stimulation, only the fast-twitch regulatory light chains (LC2f) in the EDL underwent a net increase in phosphorylation (Fig. 1B and D). No change was observed in the extent of phosphorylation of either the slow-twitch (LC2s) or fast-twitch (LC2f) myosin regulatory light chains in the soleus when stimulated the same amount of time.

The time course of phosphorylation during an isometric tetanus is shown in Fig. 2. For the EDL, the maximal extent of incorporation was reached by a 12-sec tetanus and amounted to a fractional extent of 0.51 ± 0.05 (x $\pm$ SD; n = 6).

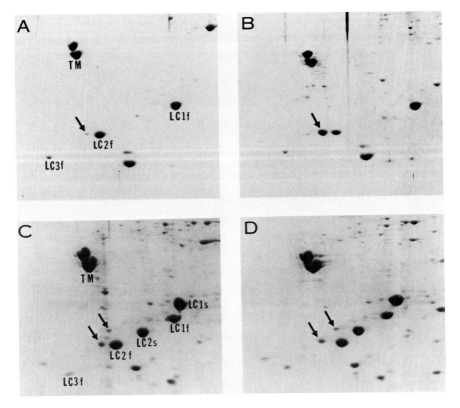

FIG. 1. Two-dimensional gel electrophoretograms of mouse fast-twitch EDL **(A,B)** and slow-twitch soleus **(C,D)** muscles. Extracts are from muscles rapidly frozen under either resting condition **(A,C)** or following a 12 sec isometric tetanus **(B,D)**. Approximately 70 μg of protein was applied to the isoelectric focusing gel. Isoelectric focusing was in the horizontal direction in the range pH 4 *(left)* to pH 6 *(right)*. The gels shown were stained with Coomassie blue R-250. Tm, tropomyosin and LC, light chains are indicated. In the EDL **(A)**, the three light chains of the fast-twitch myosin phenotype were present; these were the 22,500 (LC1f) and 16,000 dalton (LC3f) essential light chains and the 18,000 dalton (LC2f) regulatory light chain. In the mouse soleus, these fast-twitch light chains as well as two slow-twitch light chains were distinguished; these slow-twitch light chains were the 27,000 (LC1s) dalton essential light chain and the 19,000 (LC2s) dalton regulatory light chain. The *arrows* mark the phosphorylated derivative of LC2f and of LC2s. (From Crow and Kushmerick, ref. 8a, with permission.)

Further stimulation resulted in no further increase in the proportions of the phosphorylated light chains. In contrast, the fast-twitch regulatory light chains (data shown) and the slow twitch regulatory light chains (data not shown) in the soleus did not undergo any detectable change in net phosphate content on stimulation.

Subtraction of the basal level of phosphorylation from the total shows the maximal extent of phosphorylation in the EDL specifically in response to stimulation was 40% of the total LC2 content. This value corresponds to the percentage of the total cross-sectional area (36%) occupied by the fast-twitch glycolytic (FG) type fibers in the EDL (8). Thus, complete stimulus-induced phosphorylation of LC2f only in

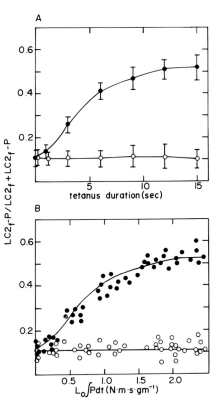

FIG. 2. The time course of phosphorylation in the fast-twitch regulatory light chains of the mouse EDL *(closed circles)* and soleus *(open circles)* muscles. Muscles were stimulated to give an isometric tetanus for the duration indicated (**A**, abscissa) and then rapidly frozen, extracted, and analyzed as described in Methods. The fractional extent of phosphorylation is expressed as the amount of phosphorylated light chain (LC2f-P) relative to total regulatory light chain content (LC2f + LC2f-P) from densitometry of Coomassie Blue stained gels. The time course of phosphorylation of the slow-twitch regulatory light chains in the soleus is not shown, but followed the same pattern of no increase in phosphorylation shown for the fast-twitch fiber types in the soleus. The *bars* indicate one standard deviation of the measurement. **A:** Extent of LClf-P formation as a function of isometric tetanus duration. Data are averaged from at least 6 muscles per point; *bars* represent ± SD. **B:** The same data are plotted individually as a function of the time-integral of tension.

FG fibers could account for our data. The remaining fibers in the EDL are fast-twitch oxidative-glycolytic (FOG) type of fibers, which are histochemically identical to the fast-twitch fibers in the soleus. These FOG fibers were not phosphorylated in the soleus in response to stimulation because no increased phosphorylation was detected in soleus. Since the FOG fibers are histochemically identical in soleus and EDL and since both muscles were stimulated identically, FOG type fibers would not be expected to be phosphorylated under comparable conditions in the EDL. The fibers that are phosphorylated must then be confined to the population of FG type fibers present in the EDL but not in the soleus.

Energetics of Isometric Contraction: A Comparison of Fast- and Slow-Twitch Muscles

The extent of high energy phosphate resynthesis during recovery ($\Delta \sim P_{rec}$) represents the total recovery metabolism and was measured with the use of the following relation: $\Delta \sim P_{rec} = 6.3 \, (XO_2) + 1.5 \, (\chi_{lac})$, where XO_2 is the suprabasal recovery oxygen consumption following a tetanus and χ_{lac} is the suprabasal lactate production during that interval. Details are published elsewhere (8). Figure 3 shows the pattern of recovery metabolism in EDL differs in two respects from

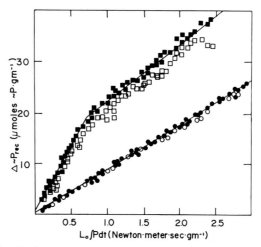

FIG. 3. The relationship between total recovery chemical input and the tension-time integral. The recovery chemical input ($\Delta \sim P_{rec}$) is expressed in μmoles $\sim$ P/g and obtained from the following equation: $\Delta \sim P_{rec} = \kappa\xi O_2 + \lambda\xi lac$, where κ and λ are stoichiometric factors equal to 6.3 and 1.5, respectively. *Open symbols* represent the contribution to the total recovery chemical input from oxygen consumption alone ($\kappa X O_2$) in the soleus *(open circles)* and EDL *(open squares)*. *Closed symbols* respresent the total recovery chemical input after the contribution from glycolytic ATP production ($\lambda X lac$) was added to the aerobic contribution in soleus *(closed circle)* and EDL *(closed squares)*. In a majority of the data points for the soleus, the contribution from lactate production was negligible and for the sake of visual clarity only the points corresponding to the total recovery input were graphed. Data from 13 soleus and EDL muscles each. Multiple determinations were performed on each muscle. Muscles were stimulated and allowed to recover for 30 min before stimulating again. (Data published in another form in Crow and Kushmerick, ref. 8.)

the soleus. First, $\Delta \sim P_{rec}$ in the fast-twitch muscle is greater for the same tension-time integral. Second, the energy cost in EDL decreases with continuous tetanization to about ½ the initial value after some 5 sec of stimulation. There is no change in slope in the soleus.

The actual splitting of PCr and other chemical changes were assessed during contraction by rapid freezing techniques. The data in Fig. 4 for soleus (upper panel) shows that $\Delta \sim P_{init}$ is limited in continuous tetanization in oxygenated muscles (open symbols). After 15 sec, an energetic steady state is achieved with high energy phosphate splitting balanced by steady oxygen consumption (data not shown). When the muscles were made anaerobic, $\sim$P resynthesis was inhibited and the increase in $\Delta \sim P_{init}$ was a linear function of tension-time integral (upper panel, closed symbols). The aerobic capacity of EDL is lower than soleus, so there was less of a difference between anaerobic $\Delta \sim P_{init}$ (lower panel, open symbols) and aerobic $\Delta \sim P_{init}$ (closed symbols).

The functional relation of $\Delta \sim P_{init}$ as a function of tension-time integral in both muscles is the same as the relationship of $\Delta \sim P_{rec}$ to tension-time integral (compare Figs. 3 and 4). When both measures of tetanic energy utilization, $\Delta \sim P_{init}$ and $\Delta \sim P_{rec}$, are compared quantitatively (Table 1), they are equal within experimental

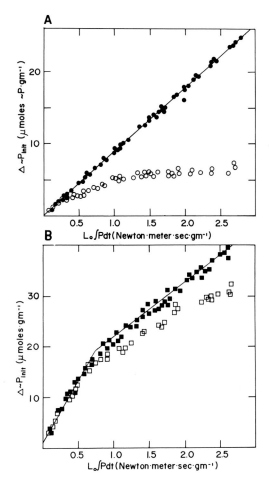

FIG. 4. The relationship between initial chemical breakdown and tension-time integral: **A:** Soleus; **B:** EDL. Units are the same as for Fig. 3. Each data point represents the initial chemical breakdown ($\Delta \sim P_{init}$) in a single muscle compared with its control from observed changes in ATP, PCr, and lactate in the manner described in Methods. The *open symbols* represent the initial chemical change assessed under aerobic conditions; the *closed circles* represent the initial chemical change assessed under anaerobic conditions. (Data published in another form in Crow and Kushmerick, ref. 8.)

error for both muscles and for all tetanic durations studied. Thus, there is no evidence of a biochemical energy imbalance in fast- or slow-twitch mammalian, in contrast to the well-known imbalance in amphibian muscles (9,15).

Light Chain Phosphorylation and the Energetics of Contraction

To summarize the results so far, the pattern of energy utilization for isometric contractions differed in two respects in the mouse EDL and soleus muscles. First, for short tetani, the initial rate of high-energy phosphate splitting during steady

TABLE 1. *Comparison of initial chemical change and recovery resynthesis in mouse muscles*[a]

Tetanus interval (sec)	Soleus		EDL	
	$\Delta{\sim}P_{init}$	$\Delta{\sim}P_{rec}$	$\Delta{\sim}P_{init}$	$\Delta{\sim}P_{rec}$
	(μmoles/newton·cm·sec)			
0–3	9.1	8.9	24.1	22.1
	($\pm$1.8)	($\pm$0.4)	($\pm$3.6)	($\pm$1.1)
3–6	8.4	8.6	19.4	19.1
	($\pm$1.4)	($\pm$0.5)	($\pm$2.1)	($\pm$1.3)
6–9	8.9	8.7	14.8	14.5
	($\pm$1.9)	($\pm$0.3)	($\pm$3.9)	($\pm$1.9)
9–12	8.6	8.6	11.8	10.5
	($\pm$1.0)	($\pm$0.2)	($\pm$3.3)	($\pm$1.2)
12–15	8.8	8.8	10.0	11.4
	($\pm$2.5)	($\pm$0.3)	($\pm$1.2)	($\pm$1.4)

Data from Kendrick-Jones et al., ref. 17.
[a]Mean values $\pm$ standard deviation obtained from nonlinear regression analysis as described in methods. Differences between $\Delta{\sim}P_{init}$ and $\Delta{\sim}P_{rec}$ are not statistically different for any tetanus intervals at the level ($p > 0.10$).

isometric contractions was about threefold higher in the EDL than in the soleus. Second, with prolonged stimulation, this rate in the EDL subsequently decreased so that after 12 sec of stimulation it had fallen to about ½ the initial rate. The reduction in the rate of energy utilization observed in the EDL followed a similar time course to stimulation-induced light chain phosphorylation. Comparable contractile activity induced no change in the rate of high energy phosphate utilization in the soleus.

The relationship between the rate of energy utilization and the fractional extent of light chain phosphorylation is shown in Fig. 5. The data show a strong correlation between increased LC2f phosphorylation and the reduction in energy cost in the EDL. In the soleus there was no detectable change in either parameter (Fig. 5, open circles). Throughout all these measurements, the tetanic force fell to not less than 80% the initial force generation in EDL and fatigue of force was less in soleus (Table 2). This relative constancy of isometric force along with the reduction in the overall energy cost for tension maintenance suggests that light chain phosphorylation is associated with a reduction in the actomyosin ATPase rate *in vivo*. This suggestion, of course, is based on the assumption that most of the reduction in energy utilization is due to reduction in the actomyosin ATPase.

Light Chain Phosphorylation and Changes in the Mechanical Properties of the EDL

Measurement of the V_{max} as a function of preceding tetanus duration tested and confirmed the hypothesis that the reduction seen in the energy cost of the EDL during prolonged stimulation was due to a reduction in the turnover rate of the

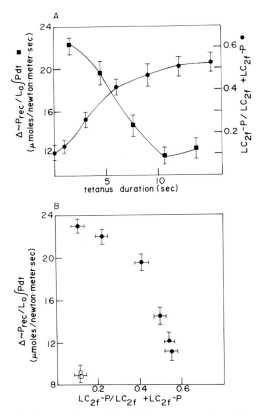

FIG. 5. The relationship between the extent of light chain phosphorylation and the energy cost for tension maintenance. **A:** The energy cost for tension maintenance in the EDL ($\Delta \sim P_{rec}$) in μmoles $\sim$P/Newton · meter · sec and the extent of light chain phosphorylation in the EDL are plotted as a function of the tetanus duration in seconds. The *bars* represent ± one standard deviation. *Closed circles,* light chain phosphorylation; *closed squares,* energy utilization rate for the EDL. **B:** The energy cost is plotted directly as a function of light chain phosphorylation. *Bars* represent ± one standard deviation. *Closed circles,* EDL; *open circles,* soleus. (From Crow and Kushmerick, ref. 8a, with permission.)

actomyosin ATPase. The values for the V_{us} measured after various durations of isometric tetanus are given in Table 2. Following short tetani, the V_{us} of the EDL was approximately three times that of the soleus studied under comparable conditions. With prolonged stimulation prior to measurement of shortening velocity, V_{us} in the EDL fell to approximately one-half its initial value. Neither brief nor prolonged stimulation induced any velocity change in the soleus.

In Fig. 6, the relationships between V_{max} and light chain phosphorylation are shown. In the EDL, the parallel reductions in the rate of energy utilization (compare Figs. 5 and 6) and of V_{us} with continued stimulation were correlated with the extent of LC2f phosphorylation (Fig. 6). In stimulations of the soleus (open symbols), no change was observed in V_{us}, or phosphorylation of either LC2f (data shown) or of LC2s (data not shown). Thus, in addition to the quantitative correlations in EDL

TABLE 2. *The relationship between light chain phosphorylation and the maximum velocity of shortening in the mouse soleus and EDL[a]*

Tetanus duration (sec)	$P_o{}^b$ (g)	Maximal velocity[c] of shortening, V_{us} (fiber lengths·s⁻¹)	$\dfrac{\text{LC2f-P}^d}{\text{LC2f-P + LC2f}}$
Soleus			
0–3	12 (±1.3)	1.88 (±0.05)	0.11 (±0.05)
6–9	11 (±1.1)	1.89 (±0.04)	0.12 (±0.06)
12–15	11 (±0.06)	1.88 (±0.03)	0.10 (±0.05)
EDL			
0–3	17 (±0.8)	5.75 (±0.13)	0.22 (±0.07)
6–9	16 (±1.2)	4.07 (±0.06)	0.45 (±0.06)
12–15	13 (±1.5)	3.21 (±0.10)	0.51 (±0.05)

[a]All measurements are expressed as the mean ± one standard deviation.
[b]The isometric tetanic force just prior to release or the termination of stimulation.
[c]The maximum velocity of shortening is measured in units of fiber lengths/sec. This parameter was measured from the velocity of unloaded shortening (V_{us}) as described in methods and by Edman (12).
[d]The fractional extent of LC2f phosphorylation measured by a densitometer of the LC2f region of gels: area of the phosphorylated form divided by the total LC2f area.

between LC2f phosphorylation, contractile energetics and mechanics, there are distinctive qualitative differences between the fast-twitch EDL and slow-twitch soleus muscles with respect to the same parameters.

DISCUSSION

There are two possible interpretations for the relationship between light chain phosphorylation and the reduction in the *in vivo* actomyosin ATPase rate of isolated fast-twitch mouse muscles reported here. One is that the correlation is a mere coincidence; that is, light chain phosphorylation and the apparent reduction in the actomyosin ATPase rate *in vivo* are parallel phenomena bearing only a fortuitous relationship to one another. The other is to postulate a causal connection between the three sets of temporally coincident results. The results presented here correlating phosphorylation with energetic and mechanical effects in EDL and the finding of no such change in soleus under similar experimental conditions support the hypothesis that light chain phosphorylation modifies the intrinsic rate of skeletal muscle actomyosin ATPase, which, in turn, leads to a reduction in the cross-bridge cycling rate *in vivo*. This postulated modification was manifested in intact muscle as a reduction in the energy cost for tension maintenance and a slowing of the velocity of shortening. It is not established that high energy phosphate splitting during contraction is equivalent quantitatively to actomyosin ATPase; some unknown fraction of the total energy cost is due to sarcoplasmic reticulum and other contraction-induced ATPases. However, V_{max} directly correlates with actomyosin ATPase (3) and a reduction in V_{max} indicates the cycling rate of cross-bridges is decreased.

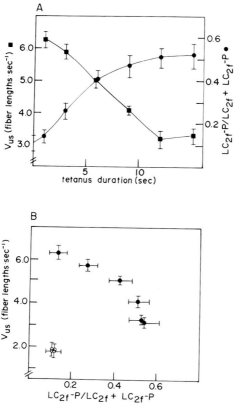

FIG. 6. The relationship between the velocity of unloaded shortening and light chain phosphorylation. Values represented are the means ± one standard deviation. **A:** The V_{us} in fiber lengths per second *(closed squares)* as well as the extent of light chain phosphorylation in the EDL *(closed circles)* are plotted as a function of the duration of tetanic stimulation in seconds. **B:** The V_{us} plotted directly against the extent of light chain phosphorylation. *Closed circles,* EDL; *open circles,* soleus.

The effects just described, that accompany phosphorylation of the regulatory light chains in the mouse fast-twitch muscles, differ in two major respects from the effects seen on phosphorylation of smooth muscle regulatory light chains. In mammalian skeletal muscles, the time course of phosphorylation occurs on a time scale at least two orders of magnitude slower than the development of tension (22,28). In contrast, phosphorylation of the light chains in smooth muscle follows a similar time course to tension development (10). Second, phosphorylation of the regulatory light chains in smooth muscle is associated with activation of the actomyosin ATPase (6,27), whereas the results reported here indicate that phosphorylation of the light chains in vertebrate (mouse) skeletal muscles is associated with a reduction in the actomyosin ATPase rate *in vivo.*

The mechanistic significance suggested by our experiments do not necessarily embrace all reported experimental results. Pemrick's (25) data indicating a decrease in the apparent K_m for actin on phosphorylation may not be relevant to the *in vivo* condition within the myofibril if the effective actin concentration is in the 10 M range, as is suggested (13). The lack of change in PCr splitting with increased phosphorylation in IAA-treated rat EDL (4) may reflect simply different experimental conditions from ours in such a way to prevent the expression of the functional role suggested here. Cooke et al. (7) reported the decrease in ATPase rate in glycerinated and thiophosphorylated rabbit psoas fibers only at 37°C and not at lower temperatures. Their results also indicate that quite specific conditions are apparently required to detect a functional expression of light chain phosphorylation. The correlation of light chain phosphorylation and posttetanic potentiation observed by Manning and Stull (22) could be explained by decreased actomyosin turnover rate. The occurrence of some degree of phosphorylation in both soleus and EDL in resting muscles indicates the potential for changing the extent of phosphorylation in the slow-twitch muscle, which may be manifested under other experimental and stimulation conditions. Finally, the phosphorylation that occurs during tension development of frog muscle (2) may also be correlated with a decreased energy cost for contraction, but specific experiments to test this point have not yet been made.

The functional significance of phosphorylation of vertebrate skeletal muscle myosin may be related to its apparent fiber-type specificity detected under our experimental conditions. Fast glycolytic fibers are fast-twitch types optimized for large power outputs (31) because of their large cross-sectional area and of their organization into large motor units. Maximum power is needed rarely and only in situations of extreme accelerations. Large power outputs occur normally only during brief portions of locomotory movements. Fast-twitch fibers have a higher energy cost for contraction than do slow-twitch fibers and, also, fast glycolytic fibers rely on their glycogen stores for high energy phosphate resynthesis so that stimulated FG fibers fatigue rapidly (5). However, high speed and maximal mechanical power is not necessary or useful for sustained forceful contractions. Phosphorylation of the regulatory light chains in these fibers and the concomitant reduction in actomyosin turnover rate (decreased ATPase and V_{us}) may be a mechanism to reduce fatigueability in large FG fibers in sustained forceful contractions.

ACKNOWLEDGMENTS

This work was supported by NIH grant AM 14485 and by a grant from the Muscular Dystrophy Association. Michael Crow was a predoctoral trainee on grant 5-T32-GM07258 and Martin Kushmerick was supported by RCDA 5-KO4-AM00178.

REFERENCES

1. Adelstein, R. S. (1980): Phosphorylation of muscle contractile proteins. *Fed. Proc.*, 39:1544–1546.
2. Barany, M. (1967): ATPase activity of myosin correlated with speed of muscle shortening. *J. Gen. Physiol.*, 50:197–216.
3. Barany, K., Barany, M., Gillis, J. M., and Kushmerick, M. J. (1980): Myosin light chain phosphorylation during the contraction cycle of frog muscle. *Fed. Proc.*, 39:1547–1551.

4. Barsotti, R. J., and Butler, T. M. (1981): Effect of myosin LC2 phosphorylation on high-energy phosphate usage in contracting skeletal muscle. *Biophys. J.*, 33:234a.
5. Burke, R. E., Levine, D. N., Tsairis, P., and Zajac, F. E. (1973): Physiological types and histochemical profiles in motor units of the cat gastrocnemius. *J. Physiol. (Lond.)*, 234:723.
6. Chacko, S., Cont, M. A., and Adelstein, R. S. (1977): Effect of phosphorylation of smooth myosin on actin activation and ca^{++} regulation. *Proc. Natl. Acad. Sci. (USA)*, 74:129–133.
7. Cooke, R., Franks, K., Ritz-Gold, C. J., and Toste, T. (1981): The function of myosin light chain phosphorylation in skeletal muscle. *Biophys. J.*, 33:235a.
8. Crow, M., and Kushmerick, M. J. (1982): Chemical energetics of slow- and fast-twitch muscles of the mouse. *J. Gen. Physiol.*, 79:147–162.
8a. Crow, M. T., and Kushmerick, M. J. (1982): Myosin light chain phosphorylation as associated with the decrease in the energy cost for contraction in fast twitch mouse muscles. *J. Biol. Chem.*, 257:2121–2124.
9. Curtin, N., and Woledge, R. C. (1978): Energy changes and muscle contraction. *Physiol. Revs.*, 58:690–761.
10. Dillon, P. R., Ackoy, M. O., Driska, S. P., and Murphy, R. A. (1981): Myosin phosphorylation and the crossbridge cycle in arterial smooth muscle. *Science*, 211:495–497.
11. Ebashi, S., Endo, M., and Ohtsuki, I. (1969): Control of muscle contraction. *Q. Rev. Biophys.*, 2:351–384.
12. Edman, K. A. P. (1979): The velocity of unloaded shortening and its relation to sarcomere length and isometric force in vertebrate muscle fibers. *J. Physiol. (Lond.)*, 291:143–159.
13. Eisenberg, E., Hill, T. L., and Chen, Y. (1980): Cross-bridge model of muscle contraction: Quantitative analysis. *Biophys. J.*, 29:195–226.
14. Hartree, W., and Hill, A. V. (1921): The regulation of the supply of energy in muscle contraction. *J. Physiol. (Lond.)*, 55:133–145.
15. Homsher, E., and Kean, C. J. (1978): Skeletal muscle energetics and metabolism. *Ann. Rev. Physiol.*, 40:93–131.
16. Kendrick-Jones, J. (1976): Myosin-linked calcium regulation. In: *Molecular Basis of Motility*, edited by L. M. G. Heilmeyer, J. C. Ruegg, and Th. Wieland, pp. 122–136. Springer-Verlag, New York.
17. Kendrick-Jones, J., Lehman, W., and Szent-Gyorgyi, A. G. (1970): Regulation in molluscan muscles. *J. Mol. Biol.*, 54:313–326.
18. Kendrick-Jones, J., Szentkiralyi, E. M., and Szent-Gyorgyi, A. G. (1976): Regulatory light chains in myosin. *Mol. Biol.*, 104:747–775.
19. Kerrick, W. G. L., Hoar, P. E., and Cassidy, P. S. (1980): Calcium activated tension: The role of myosin light chain phosphorylation. *Fed. Proc.*, 39:1558–1563.
20. Kretzschmar, K. M., and Wilkie, D. R. (1969): A new approach to freezing tissues rapidly. *J. Physiol. (Lond.)*, 202:66–67P.
21. Kushmerick, M. J., and Paul, R. J. (1976): Relationship between initial chemical reactions and oxidative metabolism for single isometric contractions of frog sartorius at 0°C. *J. Physiol. (Lond.)*, 254:711–727.
22. Manning, D. R., and Stull, J. T. (1979): Myosin light chain phosphorylation and phosphorylase activity in rat extensor digitorum longus muscle. *Biochem. Biophys. Res. Commun.*, 90:164–170.
23. Morgan, M., Perry, S. V., and Ottoway, J. (1976): Myosin light chain phosphatase. *Biochem. J.*, 157:687–697.
24. O'Farrell, P. H. (1975): High resolution of two-dimensional electrophoresis of proteins. *J. Biol. Chem.*, 250:4007–4021.
25. Pemrick, S. M. (1980): Phosphorylated L2 light chain of skeletal myosin is a modifier of the actomyosin ATPase. *J. Biol. Chem.*, 255:8836–8841.
26. Perrie, W. T., Smillie, L. B., and Perry, S. V. (1973): A phosphorylated light chain component of myosin from skeletal muscle. *Biochem. J.*, 135:151–164.
27. Small, J. V., and Sobieszek, A. (1977): Ca^{++} regulation of mammalian smooth muscle actomyosin via a kinase-phosphatase-dependent phosphorylation and dephosphorylation of the 20,000 M$_r$ light chain of myosin. *Eur. J. Biochem.*, 76:521–530.
28. Stull, J. T., and High, C. W. (1977): Phosphorylation of skeletal muscle contractile proteins *in vivo*. *Biochem. Biophys. Res. Commun.*, 77:1078–1083.
29. Szent-Gyorgyi, A. G., Szentkiralyi, E., and Kendrick-Jones, J. (1973): The light chains of scallop myosin as regulatory subunits. *J. Mol. Biol.*, 74:179–203.

30. Weeds, A. G., and Lowey, S. (1971): The substructure of the myosin molecule. II. The light chain of myosin. *J. Mol. Biol.*, 61:701–725.
31. Wuerker, R. B., McPhedran, A. M., and Henneman, E. (1965): Properties of motor units in a heterogeneous pale muscle (M. gastrocnemius) of the cat. *J. Neurophysiol.*, 28:85–99.

Basic Biology of Muscles: A Comparative Approach, edited by B. M. Twarog, R. J. C. Levine, and M. M. Dewey. Raven Press, New York © 1982.

Myocardial Adaptation to Stress from the Viewpoint of Evolution and Development

Norman R. Alpert and Louis A. Mulieri

Department of Physiology and Biophysics, University of Vermont College of Medicine, Burlington, Vermont 05405

In response to a temporary stress such as an increase in the demands of the peripheral tissues and organs for oxygen, there is an immediate increase in cardiac output. This rapid adaptation of the whole heart occurs as a result of changes in heart volume and thus sarcomere length (16,30), humoral agents or changes in autonomic balance (28), heart rate (11), ionic milieu (27), and substrate availability (26).

The heart can respond to persistent stress, such as pressure or volume overload, by increasing the size of its cells. Under these circumstances, the increase in muscle mass provides a basis for meeting the extra demands placed on the heart. This increase in mass is not necessarily pathological in that it can and frequently does occur without focal necrosis or the infiltration of the heart muscle with fibrous tissue. The changes found in the heart following the development of hypertrophy are not the arithmetic result of adding a number of identically functioning muscle components. There are significant alterations at the functional level, resulting from a molecular reorganization of the myocardial cell, which contribute to the success of the adaptation as well as to its limitations.

The type of response depends on the nature, duration, and intensity of the stress as well as the age and species involved. Following pressure overload hypertrophy of sufficient intensity, there is a decrease in the maximum unloaded velocity of shortening, an increase in force-velocity curvature, and an increase in time-to-peak twitch tension (4). Following thyrotoxic stress, there is an increase in maximum unloaded velocity of shortening, a decrease in force-velocity curvature, and a decrease in time-to-peak twitch tension (7). These changes in mechanical function involve both the contractile and excitation-contraction coupling system. They are comparable to the differences observed between frog and tortoise muscle and between fast and slow mammalian muscle. This led us to speculate that, in adapting to long-term stress, the heart uses a repertoire of functional and molecular alterations similar to those seen in frog and tortoise as well as mammalian fast and slow muscle, and thus represent fundamental functional aspects of muscle adaptation seen in evolution and development.

173

The key parameters involved in the adaptation are mechanical, chemical, and thermodynamic. Woledge (32) carried out a detailed mechanical and thermal analysis of tortoise muscle and compared the tortoise data with comparable data from frog muscle (19–21). The salient features of the mechanical performance of the two muscles: (a) a major difference in the unloaded maximum velocity of shortening, with the frog having a V_{max} 470% greater than the tortoise; (b) the curvature of the force velocity relationships as seen in the normalized force-velocity curves or in the ratio of "a" to "Po"[1] is greater in tortoise muscle than in frog muscle; and (c) the time-to-peak tension is five times longer in the tortoise (Fig. 1) than in frog.

Analogous differences are seen in Fig. 2, where the rat extensor digitorum longus (EDL) (fast) and the soleus (SOL) (slow) are compared (13,14). In the EDL, V_{max} is 150% greater, and the curvature is less as reflected by the increase in a/Po, whereas time-to-peak tension is 60% less than in the soleus.

Additional information about the subcellular reorganization responsible for the mechanical differences observed can be obtained from measurements of thermo-mechanical economy. The economy of isometric contraction is defined as isometric tension or the tension-time integral divided by the heat associated with the isometric response. From measurements of this ratio, it is clear that slow muscle develops force more economically than fast muscle. Thus the tortoise muscle is 41 times more economical than the frog muscle and the rat soleus is about 6½ times more economical than the extensor digitorum longus (Table 1) (10,17,20,21,31,32). Associated with the increase in economy there is an increase in the curvature of the force-velocity relation as seen in the decrease of the a/Po ratios for the tortoise

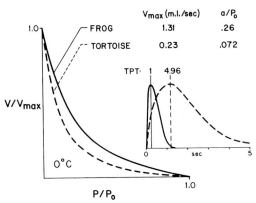

FIG. 1. Normalized force-velocity curves and isometric twitch records for frog and tortoise muscle. Unloaded shortening velocity (V_{max}), in muscle lengths per second, and the ratio, a/Po, are tabulated in the upper right hand corner. (Adapted from Hill, ref. 21, *solid line*; and Hartree, ref. 19, and Woledge, ref. 32, *dashed line*.)

---

[1]In the ratio a/Po, "a" is a constant and "Po" is isometric force at optimum length used in the standard force-velocity equation $(P + a)(V + b) = b(P_o + a)$ found in Hill (20). As "a/Po" decreases, the curvature increases.

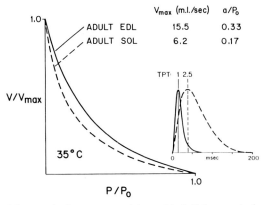

FIG. 2. Normalized force-velocity curves and isometric twitch records for rat extensor digitorum longus (EDL) and soleus (SOL) muscles. Unloaded shortening velocity (V_{max}), in muscle lengths per second, and the ratio, a/Po, are tabulated in the upper right hand corner. (Adapted from Close, refs. 13 and 14.)

TABLE 1. *Isometric economy, force-velocity curvature, and myosin ATPase in fast and slow muscles*

	Isometric tension		Myosin	Mechanical
	Isometric heat rate	a/Po	ATPase	V_{max}
Frog ÷ tortoise	0.024	3.47	15	5.6
Rat, EDL ÷ SOL	0.156	1.69	2.3	2.3

and soleus muscles in comparison with the frog and EDL muscles, respectively (13,14,17,20,21,31,32). The basis for some of these changes is an alteration in the structure of the contractile protein, myosin. As a result of these alterations, the ATPase activity (actin- or Ca^{2+}- activated) is substantially higher in the fast muscle (10,32). Finally, although normalized isometric force is virtually the same in fast and slow muscle, the maximal velocity of unloaded shortening is always faster and the time-to-peak twitch tension is shorter in the fast muscles (13,14,20,32). The functional adaptations involve changes in the contractile as well as the excitation-contraction coupling systems. The experiments described in this chapter were designed to examine the extent to which heart muscle uses the same or similar functional adaptations as those seen in evolution and development.

METHODS

The experimental plan was to produce two distinctly different models of hypertrophy in the rabbit, which represent extremes of functional and molecular reorganization of the heart muscle. Pressure-overload was produced in 1.9 kg rabbits by threading a monel metal, spiral clip around the pulmonary artery (2,18). This reduced the internal diameter of the right ventricular outflow tract by 67%. The stress on the right ventricle produced a substantial hypertrophy that remained stable for at least 2 to 10 weeks following the surgery. All the parameters studied in the preparation, namely, weight, cell size, degree of hypertrophy, mechanics, and chemistry, were generally stable during this period of time. Thyrotoxic hypertrophy was produced in 1.9 kg rabbits by 14 daily injections of 0.2 mg/kg of L-thyroxine (9).

Right ventricular intraluminal systolic pressure was significantly increased in both the pressure overloaded (P) and thyrotoxic preparations (T). Diastolic pressure was elevated from 1.2 to 2.2 mmHg in the pressure overloaded preparations (8). However, this level was not high enough to be considered an index of heart failure (Table 2). Right ventricular heart weight/body weight ratio doubled in P and T hypertrophy (Table 3) (7). Papillary muscles were increased in cross-sectional area by 78% in the P hypertrophied hearts and by 11% (n.s.) in the T hypertrophy (Table 3) (7).

To evaluate the performance of the hypertrophied myocardium, we used isolated right ventricular papillary muscle preparations and measured the detailed beat-to-beat energetics during isometric force development.

The key to making detailed myothermal measurements on papillary muscle weighing between 1 and 2 mg was the use of a low thermal capacity, planar thermopile, which has a very rapid response time. The thermopile was made by vacuum deposition of bismuth and antimony on thin mica sheets, and was then covered by a second mica sheet to provide electrical insulation (25). The thermopile was mounted in a frame so that the reference thermojunctions were in thermal contact with the jaws of the frame on each side of the center portion (Fig. 3). The measuring junctions

TABLE 2. *Right ventricular intraluminal pressure*

Pressure (mmHg)	Heart normal	Preparation	
		Pressure overload	Thyrotoxic
Systolic	17	38	31.0
Diastolic	1.2	2.2	1.2

TABLE 3. *Thyrotoxic and pressure overload hypertrophy[a]*

	RV/body weight (g/kg)	Cross-sectional area of RV papillary muscles (mm²)
N	0.31 ± 0.01	0.59 ± 0.01
Pressure overload/N	2.13[b]	1.78[b]
Thyrotoxic/N	1.97[c]	1.11 (ns)

[a]RV = right ventricle; N = normal.
[b]$p < 0.001$.
[c]$p < 0.01$.

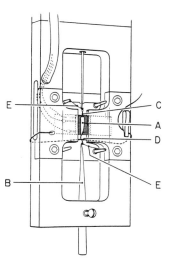

FIG. 3. The vacuum deposited bismuth and antimony thermopile assembly. *A*, papillary muscle; *B*, stationary glass hook; *C*, glass rod connected to force transducer; *D*, silk tether; *E*, stimulating electrode wires embedded in silk ties.

were located in the exposed center region of the thermopile. The papillary muscle was placed on the thermopile so that its surface was in juxtaposition to the measuring junctions at the center of the thermopile (Fig. 3A). The lower end of the papillary muscle was tied to a stationary hook (Fig. 3B) and the upper end (Fig. 3C) to a glass rod that was connected to the force transducer. The papillary muscle was kept in contact with the thermopile by using a tether (Fig. 3D) to make sure that the surface of the muscle along its whole length is apposed against the measuring junctions of the thermopile. The stimulating electrodes are 25 μm diameter wires (Fig. 3E) embedded in the silk ligatures used to fix the muscle to the stationary glass hook at the bottom and the glass rod attached to the force transducer at the top. The thermopile assembly with the papillary muscle in position was then placed in the stainless steel incubation chamber, which has glass windows in its side (25). The windows allowed for infrared heating of the papillary muscle during an experiment so that on-line calibration of the system was possible. The system was calibrated and the heat loss rate (cool-off time constant and heat capacity) were determined by a modification (25) of the Kretzschmar and Wilkie (23) method. The entire assembly was then placed in a 70 liter water bath and kept at 21°C.

The papillary muscle was allowed to equilibrate for 2 hr in normal Krebs-Ringer while being stimulated at a frequency of 0.2 Hz with the stimulus 10% above threshold. After the equilibration period, the muscle was carefully adjusted to optimum length. Muscles were rejected if the mechanical and thermal measurements were not constant for the entire period of the experiment (<5% change) and if the active to passive force ratio was less than four.

The thermal output of the muscle is made up of initial, recovery, and resting heat production. The temperature of the muscle at any given instant is determined by (a) the quantity and time course of the heat production, (b) the thermal capacity of the muscle, thermopile, and adhering solution, and (c) the conductance of the heat along the loss pathways.

Steady state, paced isometric contraction (0.2 Hz) resulted in a temperature oscillation with a mean value above the resting base line temperature (Fig. 4). The initial heat was calculated by correcting the temperature change for heat loss, which occurs during each contraction cycle, by the extrapolation procedure indicated by the upper dashed line. The dashed line was obtained by translating the cool off curve, which occurs on cessation of stimulation to permit a valid extrapolation of earlier responses. The difference between the peak of the temperature excursion and the extrapolated falling base line (I, Fig. 4) is the temperature change associated with initial heat. It is a reflection of the ATP hydrolysis linked to calcium movement and myosin cross-bridge cycling during contraction and relaxation. The corrected temperature change I was converted to the corresponding initial heat evolution by multiplication with the heat capacity of the papillary muscle and adhering Krebs solution. The infrared heating and cool-off technique was used to determine the heat capacity for each set of measurements during the experiment (25). The total activity-related heat, consisting of initial heat and recovery heat, is proportional to the shaded area (A_T) under the curve in Fig. 4. The recovery heat was calculated by the Bugnard method (12), which involves substracting an area, A_I, equivalent

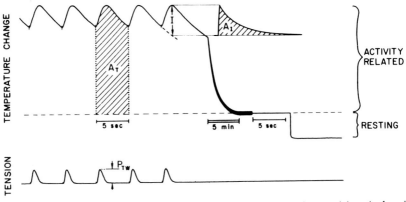

FIG. 4. A record of isometric tension *(below)* and temperature change *(above)* of a right ventricular papillary muscle stimulated repetitively at 0.2 Hz. The *horizontal dashed line*, representing the resting temperature of the muscle, is obtained by allowing the muscle to cool off to a steady base line following termination of the stimuli. This procedure is shown by the dark, 5-min cool-off *curve* following the last activity related temperature oscillation. The difference between resting temperature and ambient temperature is obtained by immersing the muscle and frame in temperature equilibrated solution. This temperature change is seen in the final excursion from the *dashed*, base-line temperature. The temperature change associated with the initial heat, I, is corrected for heat loss by translating the cool-off curve after the last stimulus to extend the cool-off curve of the previous responses (*dashed curve* below I). The initial temperature change, I, is obtained by measuring the temperature difference between this extrapolated base line and the peak temperature excursion. The area associated with the initial heat, A_{II}, is obtained by multiplying I by the cool-off time constant determined separately by the infrared heating method. The area associated with the total activity related heat, A_T, is the *shaded area* under the activity related temperature oscillation.

to the initial heat temperature rise I, from the total hatched area. This initial heat area is obtained by multiplying the initial heat temperature rise I by the cool-off time constant. From the initial heat area and recovery heat areas, the ratio of initial to recovery heat was obtained. Resting heat is obtained by measuring the temperature change that occurs when the nonstimulated papillary muscle is submerged in temperature equilibrated solution. The temperature difference that results from the immersion was multiplied by the heat loss coefficient to obtain resting heat rate.

Initial heat is a reflection of ATP splitting associated with cross-bridge cycling and calcium movement, which occur during contraction and relaxation. Recovery heat represents the energy lost in resynthesizing the ATP and creatine phosphate hydrolyzed during contraction. Resting heat reflects the energy lost in maintaining the concentration gradients and the integrity of the intracellular organelles.

The initial heat is made up of a tension dependent heat (TDH) and a tension independent heat (TIH). The TDH is associated with the ATP hydrolysis that results from myosin cross-bridge cycling. The TIH results from ATP hydrolysis linked to calcium cycling. The initial heat was partitioned into TDH and TIH components by incubating the muscle in 2.5 times normal osmotic strength mannitol-Krebs solution. This eliminated the twitch tension following stimulation and the accompanying TDH component. The remaining triggerable temperature change is the TIH (6).

Mechanical measurements on the papillary muscles were carried out using the standard methods previously described (25).

RESULTS

Mechanical Measurements

Mechanical parameters of papillary muscles from pressure overload as compared with thyrotoxic hypertrophied hearts resemble the comparisons between tortoise and frog (Fig. 1) or mammalian fast and slow muscle (Fig. 2). Mechanical V_{max} is two times as great in the thyrotoxic as compared with the pressure overload hypertrophy (Fig. 5) (29). The curvature of the force-velocity relationship is greater in the pressure overloaded preparation as indicated by the 35% decrease in the a/Po ratio. Time-to-peak tension is increased by 160%.

Initial Heat Total Heat and Thermo-Mechanical Economy

If the premise that the mechanical differences observed between thyrotoxic and pressure overload muscles are analogous to those between frog and tortoise or EDL and soleus then one would predict that the pressure overload hypertrophied heart would develop force more economically than the thyrotoxic heart and have a lower contractile protein ATPase activity. This prediction can be evaluated by comparing the thermal, mechanical, and biochemical characteristics of the hearts.

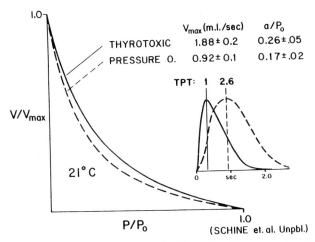

FIG. 5. Rabbit right ventricular papillary muscle. Normalized force-velocity curves and isometric force records for the pressure overload and thyrotoxic hypertrophied heart muscles. Unloaded shortening velocity (V_{max}) in muscle lengths per second, and the ratio a/Po, are tabulated in the upper right hand corner. (Adapted from Schine, ref. 29.)

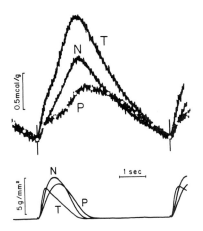

FIG. 6. Records of isometric force *(below)* and heat production *(above)* from normal (N), pressure overloaded (P), and thyrotoxic (T) right ventricular papillary muscles. (Adapted from Alpert and Mulieri, refs. 3 and 5.)

Records of isometric force are presented below and temperature change above for the normal, pressure overload, and thyrotoxic muscles in Fig. 6. In the bottom records note that the force development by the pressure overload muscle is slower and reaches a peak later than the normal muscle. The thermal counterparts of these twitches are seen in the upper trace. Although the twitch force and the tension-time integral are very nearly the same in the pressure overload preparation as in the normal, there is a significant decrease in the thermal output and the rate of heat liberation in these papillary muscles. The force development by the thyrotoxic muscle is 30% faster and reaches a peak 42% earlier than the normal muscle. In addition, the tension-time integral is significantly lower than for the normal muscle.

The thermal counterparts of the thyrotoxic isometric twitch are seen in the upper trace (Fig. 6). Although the thyrotoxic preparation has a twitch force that is 28% less than normal and abbreviated to the extent that the tension-time integral is 44% less than normal, there is a significant increase in its thermal output. The total activity-related heat production, consisting of initial and recovery heat, is lower than normal in the pressure overloaded preparation. Although in thyrotoxic preparations the total activity-related heat production is also below normal, it is 132% above the normal values when expressed per tension time integral.

Tension Dependent and Tension Independent Heat Production

The changes in the quantity and rate of initial heat production may be related to alterations in the economy of cross-bridge cycling and/or calcium cycling. To assess the separate contribution of each of these processes, activity-induced heat production is measured in normal Krebs and then in 2.5 times hyperosmotic, mannitol-Krebs solution where the twitch force is virtually eliminated (<5% Po). Records of the thermal output and isometric force for normal and hypertrophied muscles in normal Krebs and mannitol hyperosmotic Krebs solution are presented in Fig. 7. The tension dependent heat is depressed to 65% of normal in the pressure overloaded preparation and is the same as normal in the thyrotoxic hearts (Table 4). When the tension

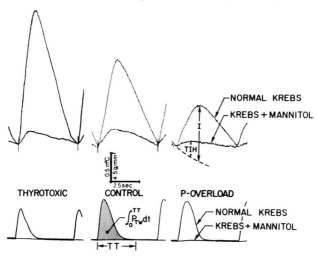

FIG. 7. Records of isometric force *(below)* and heat production *(above)* of right ventricular papillary muscles from normal *(center panel)*, thyrotoxic *(left panel)*, and pressure overload *(right panel)* hearts. The larger heat excursion *(upper tracings, upper records)* represents the initial heat production during isometric contraction and relaxation *(lower tracings, upper records)*. Initial heat consists of tension dependent heat and tension independent heat. When the papillary muscle is incubated in 2.5 × N hyperosmotic Krebs solution, an all-or-none, triggerable heat output remains. This smaller heat excursion *(upper tracings, lower records)* is the tension independent heat. Under these conditions, isometric force is virtually eliminated *(lower traces, lower records)*. The tension dependent heat is the difference between the initial heat and the tension independent heat.

TABLE 4. *Tension dependent and independent heat production*

	N	PO	T
TDH (mcal/g)	1.29	0.84	1.29
TDH/p_{TW} (μcal/g cm)	2.30	1.81	3.62
TIH (mcal/g)	0.40	0.20	0.32

From Alpert et al., ref. 8.

TABLE 5. *Ratio of recovery to initial heat*

	Recovery / initial
N	1.46
P	1.42
T	0.79[a]

[a]$p < 0.01$.

dependent heat is normalized for twitch tension, the economy of force development, (P_{TW}/TDH), in the pressure overloaded hearts is 127% of normal, whereas in the thyrotoxic hearts it is 63% of normal (Table 4). The tension independent heat is reduced to 50% of normal in the pressure overloaded hearts and 80% of normal in the thyrotoxic hearts (Table 4).

Recovery Heat

Recovery heat was measured by the Bugnard method (12). The ratio of recovery to initial heat was the same for the pressure overloaded and normal preparations. In the thyrotoxic preparation, the ratio of recovery to initial heat is 56% of normal (Table 5).

DISCUSSION

General Adaptation and Thermomechanical Economy

In the development of different species, or specific muscle types, fast and slow skeletal muscles evolved to meet specific functional needs of the animals. As shown in Fig. 8P, there is a distinct similarity in the pattern of activity seen in the pressure overloaded heart and the leg muscles of the tortoise. They both are required to work against a steady, elevated force and they do so with relatively low contraction velocity. The steady force is brought about by the constriction of the pulmonary artery in the heart, whereas in the tortoise it is a result of the mechanical disadvantage of having to support the body weight on partially flexed (20°) limbs. Figure 8T shows the similarity between the pattern of activity seen in the thyrotoxic heart and

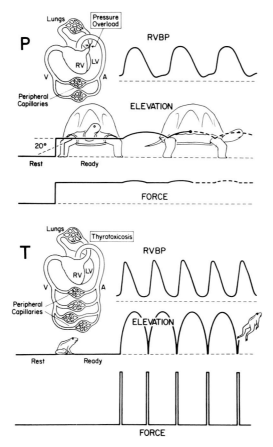

FIG. 8.   Pressure overload **(P)** and thyrotoxic hypertrophy **(T)**. RVBP is the intraluminal right ventricular pressure as a function of time. Elevation represents the vertical position of the tortoise or frog as a function of time. The force trace represents the force in the skeletal muscles of the tortoise and frog, respectively.

in the leg muscles of the frog. In both there is a requirement for rapid force development and high repetition rate due to the need to transport mass at a high velocity (blood to tissues, or, body to safety). The heart muscle adapts to different stresses in a functional manner that is dependent on the stress.

In terms of mechanical behavior, the slow muscle, in contrast to the fast muscle type, had an increased curvature of the force velocity relationship (decrease in a/Po), a decreased V_{max}, and an increase in time-to-peak tension (Figs. 1 and 2). In this regard, pressure overloaded hypertrophied heart muscle behaves like slow muscle, whereas the thyrotoxic hearts behave like fast muscle (Fig. 5). In addition, the differences between the myosin ATPase and the thermomechanical economy are qualitatively similar to those seen between fast and slow muscle and between frog and tortoise (Table 1). Similarly, the ratio of thyrotoxic to pressure overload values for twitch tension per tension dependent heat production, a/Po, myosin

ATPase, and maximum velocity of unloaded shortening are 0.5, 1.52, 2.54, and 2.62, respectively. The differences in the thermomechanical performance of the hypertrophied hearts result from adaptive changes that occur in the excitation-contraction coupling system and the contractile apparatus.

Calcium Cycling and the EC Coupling System

From measurements of the tension independent heat, it is clear that there are significant alterations in the calcium cycled during each contraction relaxation cycle. The quantity of calcium moved can be calculated from the TIH measurements (Table 4) if it is assumed that two calcium ions are transported for every ATP molecule split and that the molar enthalpy of ATP hydrolysis is 11 kcal/mole. Then, for the N, P, and T preparations, 0.072, 0.036, and 0.058 µmoles of calcium are cycled per gram of heart muscle per beat. Furthermore, this calcium is removed in 54% of the normal time in the thyrotoxic muscle and 128% of the normal time in the pressure overloaded muscle. These differences in the amount of calcium cycled and the time of removal are ready explanations for the observed alterations in the time-to-peak tension and the tension-time integral.

Cross-bridge Cycling and the Contractile Proteins

The substantial differences observed between pressure overload and thyrotoxic hypertrophy with regard to unloaded velocity of shortening, the curvature of the force-velocity relation, and myosin ATPase activity suggest that there are different types of myosin in the two preparations. Examination of this possibility indicated that the thyrotoxic heart contained primarily the V_1 isoenzyme of myosin, whereas the pressure overloaded heart contained the V_3 isoenzyme (Fig. 9) (24).

The predominance of the V_1 isoenzyme of myosin in thyrotoxic heart and the V_3 isoenzyme in the pressure overloaded heart has important implications concerning our interpretation of the kinetics of cross-bridge cycling and the economy of tension development. Force and work are produced when the globular cross-bridge head of the nonrefractory myosin molecule makes contact with an actin monomer in the thin filament and rotates from the 90° to the 45° position, moving the actin filament toward the center of the sarcomere. One recent scheme (15) for this cycle consists of the following steps: (a) ATP is rapidly hydrolyzed on the dissociated myosin (M) head ($M^*ATP \rightarrow M^{**}ADPP_i$) leaving it in an energized but refractory state; (b) a slow, rate-limiting, conformational change occurs in the myosin, transforming it from the refractory to the nonrefractory state ($M^{**}ADPP_i \rightarrow M^+_+ADPP_i$);

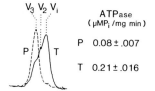

V_3 V_2 V_1

ATPase
(µMP$_i$ /mg min)

P 0.08 ± .007

T 0.21 ± .016

FIG. 9. Densitometric tracing of pyrophosphate gels of myosin from pressure overload (P) and thyrotoxic (T) rabbit hearts. The ATPase activity of the myosin is tabulated in the right hand portion of the slide. (Redrawn from Litten, et al., ref. 24.)

(c) the nonrefractory cross-bridge connects to the actin (A) filament $(M^+_+ ADPP_i + A \to AM^+_+ ADPP_i)$; (d) the contact angle between the cross-bridge head and the actin filament changes from 90° to 45°, stretching the flexible portion of the bridge and generating tension $(AM^+_+ ADPP_i \to AM\ ADPP_i)$; (e) the subsequent loss of hydrolysis products $(ADP + P_i)$ and binding of an ATP molecule to AM causes a rapid dissociation of the cross-bridge and completes the cycle $AM + ATP \to AMATP \to A + M^*ATP)$.

The main feature of this scheme is that in a single cycle of the activated contractile component, there are two distinct states, an associated, force producing state (on), and a dissociated state (off). In the associated or "on" state, force is generated when the cross-bridge head rotates from the 90° to the 45° position (d above), stretching compliant elements in the myosin molecule during the on time. In the dissociated or "off" state, the rate limiting step occurs at $M^{**}ADPP_i \to M^+_+ ADPP_i$ (b). Although each cycle of a particular bridge produces only a momentary impulse of force, the summation of these impulses from millions of randomly cycling cross-bridges gives rise to a steady force. Thus, the force developed can be analyzed in terms of the average strength of the cross-bridge (S), the duration of the on time (τ), and the rate of cross-bridge cycling (Fig. 10). The average on time, τ, for the thyrotoxic muscle, is 47% of normal, whereas the cycling rate is 156% of normal. For the pressure overloaded muscle, the on time is 154% of normal, whereas the

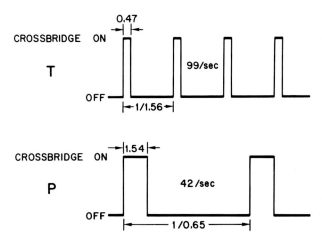

FIG. 10.   Representation of the average cross-bridge cycle during contraction and relaxation. In this analysis the on phase represents the period of connection between actin and myosin where the globular myosin head is presumed to rotate from 90° to the 45° position. The off phase represents the period in the kinetic actomyosin cycle where actin and myosin are dissociated. The average cycling rate relative to normal muscle is indicated by the frequency of appearance of the rectangular pulses and is determined by dividing the tension-dependent heat by the twitch time. The average on time relative to normal muscle, τ, is determined by dividing the tension-time integral by the tension dependent heat. All values are expressed relative to normal. The assumption is made that each cross-bridge is capable of developing the same force indicated by the height of the pulse, whether in normal, pressure overload (P), or thyrotoxic (T) hearts.

cycling rate is 65% of normal. If one assumes the myosin content of the papillary muscle to be 35 mg/g of muscle (22), then the actual cycling rate of the cross-bridge can be calculated. The average number of cross-bridge cycles per second during the twitch for the thyrotoxic and pressure overloaded hearts is 99 and 42/sec, respectively.

Optimization of Power Output

It is appealing to consider that the possibility that the differences between the frog and tortoise, EDL and soleus, and thyrotoxic and pressure overload muscles represent an adaptation involving the alterations in the relative load at which maximal power output occurs. This would be expected if the power output were sensitive to the ratio of a/Po. However, the load at which the optimum power output occurs appears to be insensitive to a/Po variations and hence to the species, muscle type or adaptation (Fig. 11) (1).

SUMMARY

Myocardial hypertrophy, which results from the increase in demand placed on the heart, does not involve the simple enlargement of the myocardial cell. Instead, the cells are restructured in a manner that depends on the nature of the stress applied. The cellular reorganization involves the contractile, EC coupling, and recovery systems. In the remodeling of each of these systems, a repertoire of functional adaptation is used which bears a remarkable similarity to the changes seen in the

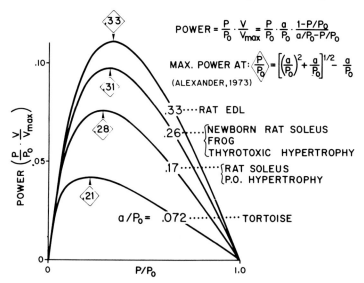

FIG. 11. Power as a function of relative load. For each muscle preparation, a/Po is indicated by the number inserted in the *solid, power curve line*. The relative force, for each muscle preparation, at which optimum power occurs is indicated within the *diamond*.

differentiation of species and the development of muscle type. Thus, in pressure overload hypertrophy, where the maintenance of persistent high pressure is a necessity, the contractile system of the muscle is slower and more economical. This is similar to the specialization seen in the tortoise or in slow muscles like the rat soleus. In thyrotoxic hypertrophy, where the demand is to move the blood at high velocity, the muscle is fast and less economical. This is similar to the specialization seen in the frog sartorius or rat EDL. The restructuring involves specific molecular changes that uniquely adapt the muscles for the tasks at hand.

ACKNOWLEDGMENTS

Supported in part by PHS RO 1 17592–05. We acknowledge the technical assistance of Robert Goulette.

REFERENCES

1. Alexander, R., McNeill, (1973): Muscle performance in locomotion and other strenuous activities. In: *Comparative Physiology*, edited by L. Bolis, K. Schmidt-Nielson, and S. M. P. Maddrell, pp. 1–21. North-Holland, Amsterdam.
2. Alpert, N. R., Hamrell, B. B., and Halpern, W. (1974): Mechanical and biochemical correlates of cardiac hypertrophy. *Circ. Res.*, 34/35(Suppl. II):71–82.
3. Alpert, N. R., and Mulieri, L. A. (1980): The functional significance of altered tension dependent heat in thyrotoxic myocardial hypertrophy. *Basic Res. Cardiol.*, 75:179–184.
4. Alpert, N. R., and Mulieri, L. A. (1981): The utilization of energy by the myocardium hypertrophied secondary to pressure overload. In: *The Heart in Hypertension*, edited by B. E. Strauer, pp. 153–163. Springer Verlag, Berlin.
5. Alpert, N .R., and Mulieri, L. A. (1982): Increased myothermal economy of isometric force generation in compensated cardiac hypertrophy induced by pulmonary artery constriction in rabbit. A characterization of heat liberation in normal and hypertrophied right ventricular papillary muscles. *Circ. Res.*, 50:491–500.
6. Alpert, N. R., and Mulieri, L. A. (1982): Heat mechanics, and myosin ATPase in normal and hypertrophied heart muscle. *Fed. Proc.*, 41:192–198.
7. Alpert, N. R., Mulieri, L. A., and Litten, R. A. (1979): Functional significance of altered myosin ATPase in enlarged hearts. *Am. J. Cardiol.*, 44:947–953.
8. Alpert, N. R., Mulieri, L. A., Litten, R. Z., Goulette, R., and Schine, L. (1979): Myothermal and enzymatic analysis of a new cardiac preparation titratable between failing and non-failing hypertrophy. *Circ. Res.*, 60:II–224.
9. Banerjee, S. K., Flink, I. L., and Morkin, E. (1976): Enzymatic properties of native and N-Ethylmaleimide-modified cardiac myosin from normal and thyrotoxic rabbits. *Circ. Res.*, 39:319–326.
10. Barany, M., and Close, R. (1971): The transformation of fast and slow muscles of myosin in cross innervated rat muscles. *J. Physiol. (Lond.)*, 213:455–474.
11. Bowditch, H. P. (1871): Uber die Eigenthümlichkeiten der Reizbarkeit welche die Muskelfasern des Herzen zeigen. *Ber. Verhandl. Sächs. Akad. Wiss.*, 23:652–689.
12. Bugnard, L. (1934): The relation between total and initial heat in single muscle twitches. *J. Physiol. (Lond.)*, 82:509–519.
13. Close, R. (1964): Dynamic properties of fast and slow skeletal muscles of the rat during development. *J. Physiol. (Lond.)*, 173:74–75.
14. Close, R. (1969): Dynamic properties of fast and slow skeletal muscles of the rat after nerve cross-union. *J. Physiol. (Lond.)*, 204:331–346.
15. Eisenberg, E., and Hill, T. L. (1978): A crossbridge model of muscle contraction. *Prog. Biophys. Mol. Biol.*, 33:55–82.
16. Frank, O. (1895): Zur Dynamik der Herzmuskels. *Ztschr. F. Biol.*, 32:570.
17. Gibbs, C., and Gibson, W. R. (1972): Energy production of rat soleus muscle. *Am. J. Physiol.*, 223:864–871.

18. Hamrell, B. B., and Alpert, N. R. (1977): The mechanical characteristics of hypertrophied rabbit cardiac muscle in the absence of congestive heart failure. *Circ. Res.*, 40:20–25.
19. Hartree, W. (1926): Heat production of tortoise muscle. *J. Physiol. (Lond.)*, 61:255–260.
20. Hill, A. (1938): The heat of shortening and the dynamic constants of shortening. *Proc. R. Soc. Lond. [Biol.]*, 126:136–195.
21. Hill, A. V. (1970): *First and Last Experiments in Muscle Mechanics*. Cambridge University Press, Cambridge, England.
22. Katz, A. (1970): Contractile proteins of the heart. *Physiol. Rev.*, 50:63–158.
23. Kretzschmar, K. M., and Wilkie, D. R. (1972): A new method for absolute heat measurements utilizing the Peltier effect. *J. Physiol. (Lond.)*, 224:18P–20P.
24. Litten, R. Z., Martin, B. J., Low, R. B., and Alpert, N. R. (1982): Altered myosin isozyme patterns from pressure overloaded and thyrotoxic hypertrophied rabbit hearts. *Circ. Res.*, 50:856–859.
25. Mulieri, L. A., Luhr, R., Trefry, J., and Alpert, N. R. (1977): Metal-film thermopiles for use with rabbit right ventricular papillary muscles. *Am. J. Physiol.*, 233:146–156.
26. Olson, R. E. (1962): The control of function of the heart. In: *Handbook of Physiology, Section 2, Circulation, Vol. 1*, edited by W. F. Hamilton and P. Dow, pp. 489–532. American Physiological Society, Washington, D.C.
27. Ringer, S. (1880–82): Concerning the influence exerted by each of the constituents of the blood on the contraction of the ventricle. *J. Physiol. (Lond.)*, 3:380.
28. Sarnoff, S. J., and Mitchell, J. H. (1962): The control of function of the heart. In: *Handbook of Physiology, Section 2. Circulation, Vol. 1*, edited by W. F. Hamilton and P. Dow, pp. 489–532. American Physiological Society, Washington, D.C.
29. Schine, L. (1982): *The Mechanical Performance of the Pressure Overloaded, Thyrotoxic and Double Treated Hypertrophied Hearts*. Masters Thesis, University of Vermont.
30. Starling, E. H. (1918): *The Linacre Lecture on the Flow of the Heart, Given at Cambridge, 1914*. Longmass, Green and Co. et al., London.
31. Wendt, I. R., and Gibbs, C. (1973): Energy production of rat of extensor digitorum longus muscle. *Am. J. Physiol.*, 224:1081–1086.
32. Woledge, R. (1968): The energetics of tortoise muscle. *J. Physiol. (Lond.)*, 197:685–707.

*Basic Biology of Muscles: A Comparative
Approach*, edited by B. M. Twarog,
R. J. C. Levine, and M. M. Dewey.
Raven Press, New York © 1982.

Chemical Energetics of Contraction in Mammalian Smooth Muscle

Thomas M. Butler, Marion J. Siegman, and Susan U. Mooers

*Department of Physiology, Jefferson Medical College, Thomas Jefferson University,
Philadelphia, Pennsylvania 19107*

Measurements of chemical energy usage in muscle during different types of contraction provide insights into the mechanism whereby chemical energy is transduced into mechanical output. Although many studies have been reported for striated muscles (9), there is a paucity of comparable information for mammalian smooth muscle. In some studies, indirect measures of ATP utilization, such as oxygen consumption and/or lactate production, were made, but these suffer from a possible lack of time resolution and the assumption of a constant coupling ratio between ATP synthesis and oxygen utilization (14,16,25). The difficulties of direct measurements of high energy phosphate usage are that respiration and glycolysis must be stopped in order to prevent ATP synthesis, and this must be accomplished without untoward effects on chemical contents, mechanical output, or a change in the relationship of chemical energy input and mechanical output.

A smooth muscle preparation suitable for the direct determination of chemical energy usage during contraction has been developed in our laboratories. The rabbit taenia coli preparation, whose mechanical properties are well defined (6,15,31), is treated with 5 mM Fluoroacetate, 0.5 mM Iodoacetate-Krebs solution under anaerobic conditions at 5°C for 30 min. This is followed by rewarming to 18°C for 3.5 min, prior to the initiation of the appropriate experimental design. We have shown that this treatment blocks respiration and glycolysis without altering the chemical composition, mechanical behavior, or ultrastructural characteristics of the muscle (7,32). The measurement of changes in nucleotide and phosphorylcreatine contents of the muscle under appropriate experimental designs gives a direct measure of total high energy phosphate usage.

CHEMICAL ENERGY USAGE DURING ISOMETRIC CONTRACTION

Figure 1 shows the ATP and phosphorylcreatine (PCr) changes during tetani of different durations at 85% L_0 compared to paired unstimulated controls. Initially there was a net breakdown of ATP with no PCr splitting. This was followed by an increase in the rate of PCr splitting with net ATP synthesis. These results for smooth

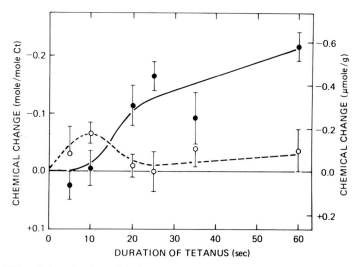

FIG. 1. ATP and phosphorylcreatine changes in the rabbit taenia coli at 18°C during isometric tetani of different durations compared to paired unstimulated controls. *Closed circles,* PCr; *open circles,* ATP. Chemical contents are expressed on the basis of total creatine (C_t) contents, which is 2.7 µmol/g wet wt, and is taken as an index of muscle mass. (From Butler et al., ref. 7.)

muscle are very different from those reported for skeletal muscle. Measurements of ATP and PCr changes during stimulation of frog skeletal muscle are consistent with the creatine kinase reaction being near equilibrium at all times (8,21). The average rate of total energy usage (ΔATP $+$ ΔPCr) was about 0.01 µmol/g · sec^{-1} for up to 60 sec of stimulation. Although there was no significant change in the average rate of total energy usage during different phases of the tetanus, there was a possibility that the rate was higher during force development; a different experimental design was required to test this (32).

Three segments were obtained from each taenia coli and adjusted to 95% L_0. One segment (A) was stimulated for 25 sec, the time required to develop maximal isometric force. The second segment (B) was stimulated for 60 sec, so that it could develop and then maintain near maximum isometric force for 35 sec. The muscles were freeze-clamped at the end of stimulation. A third segment (C), unstimulated but otherwise identically treated, was frozen 25 sec after the equilibration period at 18°C.

The average rate of energy usage during isometric force development (A–C) was 8.2 $\pm$ 0.8 mmol ~P/mol C_t · sec^{-1} and this was significantly greater than that for isometric force maintenance (B–A), 2.8 $\pm$ 0.8 mmol ~P/mol C_t · sec^{-1}. The latter number includes a component of resting energy utilization. In a separate series of experiments in which lactate production was measured under anaerobic conditions, resting metabolism was found to be 0.85 $\pm$ 0.19 mmol ~P/mol C_t · sec^{-1}. After correction for resting energy usage, the average of ~P usage during force maintenance was 2.0 $\pm$ 0.6 mmol ~P/mol C_t · sec^{-1}, and is more than fourfold lower than the average rate during initial force development.

We next addressed the question of whether the high energetic cost of force development was due to internal work done against the series elasticity. The experimental design used to test this is shown in Fig. 2. The experimental muscle was stimulated for 20 sec at 105% L_0, then quickly released to 95% L_0 and, during continued stimulation for 25 sec, active force redeveloped from 0 to 95 $\pm$ 5% P_0. The control muscle was stimulated for 20 sec at 105% L_0 and quickly frozen. In this way, the energy usage associated with the work production against the SEC that is identical to that occurring during initial force development is measured later during the tetanus. The ~P usage during force redevelopment was 0.08 $\pm$ 0.05 mol/mol C_t and is significantly less than the 0.2 $\pm$ 0.02 mol/mol C_t associated with initial force development. Therefore, the energy usage associated with work production against the series elasticity is *not* the main determinant of the high energy usage during initial force development.

If it is assumed that the major energy-utilizing process in the smooth muscle is actin-activated myosin ATPase activity, that the myosin molecule is a cross-bridge, and that all cross-bridges are cycling using 1 ATP molecule per cycle, then the average cross-bridge cycle duration can be estimated by dividing the molar myosin content by the average rate of energy utilization. Using a value of 5.4 nmol $\sim P/g \cdot sec^{-1}$ for isometric force maintenance in the taenia coli and a myosin content of 0.04 μmol/g wet wt (32), the average cross-bridge cycle duration is 7.5 sec at 18°C. A similar calculation using the energetics results of DeFuria and Kushmerick (11) and the myosin content reported by Ebashi et al. (13) for skeletal muscle, gives an average cycle duration of 0.05 sec. This 150-fold difference is most likely a reflection of the differences in intrinsic rate of actin-activated myosin ATPase activities of the two muscle types.

"Economy" of force maintenance has been defined as the force per cross-sectional area divided by the rate of energy utilization per gram wet weight. Such a calculation for the taenia coli based on the values reported here, is 700 $(N/cm^2)/(\mu mol/g/sec)$, and some 100-fold higher than that for the frog sartorius (11). This difference may be traced to the slower cross-bridge cycle time and the 40% longer myosin filament (3) of the smooth muscle, which puts more cross-bridges in parallel. The anterior byssus retractor muscle (ABRM) of *Mytilus* in the catch state has a 10-fold higher economy than the taenia coli. However, when the 10 $\times$ longer myosin filament

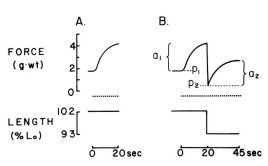

FIG. 2. Experimental design used to enable an amount of internal work equal to that done during the initial development of force to be done later in the tetanus. **A:** Control muscle, **B:** Experimental muscle. *Dotted lines* show stimulus duration. p_1 and p_2 show the passive force before and after the quick release; a_1 and a_2 are the active force before and after quick release, respectively. (From Siegman et al., ref. 31.)

length of the ABRM (20) is accounted for, the economy of force maintenance by each cross-bridge would not be different from that of the taenia coli. The unique specialization of smooth muscles showing catch may be the ability to vary their economy of contraction according to the mode of their activation. Our results have shown that mammalian smooth muscle can vary the economy of force output during the time course of a tetanus. Recent studies of mechanical properties of smooth muscle (12,34) have shown that maximum velocity of shortening can vary with duration of stimulation, and this may mean that the intrinsic rate at which cross-bridges cycle can vary. It remains to be seen how similar the classic catch state of invertebrate muscles is to the isometric force maintenance state of smooth muscle.

There is now considerable evidence that in smooth muscle, actin-activated myosin ATPase activity can be regulated by a calcium-dependent myosin light chain kinase (1,30,33). We, therefore, examined the relationship of light chain phosphorylation and ATP utilization under a variety of mechanical conditions. The time course of light chain phosphorylation during an isometric tetanus at L_0 in the rabbit taenia coli at 18°C is shown in Fig. 3. At rest, the degree of phosphorylation is 11 ± 1% (N = 6). On stimulation, there is an increase of phosphorylation to 31% at 5 sec followed by a slow but significant decline from 5 to 60 sec. The peak of phosphorylation precedes maximum force development and is approximately coincident with the maximum rate of force development. It is obvious that the relative degree of light chain phosphorylation during force development and force maintenance are not proportional to the relative rates of energy utilization discussed earlier. The magnitude of phosphorylation of the myosin could account for only 5% of the total observed energy utilization during the initial 25 sec of force development if the phosphorylated form of the myosin is not turning over. It is, however, possible that both kinase and phosphatase activity are occurring. An upper limit on the magnitude of these activities during the period of force development can be calculated from the average rate of high energy phosphate utilization of 22.1 nmol/g/sec and a myosin S_1 content of 0.08 μmol/g taenia coli. In the very unlikely event that all of the observed energy is used for kinase-phosphatase activity, then on average each S_1 would undergo a phosphorylation-dephosphorylation cycle every 3.6 sec during the period of isometric force development. A similar calculation for the period of force maintenance shows that on average each S_1 would undergo a phosphorylation-dephosphorylation cycle only every 15 sec.

ENERGY USAGE AND MYOSIN LIGHT CHAIN PHOSPHORYLATION DURING RELAXATION

The energy usage for relaxation from an isometric tetanus was determined as the difference in ~P contents in muscles stimulated for 25 sec and those stimulated for 25 sec followed by 45 sec of relaxation. After correction for resting metabolism, $\Delta \sim P$ was very low (0.016 ± 0.031 mol/mol C_1, N = 17). The time course of myosin light chain phosphorylation during relaxation is shown in Fig. 3. The degree of phosphorylation declines more rapidly than does isometric force.

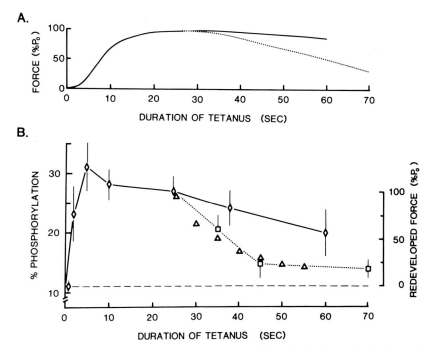

FIG. 3. Myosin light chain phosphorylation and force during stimulation and relaxation under isometric conditions. **A:** The *solid line* shows the force response to stimulation under isometric conditions, and the *dotted line* shows the force during relaxation following a 25 sec tetanus. **B:** *Diamonds*, percent phosphorylation during tetani of different durations; *squares*, percent phosphorylation during relaxation following a 25 sec tetanus; *triangles*, active state during contraction and relaxation determined as described in the text. (From Butler and Siegman, ref. 5.)

We have found that the economy of force maintenance is very high during relaxation (32). The process of relaxation from an isometric tetanus was further probed by determination of the ability of the muscle to redevelop force following a quick release at different times during relaxation. The maximum redeveloped force was taken as a measure of the relative number of cross-bridges able to cycle as progressive loss of contractile activity occurred.

Determinations of the maximum redeveloped force were made at the peak of a tetanus during stimulation and at intervals following cessation of 25 sec of stimulation. The data are shown in Fig. 3. During relaxation, the ability to redevelop force decays more rapidly than does the force maintained by the muscle. This measure of active state is consistent with the observed rate of energy utilization and phosphorylation during relaxation. The extra force observed may be due to the straining of attached, noncycling cross-bridges (31). Although energy utilization and phosphorylation are not proportional during the development and maintenance phases of the isometric tetanus, during relaxation these parameters and active state all decrease at approximately the same rate, which is faster than the rate of decrease of isometric force.

The net rate of dephosphorylation of myosin during relaxation is 0.5 nmol/g $\cdot$ sec^{-1}. This is a minimum estimate of the *in vivo* activity of the myosin light chain phosphatase. It is interesting to note that this rate is 10% of the suprabasal rate during isometric force maintenance. Therefore, if the activity of the phosphatase is not regulated during the contraction, the energy required for kinase-phosphatase activity is at least 10% of the total required during force maintenance.

CHEMICAL ENERGY USAGE DURING SHORTENING AND WORK PRODUCTION

It is well known that in skeletal muscle the total energy utilization varies with shortening and the performance of external work. We have determined the chemical energetics during isovelocity shortening in the rabbit taenia coli in order to probe the kinetics of cross-bridge cycling under shortening conditions and to determine the relationship between chemical energy input and external work production.

The taenia coli was divided into three segments, and each was either stimulated isometrically at L_0, stimulated and subjected to an isovelocity shortening, or treated as an unstimulated control. Table 1 shows the results of the two velocities studied. There is no significant difference in the energy utilization in the shortening and isometric design. At a shortening velocity near that of maximum power output, the external work production was only 4.6 KJ/mol $\sim$P used. The similarity of the energy usage during isometric force development and during isovelocity shortening from the onset of stimulation might be attributed to the large energy usage associated with the initial activation of the muscle under both conditions. If so, then any extra energy usage associated with work production would be comparatively small, resulting in a low work production per ATP used. We tested this by stimulating the muscle tetanically for 20 sec during which time maximum force developed, and then allowed it to shorten under isovelocity conditions. The rationale was that during force maintenance the rate of energy usage was low so that an increment, if it occurs, in energy usage associated with work output, could be measured. The average rates of energy utilization during shortenings at different velocities are shown in Fig. 4 with those for isometric force development, maintenance, and resting conditions. At velocities of up to 50% V_{max}, there was a threefold increase

TABLE 1. *Energy usage under isometric and shortening conditions[a]*

Duration of stimulation (sec)	Velocity of shortening (L_0/sec)	$\Delta\sim$P (mol/mol C_t)			Active work ($J \times 10^{-2}$/ mol C_t)	Corrected active work[b] ($J \times 10^{-2}$/mol C_t)
		Shortening − control	Isometric − control	Shortening − isometric		
12	0.015	−0.102 ±0.033	−0.092 ±0.042	−0.011 ±0.029	4.7 ± 0.9	5.4
35	0.005	−0.210 ±0.040	−0.154 ±0.050	−0.056 ±0.047	7.9 ± 1.0	8.9

[a]Muscles were stimulated and immediately shortened from 105 to 87% L_0. Muscles held under isometric conditions were stimulated at L_0 for the same duration as those allowed to shorten.
[b]This is corrected for work done against the series elasticity.

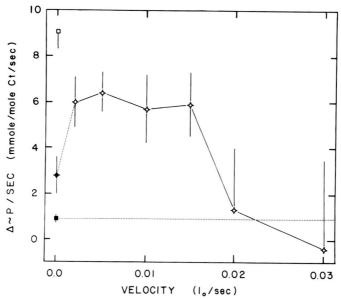

FIG. 4. Average rate of high energy phosphate usage vs velocity of shortening. Symbols represent rates of energy usage during rest *(solid square)*; isometric force development *(open square)*; isometric force maintenance *(solid circle)*; and during shortening following a 20 sec isometric tetanus *(open circles)*.

in average rate of energy usage compared to that during isometric force maintenance. This occurred with no change in the degree of myosin light chain phosphorylation. At very fast velocities, the average rate of energy usage tends to fall. During isovelocity shortenings at 7 to 50% V_{max} external work, corrected for work done by the series elasticity, was produced at 2.7 to 7.8 KJ/mol ~P used.

Why could an increase in energy usage be shown only during the period when the average rate of energy usage by the muscle was already low? The average rates of energy usage for shortenings of 0.005 L_0/sec starting from the onset of the tetanus or from the plateau were not different (6.1 ± 1.1 and 5.6 ± 0.8 mmol ~P/mol $C_t \cdot sec^{-1}$, respectively). Also, the work output per mol of ATP used in identical shortenings from the onset of the tetanus and from the plateau were similar (4.2 KJ/mol and 2.7 KJ/mol, respectively). It thus seems that the differences in the results from comparisons of isometric and shortening conditions from the onset and from the plateau of a tetanus lie in the high rate of energy usage during the development of isometric force. It is likely that the high rate of energy usage during force development is due to cross-bridge cycling, because if the muscle is allowed to shorten, this energy is transduced into work at the same energy cost as when the muscle shortens from the plateau of the tetanus. The slowing of cross-bridge cycling during isometric force maintenance is controlled by a process other than the dephosphorylation of the myosin light chain, as has been proposed by Dillon et al. (12) for the hog carotid artery.

Compared to other types of muscles, the taenia coli has a higher cost of chemical energy for work production. It does 2.7 to 7.8 KJ of work per mol of phosphagen used at velocities of 7 to 50% V_{max}. This is considerably less than that reported for striated muscles (19,26) and for ABRM of *Mytilus* (24). The high chemical energy cost for work production may not occur in all smooth muscles since Davey et al. (10) using myothermal techniques, found that the rabbit rectococcygeus muscle at 27°C had a total mechanical efficiency similar to mammalian and amphibian skeletal muscle. It should be noted, however, that the rectococcygeus muscle might be expected to be more similar to the fast striated muscles since it has both a 10-fold higher maximum velocity of shortening and rate of energy usage during isometric force maintenance than the taenia coli (10,32).

There are numerous possible causes for the high cost of work production. It may be due to the presence of an internal resistance to shortening presented by the existence of attached, but noncycling cross-bridges (12,31), or filament and/or cell misalignment may occur, resulting in work production in an axis other than that measured. Recent ultrastructural studies (described in another section, below) show that this does not occur at the muscle lengths used in these experiments. Another possibility is that cellular processes, other than contraction, use ATP to a large extent. This may be ruled out because the rate of energy utilization increases by threefold when the muscle shortens from the plateau of a tetanus; any noncontractile process that uses ATP would have to increase its activity when the muscle shortens. The inefficiency of smooth muscle in the performance of external work may be overcome to a certain extent by the presence of a large passive tension at L_0, which would enable work to be done by passive mechanical processes and thereby contribute to the specialization of reservoir function.

CHEMICAL ENERGY USAGE, FORCE, STIFFNESS, AND MYOSIN LIGHT CHAIN PHOSPHORYLATION AS A FUNCTION OF MUSCLE LENGTH

The length-tension relationship at lengths below L_0 in smooth muscle is somewhat similar to that found for skeletal muscle (23). It is not clear, however, why force does decrease at lengths shorter than L_0 in smooth muscle. At least four possible mechanisms could operate: (a) an internal dissipation of force due to compression of thick filaments at very short muscle lengths, as occurs in skeletal muscle ("internal load"), (b) a decrease in the degree of activation at short muscle lengths (27,29), (c) a structurally induced decrease in number of cross-bridges able to interact with actin, or (d) a misalignment of contractile filaments and/or cells such that force is generated in an axis other than that measured. We have probed these possibilities by determining force output, energy utilization, stiffness, degree of phosphorylation, and ultrastructure as a function of muscle length in the taenia coli.

The high energy phosphate usage during the development and maintenance of isometric force was determined in muscles at lengths of 0.5 to 1.0 L_0, where isometric force varied from 0.2 to 1.0 P_0. The experimental design included adjustment of the muscle to L_0 and stimulation to find P_0 for that muscle. The muscle was then set to the experimental length and allowed to recover for 15 min before treatment with metabolic inhibitors as described earlier. The energy usage was determined for the initial 25 sec when force is developed, for the subsequent 35 sec when force is maintained and for a total 60 sec contraction. The results for the full 60 sec contraction are shown as a function of relative force in Fig. 5. Surprisingly, the energy used does not change as force decreases at short muscle lengths. This is contrary to that observed in similar energetics measurements on skeletal muscle (17,28) and in steady state O_2 consumption measurements in smooth muscle (2,14,25).

It was possible that the energy usage did not decrease at short muscle lengths because the muscles must shorten further to develop force and thus use more energy during the initial phase of the tetanus. The energy used during the period of force maintenance may then be better related to the force produced. The energy used during maintenance of isometric force versus relative force is shown in Fig. 6. Energy usage actually increases at lengths where active force is 40 to 60% P_0. If it is assumed that the major energy utilizing process in the muscle is actin-activated myosin ATPase activity, then these data are not consistent with the concept that a decrease in actin-myosin interaction is responsible for the decrease in force at short muscle lengths in this smooth muscle.

The degree of phosphorylation of the 20,000 dalton light chain has also been measured as relative force is varied by changing muscle length. Figure 7 shows that there is a significant decrease in degree of light chain phosphorylation at 25 sec of stimulation at muscle lengths shorter than L_0. This is strong evidence that there is less calcium released on stimulation at muscle lengths below L_0 in smooth muscle. A quantitative estimate as to how this decreased activation at short muscle

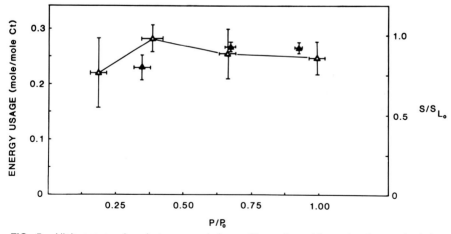

FIG. 5. High energy phosphate usage during a 60 sec isometric contraction, and relative stiffness as a function of active force output at muscle lengths at and below L_0. *Open triangles,* high energy phosphate usage; *closed triangles,* relative stiffness.

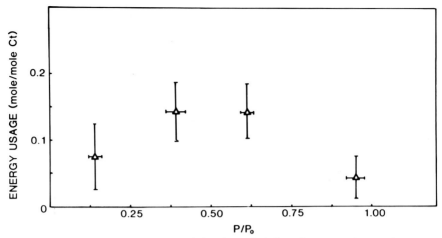

FIG. 6. High energy phosphate usage during the period of nearly constant force maintenance as a function of active force at different muscle lengths.

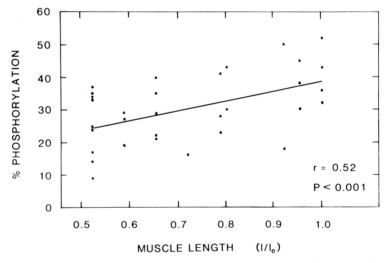

FIG. 7. Degree of light chain phosphorylation following 25 sec of stimulation under isometric conditions as a function of relative muscle length. The regression line is calculated assuming all error is in the measurement of degree of phosphorylation.

lengths contributes to the decrease in force output can be made. The degree of phosphorylation after 25 sec of stimulation at L_0 in these experiments is $39 \pm 3\%$, whereas it is $25 \pm 3\%$ at 55% L_0. In a separate set of experiments in our laboratories, P. Dillon *(unpublished results)* found that a similar decrease in degree of phosphorylation caused by lowering the calcium concentration of the bathing medium at L_0 gave only 20% of the decrease in force observed at the short muscle length. It is concluded that there is a significantly decreased calcium release at lengths below L_0 in smooth muscle, but that the magnitude of the activation can account for only 20% of the observed decrease in active force at a length of 55% L_0 in this muscle.

The observation of some decreased activation at short muscle lengths makes even more remarkable the finding that energy usage does not decrease at these lengths. It also suggests that a decreased activation may account for some of the decrease in energy usage associated with a decrease in force at short muscle lengths in studies on other smooth muscles (2,14,25).

Muscle stiffness was determined during isometric contractions at different lengths, using the method described by Bressler and Clinch (4). All measurements were made after maximum force had been developed at each length. The stiffness of stimulated muscles was corrected for resting stiffness and the results are shown in Fig. 5 as a function of relative force. Relative stiffness is always greater than relative force at muscle lengths shorter than L_0. These results are similar to those obtained by Meiss (22) who used sinusoidal length perturbations to measure stiffness in the rabbit mesotubarium. The close correlation of stiffness and energetics data compared to relative force suggests that a similar process may be responsible for stiffness and energy usage, but not force output. Some stiffness measurements were also done under conditions of lowered external calcium at L_0. Under these conditions, both energy usage and stiffness are directly proportional to force output. For a given active force, the stiffness of the muscle at short muscle lengths exceeds that of the muscle stimulated at L_0 in a low calcium medium. The fact that stiffness and energy usage are proportional no matter how external force is varied suggests that the stiffness measurements reflect the properties of the cross-bridges in smooth muscle (18,24). In summary, these data suggest that the major portion of the decrease in active isometric force at muscle lengths below L_0 is not due to a decrease in number of cross-bridges interacting with actin. Rather, it is likely that the decrease in force is due to structural complexities, such as the presence of an internal load or perhaps to filament and/or cell misalignment.

ULTRASTRUCTURE OF THE RABBIT TAENIA COLI UNDER RESTING AND STIMULATED CONDITIONS

We have done ultrastructural studies on the rabbit taenia coli at different muscle lengths under resting and stimulated conditions. In short, these studies showed that, at L_0, the filaments and cells are well aligned in the axis in which force is measured. At slack length (65% L_0), the cells are slightly misaligned at rest, but following stimulation under isometric conditions there is a good alignment. At lengths below slack length, there is considerable disorder in cells and filaments, which changes only slightly on stimulation.

The observed decrease in force and lack of change in energy usage at lengths between slack length and L_0 cannot be explained by filament and/or cell misalignment resulting in force output in an axis other than that measured. At lengths below 65% L_0, at which about 50% maximum force is generated, such misalignment probably accounts for a significant decrease in the measured force and this may play a role in the observed disassociation of force output and energy usage at very short muscle lengths.

ACKNOWLEDGMENTS

The studies reported here were supported by National Institutes of Health grant HL15835 to the Pennsylvania Muscle Institute. T. M. Butler is the recipient of a Research Career Development Award, AM00973.

REFERENCES

1. Aksoy, M. O., Williams, D., Sharkey, E. M., and Hartshorne, D. J. (1976): A relationship between Ca-sensitivity and phosphorylation of gizzard actomyosin. *Biochem. Biophys. Res. Comm.*, 68:35–42.
2. Arner, A., and Hellstrand, P. (1981): Energy turnover and mechanical properties of resting and contracting aortas and portal veins from normotensive and spontaneously hypertensive rats. *Circ. Res.*, 48:539–548.
3. Ashton, F. T., Somlyo, A. V., and Somlyo, A. P. (1975): The contractile apparatus of vascular smooth muscle: Intermediate high voltage electron microscopy. *J. Mol. Biol.*, 98:17–29.
4. Bressler, B. H., and Clinch, N. F. (1974): The compliance of contracting skeletal muscle. *J. Physiol. (Lond.)*, 237:477–495.
5. Butler, T. M., and Siegman, M. J. (1982): Physiology of smooth muscle. In: *Coronary Artery Disease*, edited by W. P. Santamore and A. A. Bove. Urban and Schwarzenberg, Baltimore.
6. Butler, T. M., Siegman, M. J., and Davies, R. E. (1976): Rigor in vertebrate smooth muscle. *Am. J. Physiol.*, 231:1509–1514.
7. Butler, T. M., Siegman, M. J., Mooers, S. U., and Davies, R. E. (1978): Chemical energetics of single isometric tetani in mammalian smooth muscle. *Am. J. Physiol.*, 235:C1–C7.
8. Canfield, P., Lebacq, J., and Marechal, G. (1973): Energy balance in frog sartorius muscle during an isometric tetanus at 20°C. *J. Physiol. (Lond.)*, 232:467–483.
9. Curtin, N. A., and Davies, R. E. (1973): ATP breakdown following activation of muscle. In: *Structure and Function of Muscle*, edited by G. H. Bourne, pp. 472–515. Academic Press, New York.
10. Davey, D. F., Gibbs, C. L., and McKirdy, H. C. (1975): Structural, mechanical and myothermic properties of rabbit rectococcygeus muscle. *J. Physiol. (Lond.)*, 248:207–230.
11. DeFuria, R. R., and Kushmerick, M. J. (1977): ATP utilization associated with recovery metabolism in anaerobic frog muscle. *Am. J. Physiol.*, 232:C30–C36.
12. Dillon, P. F., Aksoy, M. O., Driska, S. P., and Murphy, R .A. (1981): Myosin phosphorylation and the cross-bridge cycle in arterial smooth muscle. *Science*, 211:495–497.
13. Ebashi, S., Endo, M., and Ohtsuki, I. (1969): Control of muscle contraction. *Q. Rev. Biophys.*, 2:351–384.
14. Gluck, E., and Paul, R. J. (1977): The aerobic metabolism of porcine carotid artery and its relationship to isometric force energy cost of contraction. *Pfluegers Arch.*, 370:9–19.
15. Gordon, A. R., and Siegman, M. J. (1971): Mechanical properties of smooth muscle. I. Length-tension and force-velocity relations. *Am. J. Physiol.*, 221:1243–1249.
16. Hellstrand, P. (1977): Oxygen consumption and lactate production of the rat portal vein in relation to its contractile activity. *Acta Physiol. Scand.*, 100:91–107.
17. Infante, A. A., Klaupiks, D., and Davies, R. E. (1964): Length, tension and metabolism during short isometric contractions of frog sartorius muscles. *Biochem. Biophys. Acta*, 88:215–217.
18. Johansson, B., Hellstrand, P., and Uvelius, B. (1978): Responses of smooth muscle to quick load change studied at high time resolution. *Blood Vessels*, 15:65–82.
19. Kushmerick, M. J., and Davies, R. E. (1969): The chemical energetics of muscle contraction 2. The chemistry, efficiency and power of maximally working sartorius muscles. *Proc. Roy. Soc. Lond. [Biol.]*, 174:315–353.
20. Lowy, J., and Hanson, J. (1962): Ultrastructure of invertebrate smooth muscle. *Physiol. Rev.*, 42:34–47.

21. McGilvery, R. W., and Murray, T. W. (1974): Calculated equilibria of PCr and adenosine phosphates during utilization of high energy phosphate by muscle. *J. Biol. Chem.*, 249:5845–5850.
22. Meiss, R. A. (1978): Dynamic stiffness of rabbit mesotubarium smooth muscle: Effect of isometric length. *Am. J. Physiol.*, 234:C14–C26.
23. Murphy, R. A. (1976): Contractile system function in mammalian smooth muscle. *Blood Vessels*, 13:1–24.
24. Nauss, K. M., and Davies, R. E. (1966): Changes in inorganic phosphate and arginine during the development, maintenance and loss of tension in the anterior byssus retractor muscle of *Mytilus edulis*. *Biochem. Z.*, 345:173–187.
25. Paul, R. J., and Peterson, J. W. (1975): Relation between length, isometric force, and O_2 consumption rate in vascular smooth muscle. *Am. J. Physiol.*, 228:915–922.
26. Pool, P. E., Chandler, B. M., Seagren, S. C., and Sonnenblick, E. H. (1968): Mechanochemistry of cardiac muscle. II. The isotonic contraction. *Circ. Res.*, 22:465–472.
27. Rudel, R., and Taylor, S. (1971): Striated muscle fibers: Facilitation of contraction at short lengths by caffeine. *Science*, 172:387–388.
28. Sandberg, J. A., and Carlson, F. D. (1966): The dependence of phosphorylcreatine hydrolysis during an isometric tetanus. *Biochem. Z.*, 345:212–231.
29. Schoenberg, M., and Podolsky, R. J. (1972): Length-force relation of calcium activated muscle fibers. *Science*, 176:52–54.
30. Sherry, J. M., Gorecka, A., Aksoy, M. O., Dabrowska, R., and Hartshorne, D. J. (1978): Roles of calcium and phosphorylation in the regulation of the activity of gizzard myosin. *Biochemistry*, 17:4411–4418.
31. Siegman, M. J. F., Butler, T. M., Mooers, S. U., and Davies, R. E. (1976): Calcium-dependent resistance to stretch and stress relaxation in resting smooth muscles. *Am. J. Physiol.*, 231:1501–1508.
32. Siegman, M. J., Butler, T. M., Mooers, S. U., and Davies, R. E. (1980): Chemical energetics of force development, force maintenance, and relaxation in mammalian smooth muscle. *J. Gen. Physiol.*, 76:609–629.
33. Small, J. V., and Sobieszek, A. (1977): Myosin phosphorylation and Ca-regulation in vertebrate smooth muscle. In: *Excitation-Contraction Coupling in Smooth Muscle*, edited by R. Casteels, pp. 385–393. Elsevier/North Holland, Amsterdam.
34. Uvelius, B. (1979): Shortening velocity, active force and homogeneity of contraction during electrically evoked twitches in smooth muscle from rabbit urinary bladder. *Acta Physiol. Scand.*, 106:481–487.

Basic Biology of Muscles: A Comparative
Approach, edited by B. M. Twarog,
R. J. C. Levine, and M. M. Dewey.
Raven Press, New York © 1982.

Mechanics and Energetics of Contraction in Thick and in Thin Filament Regulated Muscles

Jack A. Rall

Department of Physiology, Ohio State University, Columbus, Ohio 43210

The striated adductor muscle of the scallop is responsible for its jet-propulsive swimming behavior and is of general interest because contractile force is thought to be under Ca^{2+} control via thick filaments rather than via thin filaments as in vertebrate striated muscle (12,25). The purpose of the present investigation was a comparative mechanical and energetic characterization of the isolated striated adductor muscle of the sea scallop *(Placopecten magellanicus)* and the striated semi-tendinosus muscle of the frog *(Rana pipiens)*, a more extensively studied muscle.

PROPERTIES OF THE STRIATED ADDUCTOR MUSCLE OF THE SCALLOP

The scallop, like some other molluscs, possesses the ability to swim, but unlike other molluscs, except the cephalopod, swimming is accomplished by means of jet-propulsion. Swimming movements result from the sudden expulsion of water produced by rapid, repeated adduction of the shells under the control of the striated adductor muscle (7). The rubber-like hinge ligament of the scallop contributes to its ability to sustain swimming movements by providing an elastic restoring force opening the shells during muscle relaxation, thus completing the flapping cycle (2). In this way, swimming velocities of 30 cm/sec can be attained for distances of several meters (17). After approximately 30 brief contractions of about 0.3 to 0.5 sec duration, the muscle becomes inexcitable and the shells remain closed, for up to a few hours, due to contraction of the nonstriated adductor muscle (5,8,17,27).

Molecular aspects of the striated adductor muscle have attracted attention based on the observation that contraction is under Ca^{2+} control via thick filaments rather than via thin filaments as in vertebrate striated muscle. Although it has been suggested that contraction in scallop striated muscle may be under control of both filaments (13), to date only thick filament control has been shown to be functional (24). Thick filament, or myosin-linked, regulation of contraction is mediated via the regulatory light chains of myosin (26). Because of this molecular divergence in the mechanism of Ca^{2+} control of contraction, it is worthwhile to determine to what extent other properties of scallop striated muscle differ from vertebrate striated muscle. Scallop striated muscle is almost unique among molluscs in that it is striated, with band patterns similar to those found in vertebrate striated muscle (9). Individual scallop muscle cells are ribbon-shaped, less than 2 mm long, about 0.5 to 2 μm thick and 3 to 10 μm wide (Fig. 1) (11,18,19,23). Each cell contains a single myofibril, sarcoplasmic reticulum but no transverse tubular system (Fig. 2) (11,18,19,23). Resting sarcomere lengths and thick and thin filament dimensions are similar to those found in the frog muscle (Fig. 2) (16,18). Thick filaments contain myosin and a small amount of paramyosin (approximately 5% of the total myosin) and thin filaments contain actin and tropomyosin (16,25). The amount of troponin-like protein is a small fraction of that found in vertebrate striated muscle with none detectable on thin filaments by X-ray diffraction techniques (16,26,28). Despite these observations, I bands of scallop myofibrils have been shown to contain troponin-C and troponin-I (13). The molecular architecture of this muscle seems to be highly ordered since the muscle is suited for structural studies with X-ray diffraction techniques (16,28). The cells are multiply innervated, display resting membrane potentials of about -60 mV, and are capable of producing spike potentials on stimulation (14). The muscle displays a high glycolytic potential and during contraction mobilizes ATP and arginine phosphate (rather than creatine phosphate) (8). Actomyosin ATPase activity is similar in scallop and in frog striated muscle when compared at comparable temperatures (4). The striated adductor mus-

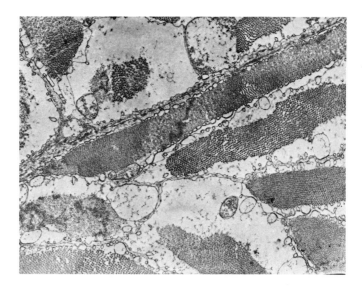

FIG. 1. Electron micrograph of a cross-section from the striated adductor muscle of the scallop. Note ribbon shaped cells of about 1 μm thickness, one myofibril per cell, and hexagonal packing of thick filaments. ×11,600.

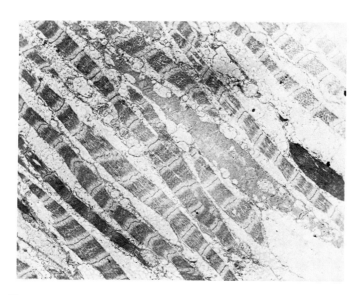

FIG. 2. Electron micrograph of a longitudinal section from the striated adductor muscle of the scallop. Note sarcomeres of about 2.6 μm length, one myofibril per cell, I bands, and A bands of about 1.7 μm length. ×2,800.

cle seems to have a contraction time that is similar to that observed in frog skeletal muscle (5,15,16), but no quantitative comparison has been made. Maximum velocity of shortening in scallop striated muscle at 14°C (3 muscle lengths/sec;) (9) seems to be similar to that for frog muscle at 0°C (2 muscle lengths/sec) (22), but this point is unsettled because of the fivefold variation in maximum velocity of shortening reported for scallop striated muscle at the same temperature (0.6 muscle lengths/sec at 14°C) (1).

MECHANICS AND ENERGETICS OF ISOMETRIC TWITCHES

Isometric mechanical properties and energy liberation, measured as heat production, were determined in isolated bundles of the striated adductor muscle of the scallop and in isolated semitendinosus muscles of the frog at 10°C (unless stated otherwise). For technical details of the dissection, mechanical and myothermal measurements, and experimental protocol, see Rall (20,21).

An example of maximum twitch force production per cross-sectional area and accompanying energy liberation is shown for scallop (Fig. 3A) and frog muscle (Fig. 3B). The energy liberation record for scallop muscle in Fig. 3A (upper panel) is shown before and after correction for stimulus heat. In this example, stimulus heat was 26% of the physiological response. Scallop muscles were quite inexcitable compared to frog muscle and large stimulating voltages were required. A similar correction for stimulus heat, not shown, would be about 2% in frog muscle. The large stimulus heat in scallop muscle may be due to difficulty in activating the

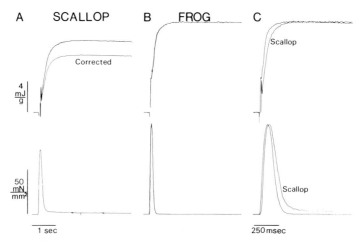

FIG. 3. Energy liberation **(top)** and force production **(bottom)** in isometric twitches of scallop **(A)** and frog **(B)** muscle. The corrected trace is the original trace minus the stimulus heat. In **C**, the thermal and mechanical records have been scaled to the same peak values so that kinetic comparisons can be more clearly visualized. The mechanical records also have been expanded in time in **C**. (From Rall, ref. 21, with permission.)

TABLE 1. *Comparison of mechanical and energetic properties of scallop and frog striated muscle*

	Scallop (N = 7)	Frog (N = 6)
Twitch force[a] (mN/mm²)	77 ± 4[b]	106 ± 7
Twitch energy liberation (mJ/g)	7.6 ± 0.5	11.1 ± 0.7
Contractile economy	10.6 ± 1.0	9.7 ± 0.4
Time to peak force (msec)	87 ± 2	86 ± 3
Time to half relaxation from peak force (msec)	104 ± 2	66 ± 5
Tetanus force[a] (mN/mm²)	132 ± 5	188 ± 9
Fatigue resistance[c]	0.51 ± 0.05	0.98 ± 0.002

From Rall (21), with permission.
[a]Twitch and tetanus force are calculated per cross-sectional area as: Force × length/ blotted mass. Length for scallop muscle was muscle length and for frog muscle was estimated fiber length (21).
[b]Values given as mean ± SE of mean.
[c]Fatigue resistance is calculated as: the ratio of the force at the last stimulus to the peak force during a tetanus.

small cells of this muscle and to the high conductivity of sea water, which would short-circuit most of the stimulus current. In this example, scallop muscle developed 29% less twitch force and liberated 37% less energy than frog muscle. On the average, scallop muscle developed significantly less twitch force per cross-sectional area while liberating significantly less energy than frog muscle (Table 1). A 5% level of significance ($p < 0.05$), based on a t-test of sample means, was accepted throughout.

In Fig. 3C, mechanical and energetic parameters have been scaled to emphasize kinetic features. In the lower panel, where peak forces have been superimposed, time to peak force was 12% longer and time to half relaxation from peak force was 40% longer in scallop muscle. On the average, time to peak force was similar in scallop and frog muscle, whereas time to half relaxation from peak force was significantly longer in scallop muscle (Table 1). Despite a statistically significant prolongation of the kinetics of relaxation in an isometric twitch, it is clear that the time course of an isometric twitch in scallop muscle is similar to that in frog muscle at the same temperature. Similar conclusions are reached when evaluating the time course of energy liberation.

It is possible to derive an estimate of contractile economy in the isometric twitch by determining the ratio of peak force development to energy liberation in the twitch (10). Interestingly, the twitch contractile economy is not significantly different in scallop and frog muscle (Table 1). The similarity of the isometric contractile economies suggests a similarity in mechanochemical conversion in scallop and frog muscle. But it should be emphasized that this conclusion depends on: (a) an unverified similarity of the *in vivo* enthalpy change for creatine phosphate and arginine phosphate hydrolysis and (b) the calculation of force per cross-sectional area for each muscle. Even though individual cells in the scallop are less than 10% of the length of the muscle (muscle length was typically 3 cm), the muscle length was employed in the force calculations (21). This procedure assumes that cells are arranged in series so that force is transferred from cell to cell in such a way that the cells effectively traverse the length of the muscle. Estimated fiber length was employed in the force calculation in frog muscle (21).

MECHANICS OF ISOMETRIC TETANIC CONTRACTIONS

An example of maximum twitch and tetanic force per cross-sectional area (for a 1.5-sec tetanus at 50 Hz stimulating frequency) is shown for scallop (Fig. 4A) and frog (Fig. 4B) muscle. In this example, the scallop muscle developed 48% less tetanus force with a similar twitch to tetanus ratio. Mean twitch to tetanus ratios were similar (Table 1). The average value of maximum tetanic force per cross-

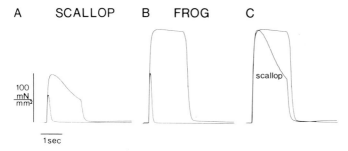

FIG. 4. Isometric twitch and tetanus records from scallop **(A)** and frog **(B)** muscle. Tetanus duration was 1.5 sec. In **C**, the tetanic contractions have been scaled to the same peak value. (From Rall, ref. 21, with permission.)

sectional area in scallop muscle was significantly less than that observed in frog muscle (Table 1). Assuming that the calculations of force per cross-sectional area are appropriate, scallop muscle generates 30% less force than frog muscle. One reason that force development may be relatively low in scallop muscle is that each cell is extremely small and contains only one myofibril (Figs. 1 and 2). Thus, throughout an equivalent cross-sectional area, scallop muscle may have a greater amount of membrane and consequently fewer myofibrils (Fig. 1). The net result would be a decreased force per cross-sectional area.

In Fig. 4C, superposition of maximum tetanus forces shows the dramatic fatigue in scallop muscle during a 1.5-sec tetanus. The average resistance to fatigue, quantitated as the ratio of force at the last stimulus to maximum tetanus force, was significantly less in scallop muscle (Table 1). Frog muscle had to be stimulated 20 times longer to produce fatigue of a comparable extent to that observed in scallop muscle after 1.5 sec of stimulation (Fig. 5). Furthermore, scallop muscle fatigues to the same or to a greater extent when stimulated at 0°C as when stimulated at 10°C (Fig. 6). The fact that the amount of fatigue was not less at a lower temperature for a tetanus of constant duration suggests that fatigue is not controlled by energy supply in scallop muscle. A muscle developing the same force for the same length of time would be expected to utilize less energy at the lower temperature. A more likely possibility is neuromuscular or excitation-contraction coupling fatigue. There is not a great deal of information available concerning scallop neuromuscular transmission but recent evidence suggests that the neurotransmitter in scallop muscle may be L-glutamate (6). Since *in vivo* the striated adductor muscle of the scallop develops and maintains force for only brief periods of time (0.3 to 0.5 sec) (5,17,27), 50% fatigue in a tetanus of 1.5 sec may not be surprising. The striated adductor muscle of the scallop is not designed to maintain force during contraction, as is the nonstriated adductor muscle, but rather to do external work.

Because of the large stimulus heat in tetanic contractions of scallop muscle, energy liberation records were considered unreliable.

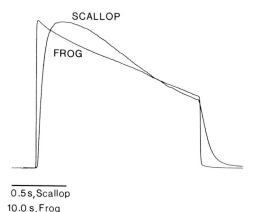

SCALLOP

FROG

0.5 s, Scallop
10.0 s, Frog

FIG. 5. Isometric tetanus force development in scallop and frog muscle. Scallop: 1.5-sec tetanus. Frog: 30-sec tetanus. Forces scaled to same peak value. Frog muscle had to be stimulated 20 times longer to produce comparable fatigue. Note differences in time base for frog and scallop.

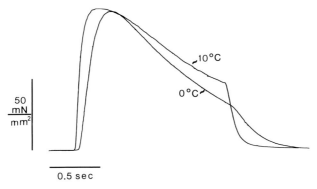

FIG. 6. Isometric tetanus force development in scallop muscle at 10 and 0°C. Same muscle, data recorded first at 10°C then at 0°C. Tetanus duration: 1.5 sec. Stimulating frequency: 50 Hz at 10°C and 20 Hz at 0°C. A stimulating frequency of 50 Hz at 0°C produced a larger peak force but more rapid fatigue. Records have not been scaled.

MECHANICS AND ENERGETICS OF POSTTETANIC TWITCHES

Experiments were designed to examine force production and energy liberation in twitches elicited at variable times after a 1.5-sec tetanus in scallop and in frog muscle. In scallop muscle, a twitch elicited after a tetanus was transiently potentiated with peak force occurring 5 to 15 sec after the tetanus and returning to control values by 3-min posttetanus (Fig. 7A, *closed circles*). Posttetanic twitch potentiation in frog muscle was considerably smaller (Fig. 7B, *closed circles*). In posttetanic twitches of scallop muscle, energy liberation increased more than peak force development increased (Fig. 7A, *open circles*). For example, even though twitch force and kinetics can be the same in scallop muscle 90 sec before and 3 min after a tetanus, Fig. 7A shows that the twitch after the tetanus liberates significantly more energy. The mean isometric contractile economy 3 min after a tetanus decreased by 25 ± 2% in scallop muscle. In comparable experiments in frog muscle (Fig. 7B), twitch force was potentiated by 4% 3 min after a tetanus, whereas energy liberation was increased by 5%. Thus, under these conditions, the average isometric contractile economy decreased by 1 ± 1% in frog muscle. After a tetanus, the scallop muscle contraction is potentiated in amount but transiently less economical. The significance and mechanism of these observations are unknown.

SUMMARY

The striated adductor muscle of the sea scallop, *P magellanicus*, and the semitendinosus muscle of the frog, *R pipiens*, exhibited certain similarities and differences in mechanical and energetic properties. Scallop and frog striated muscle exhibited similar isometric: (a) mechanical twitch kinetics, (b) twitch to tetanus ratios, and (c) contractile economies. There were significant differences between scallop and frog muscle in that scallop muscle demonstrated: (a) a 30% lower maximum isometric force, (b) an inability to maintain force during a tetanus, and

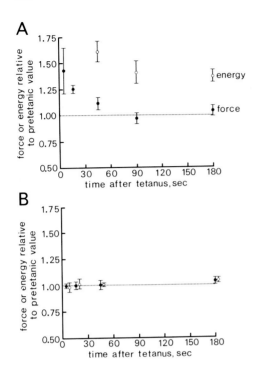

FIG. 7. Isometric twitch force and energy liberation relative to pretetanic values versus time after a 1.5-sec tetanus in scallop and frog muscle. **A:** Scallop muscle, 23 observations on 6 muscles. **B:** Frog muscle, 15 observations on 5 muscle pairs. *Open circles,* relative energy liberation; *closed circles,* relative force production both represented as mean ± SE of mean. (From Rall, ref. 21, with permission.)

(c) a transient potentiation of twitch force after a tetanus associated with a transient decrease in contractile economy.

Finally, it is clear that the present study, although contributing mechanical and energetic information on the scallop striated adductor muscle, does not consider the type of contraction normally induced by the scallop *in vivo*, a shortening, working contraction. It has been suggested (3,17) that swimming in the scallop may be most efficient when the load is high (approximately one-half of the force that can be generated) and the velocity of shortening of the striated adductor is low. This situation is somewhat different than would be predicted in frog muscle where efficiency is highest when velocity of shortening is relatively high and load is relatively low (about one-third of the maximum force that can be generated). A comparison of the mechanical efficiency versus load relation during working contractions in scallop and frog muscle might produce further interesting results.

ACKNOWLEDGMENTS

The author thanks B. Kaminer, B. Millman, A. G., Szent-Györgyi, and B. Twarog for useful discussions, J. Wells for valuable technical advice, and S. Houghton for performing the electron microscopy. Experiments on the sea scallop were conducted at the Marine Biological Laboratory at Woods Hole, MA. This investigation was supported in part by the Ohio State University College of Medicine and by United States Public Health Service grant AM-20792 from the National

Institutes of Health. The author is a recipient of Research Career Development Award 1K04NS-00324 from the National Institutes of Health.

REFERENCES

1. Abbott, B. C., and Lowy, J. (1957): Stress relaxation in muscle. *Proc. R. Soc. Lond. [Biol.]*, 146:280–288.
2. Alexander, R. McN. (1966): Rubber-like properties of the inner hinge ligament of Pectinidae. *J. Exp. Biol.*, 44:119–130.
3. Alexander, R. McN. (1979): Clams. In: *The Invertebrates*, pp. 295–315. Cambridge University Press, London.
4. Bárány, M. (1967): ATPase activity of myosin correlated with speed of muscle shortening. *J. Gen. Physiol.*, 50:197–218.
5. Bayliss, L. E., Boyland, E., and Ritchie, A. D. (1930): The adductor mechanism of Pecten. *Proc. Roy. Soc. Lond. [Biol.]*, 106:363–376.
6. Bone, Q., and Howarth, J. V. (1980): The role of L-glutamate in neuromuscular transmission in some molluscs. *J. Mar. Biol. Ass. U.K.*, 60:619–626.
7. Buddenbrock, W. V. (1911): Untersuchungen über die schwimmbewegungen and die statocysten der gattung pecten. *Sitz. Heldelber. Akad. Wiss.*, 28:1–24.
8. de Zwaan, A., Thompson, R. J., and Livingstone, D. R. (1980): Physiological and biochemical aspects of the valve snap and valve closure responses in the giant scallop *Placopecten magellanicus* II. Biochemistry. *J. Comp. Physiol.*, 137:105–114.
9. Hanson, J., and Lowy, J. (1960): Structure and function of the contractile apparatus in the muscles of invertebrate animals. In: *The Structure and Function of Muscle*, edited by G. H. Bourne, pp. 265–335. Academic, New York.
10. Hill, A. V. (1959): The relation between force developed and energy liberated in an isometric twitch. *Proc. Roy. Soc. Lond. [Biol.]*, 149:58–62.
11. Kawaguti, S., and Ikemoto, N. (1958): Electron microscopy on the adductor muscle of the scallop, *Pecten albicans. Biol. J. Okayama Univ.*, 4:191–205.
12. Kendrick-Jones, J., Lehman, W., and Szent-Györgyi, A. G. (1970): Regulation in molluscan muscles. *J. Mol. Biol.*, 54:313–326.
13. Lehman, W., Head, J. F., and Grant, P. W. (1980): The stoichiometry and location of troponin I- and troponin C-like proteins in the myofibril of the bay scallop, *Aequipecten irradians. Biochem. J.*, 187:447–456.
14. Mellon, D. (1968): Junctional physiology and motor nerve distribution in the fast adductor muscle of the scallop. *Science*, 160:1018–1020.
15. Millman, B. M. (1967): Mechanism of contraction in molluscan muscle. *Am. Zoologist*, 1:583–591.
16. Millman, B. M., and Bennett, P. M. (1976): Structure of cross-striated adductor muscle of the scallop. *J. Mol. Biol.*, 103:439–467.
17. Moore, J. D., and Trueman, E. R. (1971): Swimming of the scallop, *Chlamys opercularis (L). J. Exp. Mar. Biol. Ecol.*, 6:179–185.
18. Morrison, C. M., and Odense, P. H. (1968): Ultrastructure of the striated muscle of the scallop *(Placopecten magellanicus) J. Fish. Res. Bd. Canada.*, 25:1339–1345.
19. Morrison, C. M., and Odense, P. H. (1974): Ultrastructure of some pelecypod adductor muscles. *J. Ultrastruct. Res.*, 49:228–251.
20. Rall, J. A. (1979): Effects of temperature on tension, tension-dependent heat, and activation heat in twitches of frog skeletal muscle. *J. Physiol. (Lond.)*, 291:265–275.
21. Rall, J. A. (1981): Mechanics and energetics of contraction in striated muscle of the sea scallop, *Placopecten magellanicus. J. Physiol. (Lond.)*, 321:287–295.
22. Ritchie, J. M., and Wilkie, D. R. (1956): Muscle: Physical properties. In: *Handbook of Biological Data*, edited by W. S. Spector, p. 295. Saunders, Philadelphia.
23. Sanger, J. W. (1971): Sarcoplasmic reticulum in the cross-straited adductor muscle of the bay scallop, *Aequipecten irridians. Z. Zellforsch.*, 118:156–161.
24. Simmons, R. M., and Szent-Györgyi, A. G. (1978): Reversible loss of calcium control of tension in scallop striated muscle associated with the removal of regulatory light chains. *Nature*, 273:62–64.

25. Szent-Györgyi, A. G. (1976): Comparative survey of the regulatory role of calcium in muscle. *Symp. Soc. Exp. Biol.*, 30:335–347.
26. Szent-Györgyi, A. G., Szentkiralyi, E. M., and Kendrick-Jones, J. (1973): The light chains of scallop myosin as regulatory light chains. *J. Mol. Biol.*, 74:179–203.
27. Thompson, R. J., Livingston, D. R., and de Zwaan, A. (1980): Physiological and biochemical aspects of the valve snap and valve closure responses in the giant scallop *Placopecten magellanicus.* I. Physiology. *J. Comp. Physiol.*, 137:97–104.
28. Vibert, P., Szent-Györgyi, A. G., Craig, R., Wray, J., and Cohen, C. (1978): Changes in crossbridge attachment in a myosin-regulated muscle. *Nature*, 273:64–66.

Basic Biology of Muscles: A Comparative
Approach, edited by B. M. Twarog,
R. J. C. Levine, and M. M. Dewey.
Raven Press, New York © 1982.

Biochemical Energetics of A-Band Shortening in Contracting *Limulus* Muscle

Sandra Davidheiser and Robert E. Davies

*Department of Animal Biology, School of Veterinary Medicine,
University of Pennsylvania, Philadelphia, Pennsylvania 19104*

The striated telson muscles of the horseshoe crab *Limulus* can develop force and shorten over a wide range of sarcomere lengths from 12 to 4 μm *in vivo*. The contractile proteins of these muscles include 4 to 4.5 μm long thick filaments composed of myosin molecules surrounding a paramyosin core and 2.4 to 2.8 μm long actin, thin filaments. The observed decrease in A-band length during sarcomere shortening results initially from realignment of staggered, long thick filaments at sarcomere lengths from 12 to 7 μm (L_0) followed by actual shortening of thick filaments to 3 to 3.5 μm at sarcomere lengths less than 7 μm (3,4). These muscles provide us with a useful model for studying the effects of both sarcomere and thick filament length changes on the mechanical properties and energy requirements associated with contraction. Our investigations have been concerned with: (a) developing a suitable telson muscle preparation for studies of the energetics of contraction; (b) determining the effects of muscle length on chemical energy usage in stimulated telson muscles during isometric and isovelocity contractions; and (c) in collaboration with Drs. R. Levine and R. Kensler, investigating the effects of stimulation on A-band and thick filament length in telson muscles stretched beyond the overlap of thick and thin filaments.

METHODS

Telson levator and depressor muscle bundles (15 to 30 mg, 15 to 30 mm long, and 1 to 2 mm wide) are dissected from *Limulus* with a small piece of carapace and tendon at either end. The muscles are maintained in aerated artificial sea water, pH 7.4 at 18°C. The sarcomere length at L_0 for these muscles has been found to be 7 μm (9) and the sarcomeres are adjusted to this length on the apparatus using laser diffraction. The muscle bundle is mounted on the mechanical apparatus by attaching one end to an adjustable rod and the other end to a wire attached to a tension transducer mounted on the arm of an ergometer. The output of the transducer is displayed on a chart recorder. The bundle is adjusted to the appropriate length in relation to L_0 and allowed to recover for 10 min in a bath of artificial sea water.

213

The bath is then drained and the experimental muscle is maximally activated by passing alternating current (60 Hz, 10 to 12V) through the ends of the bundle. Resting (control) and stimulated muscles are rapidly frozen in a mixture of freon 12 and 13 ($-190°C$), ground and extracted in perchloric acid, and the neutralized extract analyzed for free and total arginine (Arg·t), adenine nucleotides, and L ($+$) lactate (2).

RESULTS AND DISCUSSION

We have found the isolated telson muscle to be suitable for mechanical and energetics studies. The bundles can be dissected from the animal without damage with intact origins and insertions. The length-tension relation is known (9) so the muscle can be adjusted to the appropriate length as required by the experimental design. Maximal activation can be achieved by either electrical field stimulation in a bath or by electrical stimulation through the ends of the muscle suspended in air. The muscle bundle can be stretched to $\simeq 1.6\ L_0$ before passive tension becomes a significant percentage of the total isometric force ($\simeq 15\%$ at 11 μm sarcomere length). The muscles can be incubated in aerated, artificial sea water at 18°C for many hours and maintain high levels of total arginine (30 to 60 μmoles/g), phosphorylarginine (PArg)/total arginine (0.84 to 0.89 mole/mole), and ATP (3 to 4 μmoles/g) (2). Since total arginine is not significantly different between control and stimulated muscles of the same animal, it can be used as a measure of relative muscle mass, thus eliminating the need for weighing small frozen samples that might thaw.

Recovery Processes

The decrease in phosphorylarginine measured during a contraction should be equivalent to the amount of ATP utilized in the contractile process providing the arginine kinase reaction maintains ATP and phosphorylarginine near equilibrium at all times, and no significant recovery of phosphorylarginine from glycolysis and oxidative phosphorylation occurs during the period of stimulation. We have found that the ATP level remains constant and that recovery of phosphorylarginine occurs within 15 min following a 30-sec isometric tetanus at L_0 under aerobic conditions at 18° (Table 1). The recovery of phosphorylarginine would only amount to $\simeq 10\%$ of the total phosphorylarginine (ΔPArg) utilized during the 30-sec tetanus calculated using the rate of recovery during the first 2-min period following the tetanus. However, this initial rate is likely to be much higher than the actual average rate during the period of stimulation since, in all muscles that have been studied, there is a lag of at least several seconds before the initiation of glycolytic and oxidative recovery processes, followed by a further delay until these processes reach maximal rates. We believe these results validate our use of telson muscles that have not been pretreated with metabolic inhibitors.

The end-products of the metabolic pathways involved in resynthesizing ATP under anaerobic and aerobic conditions in telson muscles are not known. We have

TABLE 1. *Recovery of phosphorylarginine in* Limulus *telson muscles following a 30-sec isometric tetanus at L_0 (18°C)*

	Time following tetanus (min)					
	0	2	5	15	30	60
Stimulated—Control						
ΔPArg (mole/mole Arg·t)						
Mean	−0.11	−0.08	−0.03	−0.01	+0.01	+0.02
±SE	0.018	0.026	0.015	0.016	0.013	0.013
n[a]	20	6	10	10	8	8
ΔPArg (μmole/g)[b]	−5.28	−3.84	−1.44	−0.48	+0.48	+0.96

[a]Muscles from 8 animals.
[b]48 μmoles total arginine/g wet weight.

found that lactate is not the main anaerobic end-product of glycolysis in these muscles. Tissue lactate concentrations did not increase significantly in resting muscles incubated for 2 hr in artificial sea water gassed with N_2 (anaerobic, 0.385 ± 0.035, n = 12 versus aerobic, 0.490 ± 0.035, n = 8 μmole lactate/g), and the basal rate of lactate production in anaerobic muscles determined by measuring incubation medium lactate was low (4.3 ± 1.05 nmoles/g/min, n = 10). No significant differences in lactate and ATP (experimental minus control) were measured in anaerobic muscles stimulated isometrically for 30 sec (+ 0.210 ± 0.175 μmoles lactate/g and + 0.070 ± 0.385 μmoles ATP/g, n = 7).

Oxygen utilization was measured in telson muscles in a 3-ml chamber equipped with a Clark-type O_2 electrode for monitoring the decrease of O_2 in the artificial sea water gassed with air (21% O_2). Although an increase in O_2 utilization was observed immediately following a 30-sec tetanus, the total amount of recovery O_2 above baseline was only 0.20 ± 0.032 μmoles/g, n = 10, during the next 30-min period. This amounts to 1.2 μmoles/g ATP based on an assumed maximum P/O_2 ratio of 6 and could only account for 10 to 20% of the total amount of phosphorylarginine utilized during a 30-sec tetanus. It is likely that muscular contraction in *Limulus* depends initially on a glucose-amino acid-coupled catabolic pathway for ATP synthesis similar to those described for some intertidal bivalves (8). This type of pathway does not require O_2 nor does it produce lactic acid.

Effect of Muscle Length on Isometric Force and Phosphorylarginine Usage

In frog skeletal muscle, after allowance for activation, the amount of ATP utilized during an isometric tetanus has been shown to be directly related to the developed tension and, therefore, to muscle length (6). If the change in phosphorylarginine (ΔPArg) we measure in telson muscle is related to isometric force, then less phosphorylarginine should be utilized at sarcomere lengths greater or less than L_0 where external force decreases. The results of experiments measuring external isometric force and the change in phosphorylarginine for a 30-sec tetanus at 0.3, 0.6, 1.0, 1.7, and 2.0 L_0 are summarized in Fig. 1. Force, represented by the closed circles, and energy utilization, represented by the open circles, are expressed as a fraction

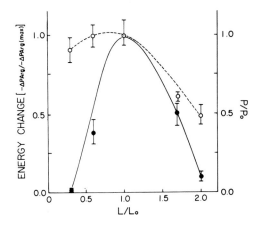

FIG. 1.   Effect of muscle length on isometric force *(closed circles)* and phosphorylarginine utilization *(open circles)* of *Limulus* telson muscle (18°C). Muscles were stimulated at either 0.3, 0.6, 1.0, 1.7, or 2.0 L_0 for 30 sec. Results expressed as a fraction of the maximum values obtained at L_0 (7 μm sarcomere length). Each *point* represents the mean ± SE of 9 to 18 muscles from a total of 14 animals.

of the values obtained at L_0. At 1.7 and 2.0 L_0 where force was either 50% or 10% of maximum, phosphorylarginine utilization was 62 and 50% of that at L_0. Extrapolating the force and Δphosphorylarginine data to the muscle length beyond L_0 corresponding to zero force, we calculated an energy usage for activation processes of 40% of the maximal energy usage. Unlike the situation in frog skeletal muscle, in *Limulus* telson muscle at lengths of 0.6 and 0.3 L_0, when external force decreased to 39% of the maximum and to virtually 0%, energy utilization was not significantly less than its maximal value at L_0. This leaves ≃ 3 to 5 μmoles/g phosphorylarginine not accounted for by external tension production.

Changes in Phosphorylarginine with Time During Isometric Tetani at L_0 and 0.5 L_0

One possible explanation for this large energy usage at short muscle lengths is that energy is required to initiate or maintain thick filament shortening. If a large energy cost is required for the shortening process, then we should observe a higher rate of energy utilization at short muscle lengths at the beginning of a tetanus. Figure 2 shows the results of experiments where we compared the changes in phosphorylarginine during isometric tetani of 5, 15, and 30 sec in muscles held at L_0 (open circles) or allowed to shorten and develop isometric force at 0.5 L_0 (filled circles). The changes in phosphorylarginine were the same in both groups despite the fact that the 0.5 L_0 muscles developed only 28% of maximum force (P_0). Therefore, initiation of thick filament shortening cannot account for all of the energy usage not associated with developed external force during a 30-sec tetanus. In a separate series of experiments, we measured phosphorylarginine utilization in two groups of telson muscles shortening equivalent distances during rapid, unloaded, isovelocity contractions at sarcomere length ranges of 11 to 7.5 and 7.5 to 4.0 μm. Shortening at a speed of ≃ 0.8 V_{max}, these muscles developed less than 5% of maximum external force, and no significant changes in energy usage from the control muscles were measured in either group. These findings also indicate that

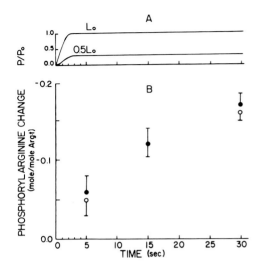

FIG. 2. Changes in force (P/P_0) **(A)** and phosphorylarginine utilization **(B)** during isometric tetani of different durations at 18°C in telson muscles at L_0 *(open circles)* and 0.5 L_0 *(closed circles)*. Each point represents the means ± SE of 12 to 21 muscles from a total of 9 animals. To convert ΔPArg/Arg·t to ΔPArg/g wet weight for this series, multiply by 53 (average μmoles Arg·t/g). Maximum isometric force (P_0) developed at L_0 (7 μm sarcomere length) was 22.46 ± 0.97 N/cm², n = 53; and at 0.5 L_0 (3 to 4 μm sarcomere length) was 6.28 ± 0.77 N/cm², n = 40.

thick filament shortening per se requires little, if any, energy. However, it is still possible that energy is required to maintain thick filaments in the shortened state.

Changes in Phosphorylarginine and Chemical Efficiency of Working Telson Muscles Shortening from Long and Short Sarcomere Lengths

The experiments summarized in Fig. 3 were designed to see whether the chemical efficiency for performing external work was lower in telson muscles contracting at short versus long muscle lengths. In these experiments, both experimental and control muscles were adjusted to a sarcomere length of either 7.5 (A), 9.0 (B), or 11.0 μm (C), and stimulated isometrically for 2 sec to allow maximum tension to develop. Control muscles were frozen after this 2-sec isometric tetanus. The experimental muscles of all three groups then shortened at a slow, constant velocity of approximately 3% V_{max} over equivalent distances to final sarcomere lengths of 4.0, 5.5, or 7.5 μm. In the left-hand side of Fig. 3, the open bars represent the work performed by each group and the lined bars represent the change in phosphorylarginine from the control muscles. Although the work performed by group A was approximately 50% that of groups B and C, the energy usage in all three groups was not significantly different. The right-hand side of Fig. 3 shows that the overall chemical efficiency, represented by external work divided by the change in energy usage, was significantly lower in telson muscles where thick filament shortening occurred. If we then subtract from the total energy usage for each group 40% for activation and other processes not involved in force generation and use a value of 42 mJ/mole ATP to represent 100% efficiency (7), then the chemical efficiency of the contractile mechanism for performing external work is 52% for the muscle groups B and C compared to 25% for group A.

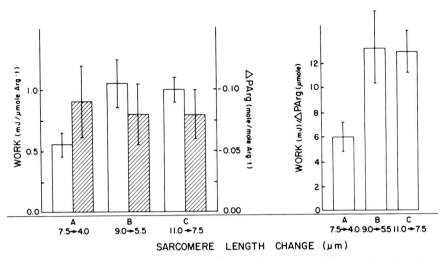

FIG. 3. External work, phosphorylarginine utilization, and chemical efficiency of *Limulus* telson muscles at 18°C. Muscles were stimulated to shorten at a constant velocity (≈ 0.03 Vmax) equivalent distances (3.5 μm sarcomere length) from starting sarcomere lengths of 7.5 **(A)**, 9.0 **(B)**, or 11.0 μm **(C)**. In the *left-hand side*, external work is represented by the *open bars* $\pm$ SE and energy usage (ΔPArg/Arg·t) by the *lined bars* $\pm$ SE; n = 17, 13, and 10 for groups **A**, **B**, and **C**, respectively. In the *right-hand side*, external work is divided by the total energy usage for groups **A**, **B**, and **C**, and represents the overall chemical efficiency.

Thick Filament Shortening at 2.0 L_0

Maximal shortening of isolated, long thick filaments from telson muscles has been shown to occur *in vitro* in the presence of ATP when Ca^{+2} was increased to 10^{-5} M, the concentration that would occur during activation (1). Therefore, it seemed likely that thick filaments should shorten when these muscles were stimulated isometrically at any sarcomere length unless, at sarcomere lengths between 7 and 12 μm, interaction with actin filaments prevented thick filament shortening. We, in collaboration with Drs. Rhea Levine and Robert Kensler, were interested in finding out what would happen to thick filament lengths when telson muscle bundles, stretched beyond thin-thick filament overlap to a length of 2.0 L_0, were isometrically stimulated. Telson muscle bundles were tied to wooden sticks at L_0 or 2.0 L_0. The control 2.0 L_0 muscles were placed in a relaxing solution containing 0.1 M NaCl, 5 mM $MgCl_2$, 1 mM DTT, and 4 mM EGTA for isolation of thick filaments (1). The experimental muscles were stimulated isometrically for 30 sec at L_0 or 2.0 L_0 and then placed immediately in the relaxing solution. The results of these experiments are summarized in Table 2. Thick filaments from unstimulated muscles at 2.0 L_0 and from the stimulated muscles at L_0 were significantly longer than thick filaments isolated from the muscles stimulated at 2.0 L_0. Therefore, thick filament shortening can also be demonstrated in telson muscles at very long sarcomere lengths beyond overlap of the Z-line-linked thin, actin filaments and the myosin-paramyosin thick filaments. Figure 4 shows a phase-contrast micrograph

TABLE 2. *Length measurements of isolated thick filaments from electrically stimulated* Limulus *telson muscles at L_0 and 2.0 L_0[a]*

	Unstimulated (2.0 L_0)	Stimulated	
		(L_0)	(2.0 L_0)
Thick filament length (μm)			
Mean	4.05	4.20	3.60
SD	0.54	0.47	0.41
SE	0.040	0.072	0.032
n	182	43	160

[a]These experiments were done in collaboration with R. Levine and R. Kensler.

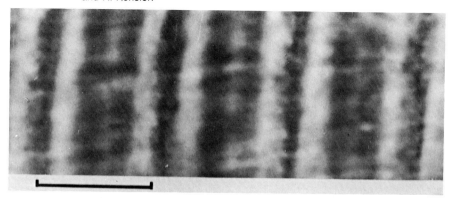

FIG. 4. Phase-contrast micrograph of an unstimulated *Limulus* telson muscle bundle with long sarcomeres ($\approx$11 μm) and long A-bands ($\approx$5.0 μm). Scale = 10 μm.

of a resting telson muscle bundle with long sarcomeres measuring approximately 11 μm and long A-bands of approximately 5.0 μm. Although we do not have pictures of the stimulated muscles from these experiments (Table 2), we do have a phase-contrast micrograph (Fig. 5) of a telson muscle bundle that was stimulated in high K^+ (400 mM) artificial sea water, allowed to shorten isotonically, and then placed in relaxing solution for several hours. Under these conditions, the sarcomeres are long (approximately 12 μm) but the A-bands are now short (approximately 3.5 μm), indicating that the thick filaments that had realigned and shortened during stimulation remained in this state despite relengthening of the sarcomeres that occurs in the relaxing solution (5). We conclude from these results that, in unstimulated telson muscles, sarcomere lengths are always greater than 6 μm and thick filaments are long. In stimulated muscles, at all lengths, thick filaments have the ability to shorten; however, this can only occur at lengths below 7 μm, or beyond 12 μm when overlap is minimal. At lengths of 2.0 L_0, in unstimulated muscles, the long thick filaments are staggered, suggesting the possibility of some actin-myosin con-

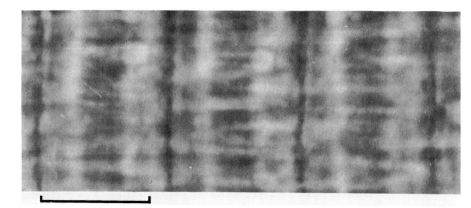

FIG. 5. Phase-contrast micrograph of a *Limulus* telson muscle bundle that had been stimulated to shorten isotonically in high K$^+$ (400 mM) artificial sea water and then placed in relaxing solution for several hours (scale = 10 μm). Sarcomeres are long (≈12 μm) and A-bands are short (≈3.5 μm).

nections at one end of the thick filament. In stimulated muscles, the thick filaments shorten and could either remain in a staggered configuration or realign in the center of the sarcomere. The ability of these muscles to generate large amounts of external tension at sarcomere lengths above 10 μm has not been explained. Even though thick filament staggering allows for some actin-myosin overlap in a half sarcomere, there must still be some additional connection to transmit force from Z-line to Z-line in these unusual muscles. As has been suggested (9), these connections could be between the thick filaments and the Z-lines, or between the thick filaments themselves either directly or through some Z-line-linked actin filaments that are longer than 2.8 μm, or through free actin filaments in the H zone.

CONCLUSIONS

In summary, we have found a large chemical energy usage not associated with external force (Figs. 1 and 2) or work production (Fig. 3) in living *Limulus* telson muscles contracting at sarcomere lengths below 7 μm when thick filaments shorten. This large expenditure of energy is probably not a result of the initiation of thick filament shortening, nor is it likely to be due entirely to maintaining thick filaments in the shortened state, since we have not observed a high rate of energy usage during isometric contractions of telson muscles stretched to a length of 2.0 L$_0$, where thick filament shortening should also occur (Table 2). It is possible that most of this energy usage is due to splitting of ATP by cross-bridges unable to generate external force either as a result of an internal force opposing shortening, or of structural changes that might occur in the shortened thick filament. Measurements of V_{max} and instantaneous stiffness at short muscle lengths would provide us with the necessary information to decide among these hypotheses, and these experiments are planned for the near future.

ACKNOWLEDGMENTS

This work was supported by NIH grant HL 15835 to the Pennsylvania Muscle Institute.

REFERENCES

1. Brann, L., Dewey, M. M., Baldwin, E. A., Brink, P., and Walcott, B. (1979): Requirements for *in vitro* shortening and lengthening of isolated thick filaments of *Limulus* striated muscle. *Nature*, 279:256–257.
2. Davidheiser, S., and Davies, R. E. (1982): Energy utilization by *Limulus* telson muscle at different sarcomere and A-band lengths. *Am. J. Physiol.*, 242:394–400.
3. de Villafranca, G. W. (1961): The A and I band lengths in stretched or contracted horseshoe crab skeletal muscle. *J. Ultrastruct. Res.*, 5:109–115.
4. Dewey, M. M., Levine, R. J. C., and Colflesh, D. (1973): Structure of *Limulus* striated muscle. The contractile apparatus at various sarcomere lengths. *J. Cell Biol.*, 58:574–593.
5. Dewey, M. M., Walcott, B., Colflesh, D., Terry, J., and Levine, R. J. C. (1977): Changes in thick filament length in *Limulus* striated muscle. *J. Cell Biol.*, 75:366–380.
6. Infante, A. A., Klaupiks, D., and Davies, R. E. (1964): Length, tension and metabolism during short isometric contractions of frog sartorius muscles. *Biochim. Biophys. Acta*, 88:215–217.
7. Kushmerick, M. J., and Davies, R. E. (1969): The chemical energetics of muscle contraction II. The chemistry, efficiency and power of maximally working sartorius muscles. *Proc. R. Soc. Lond. [B]*, 174:315–353.
8. Simpson, J. W., and Awapara, J. (1966): The pathway of glucose degradation in some invertebrates. *Comp. Biochem. Physiol.*, 18:537–548.
9. Walcott, B., and Dewey, M. M. (1980): Length-tension relation in *Limulus* striated muscle. *J. Cell Biol.*, 87:204–208.

Basic Biology of Muscles: A Comparative Approach, edited by B. M. Twarog, R. J. C. Levine, and M. M. Dewey. Raven Press, New York © 1982.

Is the Steric Model of Tropomyosin Action Valid?

C. E. Trueblood, T. P. Walsh, and A. Weber

Department of Biochemistry and Biophysics, University of Pennsylvania, Philadelphia, Pennsylvania 19104

The concept of calcium regulation of muscle activity became popular in the early fifties when earlier experiments with calcium-injected muscle fibers were complemented by experiments showing the effect of calcium on muscle extracts and on what are now called skinned muscle fibers. It took nearly another 10 years until it was accepted that calcium acts by binding to the myofibrils. After that, progress was rapid. Ebashi et al. (7) discovered troponin, the protein responsible for the switch from relaxation to contraction. He showed that actomyosin, which in the pure state always contracts with adenosine triphosphate (ATP), acquires the ability to be relaxed in the presence of ATP when the regulatory proteins troponin and tropomyosin are bound to the actin filament. Calcium binding to troponin then initiates contraction. The existence of a second mechanism of calcium control was discovered in Szent-Gyorgyi's laboratory (20): in molluscs, calcium causes contraction by binding to myosin.

By 1972, at the time of the Cold Spring Harbor Symposium on Muscle, the following picture of regulation had been developed. One troponin-tropomyosin complex regulates seven actin molecules. Of the two components of the troponin-tropomyosin complex, tropomyosin is in direct contact with each of the seven actin molecules. The heads and tails of the 42 nm long tropomyosin molecules overlap, forming one long strand.

Each strand of the actin filament is associated with one such tropomyosin strand. Troponin has three subunits, TN-I, TN-C, and TN-T, which carry out separate functions. TN-I is responsible for the inhibition of the actin-myosin interaction during relaxation; TN-C, when it becomes calcium saturated, turns off the inhibitory action of TN-I; TN-T attaches TN-I and TN-C to tropomyosin.

When relaxed muscle contracts, tropomyosin changes its position on the actin filament according to X-ray diffraction data (10,16). Such interpretation of the data was confirmed in subsequent years by diffraction studies of electron micrographs of actin paracrystals containing regulatory proteins (8,47). Contraction is always associated with a change in tropomyosin position whether contraction is initiated by calcium or by myosin forming rigor complexes (nucleotide-free actomyosin) with actin in the absence of calcium (10).

On the basis of these X-ray diffraction data, Huxley (16) and Haselgrove (10) and, independently, Parry and Squire (34) proposed what has become known as the steric model of regulation. Modeling of X-ray diffraction data suggested to them that, in relaxed muscle, tropomyosin binds to the actin filament at the periphery of the filament, away from the groove between the two strands of the actin filament. According to the three-dimensional reconstructions of Moore et al. (26), this is also the position of the myosin binding sites. Thus, tropomyosin would block myosin binding to actin in the absence of calcium (Fig. 1). It was shown in subsequent years that tropomyosin is held in the peripheral position by TN-I, which binds to actin in the absence of calcium (37). Calcium ions induce a conformational change in TN-C, causing it to bind more strongly to TN-I (in the presence of calcium, the complex TN-C-TN-I persists even in 6 M urea (35) and at the same time causing TN-I to dissociate from actin. Tropomyosin then moves closer to the groove, permitting myosin attachment to the actin filament (Fig. 1).

Although occasionally challenged, the steric model of regulation has been widely accepted since 1972. In this chapter, we shall critically examine whether 10 years later these ideas can still be the structural basis for the regulatory and cooperative phenomena that are properties of the regulated actin filament.

THE STRUCTURE OF THE REGULATED ACTIN FILAMENT

Direct steric blocking of the myosin binding site by tropomyosin requires that tropomyosin and myosin occupy the same side of each actin strand. It was not known until recently whether that actually is the case because it was not known on which side tropomyosin is positioned. An electron micrograph of an isolated regulated actin filament (that is, an actin filament containing tropomyosin and troponin) gives no clue of its polarity, as distinct from an S-1 decorated actin filament where the arrowhead pattern of the bound myosin marks the polarity. About two years

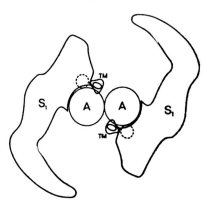

FIG. 1. Cross-section through a decorated actin filament, as visualized by Huxley (16), showing tropomyosin in the blocking position *(broken line)* in the absence of calcium and in the position near the groove in its presence.

ago, Seymour and O'Brien (39) solved that problem with diffraction patterns from electron micrographs of thin filaments still attached to the Z-band. Their data suggest that tropomyosin is on the side opposite to where the three-dimensional reconstructions of Moore et al. (26) had placed myosin.

The definition of the tropomyosin position was soon followed by a readjustment of the location of myosin. Because of the advances in computing and low-dose electron microscopy, a new study of the myosin decorated actin filament had been initiated by Taylor and Amos (42) in Huxley's laboratory. Although the density maps obtained by low-dose electron microscopy were similar to those by Moore and colleagues, the assignments given to the various parts of the density map were drastically altered. The density previously attributed to actin has now been assigned to myosin. As a result of this new interpretation, which is now accepted by most other laboratories, myosin occupies a position on the actin strand opposite to the original assignment by Moore and colleagues and on the same side as tropomyosin. Thus, the basic requirement for the steric model is fulfilled.

NEW INFORMATION ON MYOSIN BINDING SITES

Since the steric model was first formulated, new information has been obtained about the myosin binding sites on actin. According to the latest findings from Kassab's laboratory (27), each myosin molecule, when it forms a rigor complex with actin, is bound to two actin molecules.

If myosin binds to two actin molecules of the same strand, the steric model of regulation does not require any modification. That is not necessarily so if myosin binds to both actin strands. In the first case, the fact remains that myosin is bound only to the periphery of the actin filament. In the second case, the second myosin binding site may be near the groove of the actin filament, raising the possibility that tropomyosin, after it has moved to the groove, still interferes with myosin binding, that is, myosin binding to a site near the groove. It will be very interesting to learn where that second myosin binding site is located and whether nucleotide-containing myosin heads also bind to more than one actin molecule.

A few years ago we raised the question of whether calcium-free troponin-tropomyosin equally interferes with the actin binding of nucleotide-free and nucleotide-containing myosin (32,33). We compared the binding of nucleotide-free S-1 (myosin heads) and S-1 containing the ATP analogue AMPPNP (5'adenylyl imidodiphosphate). We found that there was not just a single binding site for S-1 on actin that was blocked by tropomyosin but that instead tropomyosin blocks binding of the nucleotide-free S-1 more strongly than binding of S-1 containing AMPPNP. The rather slight inhibition of S-1-AMPPNP binding was intriguing because S-1-AMPPNP attached to actin has often been considered to resemble the state of the myosin bridge in muscle immediately after attachment to actin, an attachment that was thought to be prevented by calcium-free troponin-tropomyosin. Instead, the formation of an intermediate occurring presumably later in the sequence of attached myosin states is inhibited more strongly. Next, Chalovich et al. (4), in Eisenberg's laboratory, and Wagner and Giniger (46) measured directly the actin binding of S-1-ADP·P and found that, contrary to generally held opinion, attachment of S-1-ADP·P to actin is not inhibited at all by calcium-free troponin-tropomyosin. Again, formation of a later intermediate of the attached states is strongly inhibited instead. The later intermediate studied by Greene and Eisenberg (9) was acto-S-1-ADP, rather than nucleotide-free acto-S-1. In the absence of calcium, the apparent actin binding constant of S-1-ADP is so low that it cannot be measured accurately. The equilibrium constant (L = 600) governing the distribution of tropomyosin segments between the periphery and the groove was measured directly by Trybus and Taylor (43) who used fluorescently labeled TN-I, which changes emission, presumably in response to dissociation of troponin from actin.

It appears, then, that actin does not possess only one single binding site for myosin which can be blocked by tropomyosin. This removal of calcium did not affect the actin binding constant of S-1-ADP·P (4) while lowering that of S-1-AMPPNP twofold (32) and reducing the actin binding constants of S-1 and S-1-ADP by about 10 (32) and more than 100-fold (43) respectively. Tropomyosin is most effective in preventing the formation of the myosin attached state containing ADP after dissociation of the gamma phosphate of ATP; and tropomyosin is completely ineffective in blocking the attachment of S-1-ADP·P and, possibly, S-1-ATP (Fig. 2). Thus, in terms of the steric model, there is no overlap at all between the tropomyosin domain and the first attachment site of S-1-ADP·P. During contraction, one visualizes myosin to move from its first attachment site on actin to other sites on actin, so that progression through the series of intermediate states would be associated with progression through different myosin binding sites on actin. Overlap between the tropomyosin domain and the myosin binding site would be maximal for the site associated with the intermediate state following dissociation of the gamma phosphate of ATP. Since the equilibrium constant for this complex is not known, one cannot decide whether the overlap would be partial or total. For

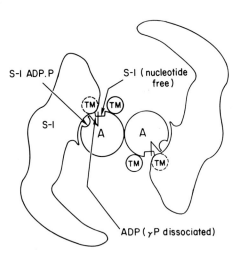

FIG. 2. Similar cross-section as in Fig. 1 with 3 binding sites added for myosin in different intermediate states: S-1-ADP *(triangles)*, nucleotide-free S-1 *(rectangle slightly overlapping with triangle)*, and S-1-ADP·P *(half circle)*. In the absence of calcium, tropomyosin *(broken line)* blocks the binding site for S-1-ADP completely, the binding site for nucleotide-free S-1 to about 15%, and the binding site for S-1-ADP·P not at all.

practical purposes, it is only necessary that the equilibrium constant for this intermediate state would be low enough to prevent its occurrence in the absence of calcium. Thus, relaxation is apparently due to the inability of the first attached state of myosin to proceed to those intermediate states that are responsible for tension development. A block at any point in the cycle will inhibit shortening or actin activated ATP turnover. The stiffness of muscle fibers in the relaxed state is very low. That is compatible with myosin attachment to actin, according to Schonberg's (38) calculations if the rate of bridge association and dissociation is very fast. Bridge attachment during relaxation would explain some resting tension phenomena that have puzzled physiologists for a long time (11,21). The attachment of bridges during relaxation would have to be different from the attachment during contraction, which precedes tension development, as postulated by Huxley and Simmons (14,15), since the latter attached state is characterized by high stiffness. It does not necessarily follow from the *in vitro* experiments, however, that the number of attached bridges (albeit noncycling) is the same during relaxation and contraction (i.e., unloaded shortening) since the *in vitro* experiments were carried out in a soluble S-1-actin system. During the *in vitro* experiment, the frequency of collision between the active sites of S-1 and actin was the same in the absence and presence of calcium (since the concentrations of S-1 and actin were the same), whereas we have no information about the frequency of collision in the fiber under either condition, nor whether it is likely to be the same during relaxation and contraction. The rate of effective collisions could be lower in the absence of calcium either if calcium binding to myosin should affect the attitude of myosin heads, or if bridge cycling during contraction should have a positive cooperative effect on the attitude of neighboring bridges.

After ADP has dissociated, blocking of myosin attachment by tropomyosin is considerably reduced again: although calcium removal causes about a 10-fold reduction of the equilibrium constant, that reduced value is still about 10^{-6} M, that is, steric interference by tropomyosin resulted in no more than a 15% decrease in the actin affinity for myosin (32). More information on myosin binding is needed to learn whether that decrement is so low because, after ADP release, myosin binds to a second actin molecule at a site near the groove outside the tropomyosin domain.

LOW LEVEL ACTIN-MYOSIN INTERACTION IN THE PRESENCE OF CALCIUM

The addition of calcium causes a 10- to 20-fold increase in the ATPase activity of regulated actomyosin, presumably associated with a movement of tropomyosin away from the periphery of the actin filament. Nevertheless, as long as the saturation of the regulated actin filament with myosin is very low, 1% or less, the rate of ATP hydrolysis (per actomyosin complex) remains about 50% lower than that of tropomyosin-free actomyosin (1,28). When the saturation with myosin increases, the ATPase activity goes up, eventually to values that are significantly higher than those of tropomyosin-free actomyosin (1,29). This modulation of ATPase activity from rates lower to rates higher than those of pure actomyosin is due to the presence of tropomyosin alone (6,31) and has been observed in the absence and presence of troponin (1,29,31).

Troponin affects the tropomyosin dependent modulation of ATPase activity least when regulated actin filaments are reconstituted with troponin-tropomyosin added to actin as the complex isolated from muscle (Fig. 3C). In that case, one might assume that troponin passively follows tropomyosin to which it is bound through TN-T. For reasons we do not yet understand, the addition of column-purified troponin to tropomyosin-actin usually raises the depressed ATPase activity to the same level or above that characteristic for the nonpotentiated state or for pure actin

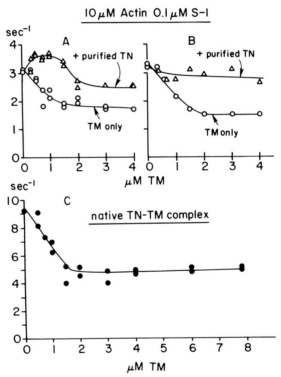

FIG. 3.   The effect of troponin on the nonpotentiated ATPase activity of tropomyosin-acto-S-1. Three variations of the effect of troponin, using three different troponin preparations. **A:** Reconstituted troponin-tropomyosin (from column-purified components), as distinct from tropomyosin alone, raised ATPase activity of acto-S-1 above that with pure actin when troponin-tropomyosin were present in concentrations up to saturation of actin (1.5 μM), but at higher concentrations, the complex inhibited ATPase activity, although not as much as did tropomyosin alone. **B:** Column-purified troponin raised ATPase activity to the same level as that of pure actin but not above; the effect was independent of the troponin-tropomyosin/actin ratio. **C:** Inhibition of ATPase activity by the native troponin-tropomyosin complex (as distinct from the reconstituted complex), isolated as such. Note that inhibition of ATPase activity is comparable to that with tropomyosin alone. Actin and S-1 as indicated on Fig., 10 mM imidazole, pH 7.0, 1 mM Mg in excess over 2 mM MgATP, 0.1 mM CaCl₂.

even when the S-1 concentration is as low as 0.1 μM with the ratio of S-1/actin equaling 1/100 (Figs. 3A and B).

According to the literature, the nonpotentiated state, that is, depressed ATPase in the presence of calcium [due to an increase in the K_m for actin (28)], can only be caused by rabbit tropomyosin and not by tropomyosin isolated from smooth muscle. We have found, on occasion, a rabbit tropomyosin preparation that did not depress ATPase, and in collaboration with S. Chacko, we have observed some depression of ATPase by fresh tropomyosin from the smooth muscle, which disappeared on aging of the tropomyosin.

ATPase rates lower than those of pure actomyosin may be explained by partial blocking of myosin binding sites that persists even in the presence of calcium. If that were true, the inhibition should disappear when the actin concentration becomes infinite, that is, the inhibition should be due to an increase in the K_m for actin while V_{max} should be the same as that for pure acto-S-1. That was indeed observed by Murray et al. (28). They explained the partial blocking on the basis of two assumptions. First, they suggested that the preference of tropomyosin for actin over the medium is rather unspecific, that is, that tropomyosin does not have a specific binding site on actin but binds about equally well to all parts of the actin strand between the periphery and the groove. They based that assumption on Wegner's (49,50) calculated low value for the affinity of tropomyosin for actin, confirmed by recent more direct experiments (see below). Second, they assumed that tropomyosin, which is not an inflexible rod (36), may meander from the peripheral blocking sites to the groove and back in bends as broad as necessary to minimize the resistance against bending (see below) (Fig. 4). In that case, a fraction of the same myosin binding sites that had been completely blocked during relaxation would remain blocked in the presence of calcium, provided that S-1, because of its low concentration, were unable to successfully compete with tropomyosin for these sites. The results of the binding studies by Greene and Eisenberg (9) are compatible with the suggestion of partial blocking of binding sites: when the S-1 concentration was as low as during the ATPase measurements, the binding of S-1-ADP to actin was inhibited. Under the assumption that the actin molecules that can associate with S-1-ADP have a high binding constant, with a value twice that for binding to pure actin, Green and Eisenberg calculated a value of five for the ratio of inhibited (blocked in terms of the steric model) and free sites (12). This ratio is higher than our value of about two. However, if the assumption of a high binding constant for the free actin sites should be incorrect, as suggested by the additional parameter of an unchanged V_{max} in our experiment, that equilibrium constant in favor of the blocked sites would become smaller and closer to our value.

Murray et al. (32) did not observe any inhibition of the actin binding of nucleotide-free S-1 in the presence of calcium. Considering that during relaxation binding of nucleotide-free S-1 was less well correlated with ATPase inhibition than binding of S-1-ADP, and taking into account that these authors used higher S-1 concentrations during their binding experiments than during their ATPase experiments, these

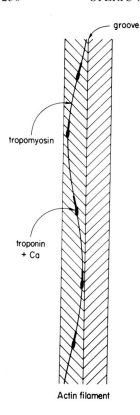

groove

tropomyosin

troponin
+ Ca

Actin filament

FIG. 4. Diagram of tropomyosin meandering between periphery and groove of the actin filament when troponin is calcium saturated and little or no S-1 is attached to the actin filament. Separation between individual actin molecules is indicated by the *parallel lines*, a notation first used by Wegner (50).

binding studies are not incompatible with the assumption of partial blocking of S-1 binding sites.

If the steric model is extended to explain the depression of ATPase in the presence of calcium in the manner just suggested (that is, blocking of peripheral sites), the tropomyosin strand cannot be positioned in the center of the groove when no or very little S-1 is attached to actin. The overall tropomyosin position as calculated from diffraction data by X-ray and of electron micrographs then would be a position averaged over the changing positions of the tropomyosin segments. The recent finding in Kassab's laboratory (27) suggests another possibility, that is, blocking of sites near the groove rather than peripheral sites if, during ATP turnover, myosin heads should interact simultaneously with both actin strands. In that case, partial blocking of myosin binding sites might be associated with a tropomyosin position near or at the center of the groove.

HIGH LEVEL ACTIN-MYOSIN INTERACTION—POTENTIATION

At high concentrations of S-1, ATPase activity (1,29) and the equilibrium constants for the formation of both the ADP containing and the nucleotide-free attached states of S-1 (after phosphate release) are increased well above the values found in the absence of tropomyosin (9,10). Furthermore, the apparent binding constant of tropomyosin for actin has been shown to be raised by a high extent of saturation with nucleotide-free S-1 (5). The transition to this potentiated state of the actin filament can take place in the presence and absence of calcium but requires a higher concentration of S-1 in its absence (9). Both ATPase activity and the apparent actin binding constant of S-1-ADP are raised, in a cooperative manner, by a factor of two or three (12,29). Obviously, that cannot be explained by a steric model. It is not known whether this change in actin cofactor activity is also associated with a change in tropomyosin position, for example, from a possible dispersed distribution of segments between periphery and groove (just described) entirely to the center of the groove.

When Haselgrove (10) compared the change in actin layer lines between electrically stimulated muscle stretched beyond overlap (no myosin attachment) and muscle going into rigor at normal length, he observed no difference. However, it is not certain whether the measurement is sensitive enough to pick up a small difference in tropomyosin position even if it exists.

It is not yet known whether all attached S-1 states can cause potentiation. So far it has been shown that the potentiated state measured by some or all of the above mentioned criteria can be induced by nucleotide-free S-1 (all criteria), S-1-ADP ("high" binding), and S-1-AMPPNP (potentiated ATPase activity) (31,33). Experiments with S-1-AMPPNP so far have only been carried out in the presence of calcium, whereas nucleotide-free S-1 (2) and S-1-ADP have been shown to cause potentiation both in the absence and presence of calcium. Activation of contraction of relaxed muscle at very low ATP concentrations is presumably caused by the potentiating effect of nucleotide-free myosin heads. We did not observe, within the scatter of our data, increased actin binding of S-1-AMPPNP during the potentiated state (as compared with binding to pure actin or to calcium-activated, nonpotentiated actin). That is thermodynamically possible if one assumes that S-1-AMPPNP binding to nonpotentiated actin has negative cooperativity, that is, that the binding constant would decrease with increasing saturation with S-1-AMPPNP if it could be arranged that the actin filament remained in the nonpotentiated state instead of escaping into the potentiated state. Postulating such negative cooperativity for the nonpotentiated state is not necessarily in contradiction with our description above, that is, partial blocking by tropomyosin of actin molecules that otherwise behave like actin molecules in a pure actin filament. Preliminary data are suggestive of a negative cooperativity of S-1 binding to pure actin filaments (29).

SUMMARY OF THE THREE STATES OF THE REGULATED ACTIN FILAMENT

We postulate that the regulated actin filament exists in three states and that the first two can be explained by steric hindrance. In the first state, in the absence of calcium, the formation of one of the major bridge attachments (associated with high stiffness) is—for all practical purposes—completely blocked by tropomyosin.

In the second state, in the presence of calcium and very low S-1 concentrations, a fraction of the actin sites involved in ATP turnover are blocked, whereas the cofactor activity of the unblocked actin molecules is the same as that of pure actin, free of tropomyosin. The blocked sites could be the same sites that are completely blocked during relaxation but they need not be. They could even be sites on the other actin strand if several of the myosin attached states include binding to the second strand. (We refer to this state as nonpotentiated.) In the third state, in the presence of very high S-1 concentrations, the cofactor activity of tropomyosin-actin and the affinity of tropomyosin-actin for myosin free of nucleotide and for myosin containing bound ADP is greatly increased over the affinity of pure actin for myosin. This state, called "potentiated" by us and "high" by Greene and Eisenberg (9), apparently involves conformational changes in one or more proteins.

We differ from Greene and Eisenberg (9) and Trybus and Taylor (43) by postulating the nonpotentiated state as a separate distinct state. They describe their binding curves of S-1-ADP to actin by a cooperative transition from the resting state (in the absence of calcium), or a mixture of the resting and potentiated state (in the presence of calcium and very low S-1 concentrations) to the "potentiated" or "high" state. Our contention that there must be a distinct nonpotentiated state (on-state in earlier publications (49)) is based on the additional parameter provided by a V_{max} measurement.

TRANSITION BETWEEN STATES

The transitions from the relaxed to the calcium-activated state of the actin filament with very low S-1 attachment and from that to the potentiated state are cooperative. The cooperativity of the calcium activation cannot be explained by simultaneous binding of two or even four calcium ions to each troponin C (30). They require additional cooperative interactions, for example, interactions between neighboring tropomyosin molecules. For a conformational rather than a steric mode of regulation, such interactions could be transmitted through conformational changes of neighboring actin molecules. It would not be necessary for the tropomyosin molecules to be bound to each other, forming a continuous strand. By contrast, in the framework of the steric model without additional conformational changes, cooperativity due to nearest neighbor tropomyosin interactions must be based on the fact that all tropomyosin molecules are interconnected, so that single tropomyosin molecules cannot individually change positions on the actin filament. That means the cooperativity should disappear on disruption of the tropomyosin strand, for example, after removal of the overlapping ends of tropomyosin by carboxypeptidase A treatment (19,23). Before describing the results of such experiments, we should like to discuss the following points: first, the evidence that actin-bound tropomyosin forms a cohesive strand; second, the existing data on actin binding of tropomyosin; and third, the ability of the tropomyosin strand to bend, that is, to what extent segments of the tropomyosin strand can be displaced from the periphery to the groove and vice versa.

EVIDENCE FOR A COHESIVE TROPOMYOSIN STRAND

Tropomyosin is a coiled coil, 42 nm long (13,25) as compared to the 38.5 nm of the one-half helical repeat formed by seven actin molecules, with carboxy- and amino ends that can bind to each other (18,40). As a result of these intermolecular associations, tropomyosin solutions are of high viscosity at a salt concentration below 0.1 M. With tropomyosin free in solution, these interactions disappear and the viscosity reaches a minimum at salt concentrations above 0.3 M (44), sggesting the importance of electrostatic interactions; the carboxy terminal end contains two aspartates and the amino terminal end the unusual sequence of three consecutive lysines (18). The following evidence suggests that these head-tail interactions also take place when tropomyosin is bound to the actin filament. Without overlap, the sum of the lengths of the tropomyosin molecules bound to actin would be longer than the actin filament. Binding between the overlapping ends of tropomyosin molecules is suggested by the high degree of positive cooperativity of tropomyosin binding to actin, attributed by Wegner (50) to the binding sites between overlapping ends that become increasingly available with increasing tropomyosin saturation of the actin filament. Wegner's conclusions are confirmed by our recent binding measurements, which show a loss of cooperativity and a reduction in the apparent binding constant after the overlapping carboxylends of tropomyosin have been removed by digestion with carboxypeptidase A. Carboxypeptidase A rapidly removes all eleven amino acids up to the two successive glutamic acid residues, provided prior to digestion all phosphate has been removed from the penultimate serine (23). About 40 to 60% of freshly extracted α-tropomyosin from rabbit heart, the preparation used by us, usually is phosphorylated.

A strand of fixed length fits flatly on the actin helix, that is, binds to identical sites on each actin molecule, only at a single distance from the groove (34). Tropomyosin cannot move further away from the groove unless it becomes longer either by stretching or by readjusting the overlap; when tropomyosin moves into the groove, it either must become shorter or it must bend, so that it would no longer bind to identical sites on each actin molecule.

TROPOMYOSIN-ACTIN BINDING

Information about the interaction between tropomyosin and actin is quite limited. The actin binding constant of an isolated native tropomyosin molecule (subtracting the contribution from the binding of tropomyosin molecules to each other) has been calculated by Wegner (50) to be between 10^4 to 10^5 M^{-1} at 37°C and was recently measured by us for carboxypeptidase A treated tropomyosin to be 0.5 to $2 \cdot 10^5$ M^{-1} at 25°C. However, it is not known whether all tropomyosin segments have a similar affinity for actin, which would translate into an equilibrium constant of 5 to 6 M^{-1} for each complex between one actin and a tropomyosin segment (33), or whether, at the other extreme, there is one single specific actin binding site on each tropomyosin molecule. In either case, actin binding of some or all tropomyosin segments is so weak that segments of tropomyosin should be dissociated from their binding sites for a significant fraction of the time. Dissociation from actin of segments of the tropomyosin strand would be limited, however, by resistance to the bending necessary to displace part of the strand from its neighboring segments.

Diffraction patterns of electronmicrographs of paracrystals from reconstituted tropomyosin-actin filaments have been interpreted to indicate that tropomyosin is placed near the groove (47). That would suggest that tropomyosin from vertebrate skeletal muscle has a greater affinity for sites near the groove than for peripheral sites of actin filaments from vertebrate skeletal muscle. We do not know whether tropomyosin may also have one or more binding sites for the periphery of the actin filament, which would hold the segment between TN-I molecules in a blocking position during the relaxed state of the actin filament. The existence of such binding sites is suggested for mollusc tropomyosin by the similarity of the X-ray diffraction pattern of relaxed mollusc muscle (45) to that of relaxed vertebrate skeletal muscle, although the myosin regulated mollusc muscles do not have (or have only very little) troponin (22). If the interpretation of the various diffraction patterns is correct, mollusc tropomyosin prefers the peripheral sites until displaced by calcium-activated myosin, whereas vertebrate skeletal tropomyosin prefers sites nearer the groove.

Troponin, in the absence of calcium, is responsible for the peripheral position of tropomyosin because TN-I binds to actin. It is not known whether, in addition, calcium-free troponin alters tropomyosin binding to actin, for example, through an effect of TN-T on tropomyosin conformation. TN-T is associated with one-third of the length of each tropomyosin molecule (24). However, it has been shown that calcium saturated troponin can increase actin binding of native tropomyosin (50) and, according to our recent findings, of carboxypeptidase-treated tropomyosin, raising the apparent actin binding constant of carboxypeptidase-treated tropomyosin by a factor of about 10 to 20. Nevertheless, calcium-saturated troponin may not always have this effect for the following reason. The increase in tropomyosin binding was caused by column-purified troponin. As mentioned above, calcium-saturated column-purified troponin raises the cofactor activity of tropomyosin-actin under conditions of very low S-1 attachment to a value higher than that of pure actin, whereas the effect of the native calcium saturated troponin-tropomyosin complex on actin is similar to that of tropomyosin alone (Fig. 3). A comparison between actin binding of the native and the reconstituted troponin-tropomyosin complex has not yet been made.

BENDING OF THE TROPOMYOSIN STRAND BETWEEN PERIPHERY AND GROOVE

Wegner and Walsh (50) recently carried out a series of experiments designed to show to what extent there is resistance in the tropomyosin strand against displacement of individual segments from the periphery to the groove and vice versa, thereby bending the strand. They measured tropomyosin binding to actin under three conditions: When all tropomyosin molecules of the strand were bound to the periphery, held there by the complex TN-T-TN-I; when all tropomyosin molecules were bound to the groove, using the calcium-saturated troponin-tropomyosin complex; and when the binding sites for the tropomyosin strand were presumably mixed because some of the tropomyosin was complexed with calcium-saturated troponin and some with TN-T-TN-I (without TN-C, TN-I remains bound to actin in the presence of calcium). In the last case, they added different constant amounts of tropomyosin complexed with TN-T-TN-I to actin and titrated the remaining free actin with increasing amounts of tropomyosin complexed with calcium-saturated troponin. Modeling of the resulting data suggests that the two different troponin-tropomyosin complexes do not bind randomly to actin but that identical complexes tend to associate with each other in short stretches, that is, that on the average, only at every fourth molecule a tropomyosin containing calcium-saturated troponin is placed next to a tropomyosin containing TN-T-TN-I. Accepting that, despite the somewhat unphysiological conditions of the experiment, tropomyosin complexed to TN-T-TN-I was at the periphery and tropomyosin complexed to calcium-saturated troponin near the groove, the data suggest that the formation of bends in the tropomyosin strand requires energy. Consequently, the occurrence of bends is reduced. Short stretches of several tropomyosin molecules are bound to either groove or periphery rather than randomly distributed between the two different positions.

This resistance against partial displacement of the tropomyosin strand may be the basis for the cooperativity of the transitions from the relaxed to the nonpotentiated or potentiated state of the regulated actin filament. It cannot be stated whether the measured resistance against partial displacement quantitatively accounts for the observed cooperativity because the experiments by Wegner and Walsh were, for technical reasons, conducted under different conditions (0.3 M KCl and 38°C) than measurements of S-1 binding to actin (12) or of actin activation of ATPase activity (29,30).

For the same reason, it is not known whether the resistance against the formation of bends may rule out that meandering distribution of tropomyosin segments that we suggested as the basis for the depression of ATPase activity in the nonpotentiated state. Kinetic considerations make it desirable that individual tropomyosin segments can be displaced. Tropomyosin movement from the periphery to the groove and vice versa on addition or removal of calcium would be a slow process if it required the simultaneous dissociation from the actin filament of all binding sites of the whole tropomyosin strand. Allowing segments containing only a few binding sites to move independently from the rest of the tropomyosin strand would accelerate the rate of tropomyosin movement, commensurate with the frequency of fast contraction-relaxation cycles, such as the 15-msec cycle of the cricothyroideus muscle of the bat.

THE EFFECT OF DISRUPTING THE TROPOMYOSIN STRAND ON THE COOPERATIVITY OF THE TRANSITIONS FROM REST TO ACTIVITY

The transition from the relaxed to the potentiated state can be achieved in the absence of calcium by increasing the S-1 concentration to high levels (9). The transition is highly cooperative when S-1-ADP binding is measured as a function of increasing concentrations of S-1-ADP. The cooperativity has been analyzed by Hill et al. (12) in terms of nearest neighbor interactions: it costs energy when neighboring tropomyosin-actin segments are not in the same state, that is, when an inhibited and a potentiated segment are next to each other. Hill's analysis is based on a two-state model whereas our ATPase dates suggest the existence of free states. We would like to know to what extent these nearest neighbor interactions are associated with the displacement of tropomyosin from the periphery rather than with the conformational changes underlying the transition to the "high" or potentiated state. Nearest neighbor interactions associated with the latter are indicated by the cooperativity of S-1 binding to actin in the presence of calcium. This cooperativity is much weaker than in the absence of calcium and has been described quite well by a simple two-state model without nearest neighbor interactions (9). Thus, it appears that the nearest neighbor interactions in the absence of calcium are largely associated with the displacement of tropomyosin from the periphery to the groove.

That is in agreement with our observation that a high degree of cooperativity attends the transition from the relaxed to the nonpotentiated state when it is measured directly as the activation of ATPase activity by the addition of calcium under conditions of very low S-1 attachment (29). That cooperativity cannot be explained by calcium binding to troponin for the following reasons. ATPase activity is activated when both of the calcium-specific sites of TN-C are saturated with calcium (2), that is, the two low affinity sites out of a total of four calcium binding sites. Kinetic measurements have shown that the calcium-induced conformational change of TN-C is not impeded when the high affinity sites contain magnesium instead of calcium (17). Since calcium binding to the specific sites is not positively cooperative (3), the only cooperativity due to calcium binding is that associated with the requirement for simultaneous binding of two calcium ions per troponin. However, that does not explain an increase in ATPase activity from 10 to 90% of maximal over only a seven- or ninefold range of calcium concentrations (30). That degree of cooperativity must be caused by interactions between nearest neighbors (12,41). In the framework of the steric model without conformational changes, such interactions require binding of tropomyosin molecules to each other, that is, the formation of a tropomyosin strand, and disappearance of the interactions on disruption of the strand. When we tested the effect on cooperativity of disrupting the tropomyosin strand by reconstituting regulated actin with discontinuous tropomyosin, using carboxypeptidase A-treated tropomyosin and column-purified troponin, we observed no reduction in the cooperativity (Fig. 5). The ATPase activity increased from 10 to 90% of maximal over the same seven- to ninefold range of calcium concentrations as with a native continuous tropomyosin strand. This result differs from that of Tawada et al. (41), who measured the effect of tropomyosin digestion on superprecipitation, a semiquantitative measurement. Our result cannot be attributed to contamination with native tropomyosin because SDS gels and analysis of the solubilized amino acids indicated complete digestion. Furthermore, to ensure removal

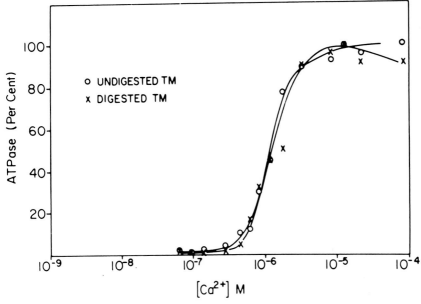

FIG. 5. The effect of carboxypeptidase A treatment of tropomyosin on the calcium dependence of the ATPase activity of regulated acto-S-1. Conditions as in Fig. 3, including 20 μM actin and 0.1 μM S-1.

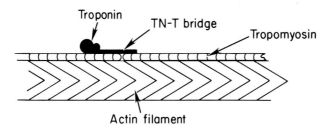

FIG. 6. Diagram showing how TN-T may join two nonoverlapping tropomyosin molecules.

of native tropomyosin below the limit of detection, we extracted the digested preparation with enough actin filaments to remove a potential 15% contamination of native tropomyosin. Viscosity measurements indicated that head-tail interactions in solutions had been abolished, and the fall in the apparent actin binding constant— and, more unambiguously, the loss of cooperativity of tropomyosin binding to actin—suggest that head-tail interaction of actin-bound tropomyosin had also been abolished. One possible explanation compatible with purely steric regulation would be the formation of a tropomyosin strand after the addition of troponin through the action of TN-T (Fig. 6). It is possible that TN-T binds tropomyosin molecules to each other since its terminal cyanogen bromide fragment is capable of binding to the amino and carboxy terminal ends of tropomyosin (23). So far evidence for or against this notion is still incomplete: the addition of troponin to digested tropomyosin in solution did not raise viscosity indicating that troponin did not cause polymerization of tropomyosin free in solution. That however does not exclude the possibility that troponin causes the polymerization of actin-bound carboxypeptidase A treated tropomyosin. The ensuing neighbor-neighbor interactions of the troponin-tropomyosin complex would then restore cooperativity to the actin binding of carboxypeptidase A treated tropomyosin and increase the apparent binding constant. Calcium-saturated troponin does increase the apparent binding constant of carboxypeptidase A treated tropomyosin about 10 to 20-fold. However, we do not yet know whether troponin restores the cooperativity of tropomyosin binding to actin that had been lost after removal of the terminal amino acids. The increase in the apparent binding constant, by itself, does not indicate tropomyosin polymerization since column-purified calcium-saturated troponin also increased the actin binding constant of the already-polymerized native tropomyosin strand (50). If troponin should not restore the cooperativity of tropomyosin binding to actin, one would have to conclude that troponin-tropomyosin is not reassembled into a continuous strand. In that case the lack of any effect of carboxypeptidase A treatment of tropomyosin on the calcium activation of acto-S-1 ATPase would mean that the transition from relaxation to activation is primarily conformational and not steric.

In that context one may be curious whether the carboxypeptidase A treatment has any effect on the transition from the non-potentiated to the potentiated state, a transition which must be primarily conformational. In two out of five experiments this transition was cooperative with native tropomyosin (Hill coefficient near two) and became non-cooperative after carboxypeptidase A treatment of the tropomyosin. In later experiments the cooperativity was very low or absent (in the range of S-1 concentrations studied) even before carboxypeptidase A treatment of the tropomyosin. That suggests that factors other than the presence of the carboxy terminal amino acids are important for either the degree of cooperativity of the transition or the range of S-1 concentrations where the transition occurs.

CONCLUSION

Most of the new evidence on tropomyosin and myosin binding to regulated actin is compatible with a steric model of regulation, that is, inhibition of ATP turnover and tension development because the tropomyosin strand blocks the formation of intermediate attached states of the myosin heads without any contribution from conformational changes of tropomyosin or actin. The only evidence that may lead to a modification of this view is the lack of change in the cooperativity during the activation of ATPase by calcium after disruption of the tropomyosin strand. Nearest neighbor interactions that persist when the tropomyosin molecules are not bound to each other must be between the underlying actin molecules. In other words, binding must be unfavorable between actin molecules associated with tropomyosin molecules near the groove and actin molecules with tropomyosin near the periphery (the tropomyosin molecules possibly also differing in configuration), implying different conformations of the two sets of actin molecules. Although that would not eliminate steric blocking, it would rule it out as the only regulatory phenomenon. Therefore, it will be interesting to obtain clear proof whether the tropomyosin strand is disrupted in regulated actin that has been reconstituted from carboxypeptidase A-treated tropomyosin and purified troponin.

REFERENCES

1. Bremel, R. D., Murray, J. M., and Weber, A. (1972): Protein—Protein interactions in the thin filament during relaxation and contraction. *Cold Spring Harbor Symp. Quant. Biol.*, 37:267–275.
2. Bremel, R. D., and Weber, A. (1971): Regulation of contraction and relaxation in the myofibril. In: *Contractility of Muscle Cells and Related Proteins*, edited by R. J. Podolsky, pp. 37–53. Prentice Hall, Englewood Cliffs.
3. Bremel, R. D., and Weber, A. (1972): Cooperation within actin filament in vertebrate skeletal muscle. *Nature (New Biol.)*, 238:97–101.
4. Chalovich, J. M., Chock, P. B., and Eisenberg, E. (1981): Inhibition of actomyosin ATPase activity without inhibition of myosin binding to actin. *J. Biol. Chem.*, 256:575–578.
5. Eaton, B. L. (1976): Tropomyosin binding to F-Actin induced by myosin heads. *Science*, 192:1337–1339.
6. Eaton, B. L., Kominz, D. R., and Eisenberg, E. (1975): Correlation between the inhibition of the acto-heavy meromyosin ATPase and the binding of tropomyosin to F-Actin: Effects of Mg^{2+}, KCl, troponin I, and troponin C. *Biochemistry*, 14:2718–2725.
7. Ebashi, S., Ebashi, F., and Kodama, A. (1967): Troponin as the Ca^{++}-receptive protein in the contractile system. *J. Biochem.*, 62:137–138.
8. Gillis, J. M., and O'Brien, E. J. (1975): The effect of calcium ions on the structure of reconstituted muscle thin filaments. *J. Mol. Biol.*, 99:445–459.
9. Greene, L. E., and Eisenberg, E. (1980): Cooperative binding of myosin subfragment-1 to the actin-troponin-tropomyosin complex. *Proc. Natl. Acad. Sci.*, 77:2616–2620.
10. Haselgrove, J. C. (1972): X-ray evidence for a confumational change in the actin containing filaments of vertebrate striated muscle. *Cold Spring Harbor Symp. Quant. Biol.*, 37:341–352.
11. Hill, D. K. (1968): Tension due to interaction between the sliding filaments in resting striated muscle. The effect of stimulation. *J. Physiol.*, 199:637–684.
12. Hill, T. L., Eisenberg, E., and Greene, L. (1981): Theoretical model for the cooperative equilibrium binding of myosin subfragment-1 to the actin-troponin-tropomyosin complex. *Proc. Natl. Acad. Sci.*, 77:3186–3190.
13. Hodges, R. S., Sodek, J., and Smillie, L. B. (1972): Tropomyosin: Amino acid sequence and coiled-coil structure. *Cold Spring Harbor Symp. Quant. Biol.*, 37:299–310.
14. Huxley, A. F. (1974): Review lecture: Muscular contraction. *J. Physiol.*, 243:1–42.
15. Huxley, A. F., and Simmons, R. M. (1972): Mechanical transients and the origin of muscle force. *Cold Spring Harbor Symp. Quant. Biol.*, 37:669–680.
16. Huxley, H. E. (1972): Structural changes in the actin and myosin containing filaments during contraction. *Cold Spring Harbor Symp. Quant. Biol.*, 37:361–376.
17. Johnson, J. D., Charlton, S. C., and Potter, J. D. (1979): A fluorescence stopped flow analysis of Ca^{2+} exchange with troponin C. *J. Biol. Chem.*, 254:3497–3502.
18. Johnson, P., and Smillie, L. B. (1975): Rabbit skeletal α-tropomyosin chains are in register. *Biochem. Biophys. Res. Commun.*, 64:1316–1322.
19. Johnson, P., and Smillie, L. B. (1977): Polymerizability of rabbit skeletal tropomyosin: Effects of enzymic and chemical modifications. *Biochemistry*, 16:2264–2269.
20. Kendrick-Jones, J., Lehman, W., and Szent-Gyorgyi, A. G. (1970): Regulation in molluscan muscles. *J. Mol. Biol.*, 54:313–326.
21. Lannergren, J., and North, J. (1973): The effect of bathing solution tonicity on resting tension in frog muscle fibers. *J. Gen. Physiol.*, 62:737–755.
22. Lehman, W., and Szent-Gyorgyi, A. G. (1975): Distribution of actin control and myosin control in the animal kingdom. *J. Gen. Physiol.*, 66:1–30.
23. Mak, A. S., and Smillie, L. B. (1981): Structural interpretation of the two-site binding of troponin on the muscle thin filament. *J. Mol. Biol.*, 149:541–550.

24. Mak, A. S., Smillie, L. B., and Barany, M. (1978): Specific phosphoralation at serine-283 of α-tropomyosin from frog skeletal and rabbit skeletal and cardiac muscle. *Proc. Natl. Acad. Sci.*, 75:3588–3592.

25. McLachlan, A. D., Stewart, M., and Smillie, L. B. (1975): Sequence repeats in α-tropomyosin. *J. Mol. Biol.*, 98:281–291.

26. Moore, P. B., Huxley, H. E., and De Rosier, D. J. (1970): Three-dimensional reconstruction of F-actin, thin filaments and decorated thin filaments. *J. Mol. Biol.*, 50:279–295.

27. Mornet, D., Bertrand, R., Pantel, P., Audemard, E., and Kassab, R. (1981): Structure of the actin-myosin interface. *Nature*, 292:301–306.

28. Murray, J. M., Knox, M. K., Trueblood, C. E., and Weber, A. (1980a): Do tropomyosin and myosin compete for actin sites in the presence of calcium? *FEBS. Lett.*, 114:169–173.

29. Murray, J. M., Knox, M. K., Trueblood, C. E., and Weber, A. (1982): The potentiated state of the tropomyosin-actin filament and nucleotide containing myosin subfragment-1. *Biochemistry (in press).*

30. Murray, J. M., and Weber, A. (1980): Cooperativity of the Calcium switch of regulated actomyosin system. *Mol. and Cell. Biochem.*, 35:11–15.

31. Murray, J. M., Weber, A., and Bremel, R. D. (1975): Could cooperativity in the actin-filament play a role in muscle contraction. In: *Calcium Transport in Contraction and Secretion*, edited by E. Carafoli, W. Clementi, W. Drabikowski, and A. Margreth, pp. 489–496. Elsevier/North Holland, Amsterdam, New York.

32. Murray, J. M., Weber, A., and Knox, M. K. (1981): Myosin subfragment-1 binding to relaxed actin filaments and steric model of relaxation. *Biochemistry*, 20:641–649.

33. Murray, J. M., Weber, A., and Wegner, A. (1980b): Tropomyosin and the various states of the regulated actin filament. In: *Regulatory Mechanisms of Muscle Contraction*, edited by S. Ebashi, M. Endo, and K. Maruyama, pp. 221–236. Springer-Verlag, Berlin.

34. Parry, D. A. D., and Squire, J. M. (1973): Structural role of tropomyosin in muscle regulation: Analysis of the X-ray diffraction patterns from relaxed and contracting muscles. *J. Mol. Biol.*, 75:33–55.

35. Perry, S. V., Cole, H. A., Head, J. F., and Wilson, F. J. (1972): Localization and mode of action of the inhibitory protein component of the troponin complex. *Cold Spring Harbor Symp. Quant. Biol.*, 37:251–262.

36. Phillips, G. N., Jr., Lattman, E. E., Cummins, P., Lee, K. Y., and Cohen, C. (1979): Crystal structure and molecular interactions of tropomyosin. *Nature*, 278:413–417.

37. Potter, J. D., and Gergely, J. (1974): Troponin, tropomyosin, and actin interactions in the Ca^{2+} regulation of muscle contraction. *Biochemistry*, 13:2697–2703.

38. Schonberg, M. (1981): The stretch response of resting and activated skeletal muscle. *Fed. Proc.*, 39:1729.

39. Seymour, J., and O'Brien, E. J. (1980): The position of tropomyosin in muscle thin filaments. *Nature*, 283:680–683.

40. Stewart, M. (1975): Tropomyosin: Evidence For No Stagger Between Chains. *FEBS. Lett.*, 53:5–7.

41. Tawada, Y., Ohara, H., Ooi, T., and Tawada, K. (1975): Nonpolymerizable tropomyosin and control of the superprecipitation of actomyosin. *J. Biochem.*, 78:65–72.

42. Taylor, K. A., and Amos, L. A., (1981): A new model for the geometry of the binding of myosin crossbridges to muscle thin filaments. *J. Mol. Biol.*, 147:297–324.

43. Trybus, K. M., and Taylor, E. W. (1980): Kinetic studies of the cooperative binding of subfragment 1 to regulated actin. *Proc. Natl. Acad. Sci.*, 77:7209–7217.

44. Vibert, P. J., Haselgrove, J. C., Lowy, J., and Poulsen, F. R. (1972): Structural changes in actin containing filaments of muscle. *Nature (New Biol.)*, 236:182–183.

45. Wagner, P. D., and Giniger, E. (1981): Binding of Myosin Subfragment One to F-Actin and Regulated Actin in the Presence of ATP. *Biophys. J.*, 33:232a.

46. Wakabayashi, T., Huxley, H. E., Amos, L. A., and Klug, A. (1975): Three-dimensional image reconstruction of actin-tropomyosin complex and actin-tropomyosin-troponin T-troponin I complex. *J. Mol. Biol.*, 93:477–497.

47. Walsh, T. P., and Wegner, A. (1980): Molecular control mechanisms in muscle contractions. *Acta Biochim. Biophys.*, 626:79–87.

48. Weber, A., and Murray, J. M. (1973): Effect of the state of oxidation of cysteine 190 of tropomyosin on the assembly of the actin-tropomyosin complex. *Physiol. Rev.*, 53:612–673.

49. Wegner, A. (1979): Equilibrium of the Actin-tropomyosin interaction. *J. Mol. Biol.*, 131:839–853.
50. Wegner, A., and Walsh, T. P. (1981): Interaction of Tropomyosin-troponin with Actin Filaments. *Biochemistry*, 20:5633–5642.

Basic Biology of Muscles: A Comparative
Approach, edited by B. M. Twarog,
R. J. C. Levine, and M. M. Dewey.
Raven Press, New York © 1982.

Comparative Aspects of the Regulation of Contraction in Vertebrate Muscle

*Samuel V. Perry, *H. A. Cole, *R. J. A. Grand, and
**B. A. Levine

*Department of Biochemistry, University of Birmingham, Birmingham B15 2TT United
Kingdom; and **Inorganic Chemistry Laboratory, University of Oxford,
Oxford OX1 2QR United Kingdom

The increase in the intracellular Ca^{2+} concentration when muscle is stimulated initiates changes in the myosin and the actin filaments that lead to contraction. Both pathways, which are summarized schematically in Fig. 1, involve Ca^{2+} binding and phosphorylation of protein components of the systems. It is now becoming clear that excitation-contraction coupling in all muscle types depends on the latter two processes to varying extents. In this chapter we wish to examine the features of these processes that are common to the regulation of contraction in all vertebrate muscle systems and those that have evolved to accommodate the special features of excitation-contraction coupling in different muscle types.

A FILAMENT REGULATION

The Role of Myosin Light Chain Kinase in the Regulation of the Contraction of Striated Muscle

The capacity to phosphorylate the P light chain of myosin is greatest in fast skeletal muscle, and, presumably, the process therefore has special significance for

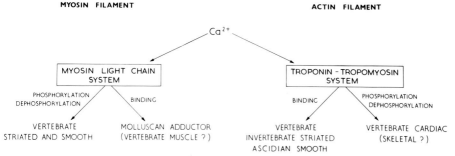

FIG. 1. Scheme illustrating the involvement of Ca^{2+} binding and phosphorylation in the regulation of contractile activity in muscle.

the function of this muscle. From the study of the properties of the phosphorylation-dephosphorylation system in rabbit white skeletal muscle, some general conclusions can be made as to what its role may or may not be. Although there is a brief report in the literature that the enzyme in rabbit skeletal muscle may exist in a higher molecular weight form (47), this has not as yet been confirmed. The best characterized preparation consists of a monomer of molecular weight about 77,000 (37). Most of the enzyme is found in the sarcoplasm, although a significant fraction is found associated with isolated myofibrils and myosin from rabbit fast muscle. From the amounts of myosin light chain kinase and ATPase present and the specific activities of the isolated enzymes from fast skeletal muscle of the rabbit, it can be calculated that the capacity of the myofibril to hydrolyse ATP in a given weight of muscle is about 50 times as great as the maximum expected rate of phosphorylation of the P light chain. Although it may be dangerous to extrapolate from such facts to the situation *in vivo*, this estimation, coupled with the fact that the myosin light chain phosphatase activity per gram of tissue is much less than that of the kinase, makes it very unlikely that P light chain phosphorylation and dephosphorylation can take place within a single cross-bridge cycle in fast skeletal muscle. The extreme specificity of the kinase and phosphatase, their wide distribution in most myosin based contractile systems, and the fact that the kinase is activated at very similar Ca^{2+} concentrations to those that result in activation of the myofibrillar ATPase (37), strongly suggest a specific role in the contractile process. From the *in vitro* studies, the amount of phosphorylation in fast skeletal muscle would be expected to increase after repeated contractions and then slowly fall. These, indeed, have been the results reported in frog (1), rat (44), and mouse (25). The latter two groups have correlated P light chain phosphorylation with posttetanic potentiation and with a reduced velocity of shortening, respectively. Butler and Siegman (2), on the other hand, were not able to correlate velocity of shortening with light chain phosphorylation in the rat. All these studies indicate that the P light chain is partially or only slightly phosphorylated *in vivo* in resting intact muscle. It may be significant that in the original study on myosin phosphorylation (31) it was reported that when rabbit biceps femoris muscle was homogenized in 6 M guanidine chloride immediately after death of the animal, the P light chain was found to be fully phosphorylated. If the more recent results represent the true state of P light chain phosphorylation *in vivo*, this would suggest that the findings of Perrie et al. (31) were due to phosphorylation stimulated during excision of the rabbit biceps femoris muscle.

In the heart, the activity of myosin light chain kinase per unit weight of tissue is much lower than in skeletal muscle (11) and much more comparable to that of the phosphatase. Thus, it might be expected, as the two enzymic activities are more finely balanced in cardiac muscle, that changes in phosphorylation would be likely to occur under physiological conditions. The original studies carried out by direct phosphate analyses on the isolated light chain fractions suggested that they were fully phosphorylated in normal heart and were dephosphorylated under a variety of conditions (13). Electrophoretic examination did not confirm this result and indicated that little change could be detected in the phosphorylation state (34). A number of investigators have now confirmed that little or no change in the state of phosphorylation of the P light chain is associated with the inotropic effect of adrenaline on the mammalian heart (18,20, but cf. 24).

In a recent detailed analysis of phosphorylation in cardiac muscle by the application of two dimensional gel electrophoresis to whole extracts of perfused rabbit heart, it was concluded that, in the resting animal, the P light chain was about 25% phosphorylated (49). If phosphorylation is not random, then this implies that about a quarter of the myosin molecules are fully phosphorylated or about one-half contain one phosphorylated head. On treatment with adrenaline, resulting in an increase in developed force, the extent of myosin phosphorylation increased only a few percent. A similar increase was obtained by making the heart anoxic and it is possible that the slight increase in P light chain phosphorylation was unrelated to the intervention with adrenaline but was the consequence of the anoxia produced by the increased metabolic demands associated with the inotropic effect. In contrast to the observations in myosin, the electrophoretic pattern of troponin I changed in a manner indicating that significant phosphorylation had occurred, as has been reported earlier (10,42).

The rise in phosphorylation of the troponin I lagged behind the rise in force produced by adrenaline intervention, indicating that the process could not be responsible for the force increase (Fig. 2). Thus, in the perfused heart, the phosphorylation of troponin I would appear to be clearly a consequence of activation of cyclic AMP-dependent protein kinase, whereas the myosin light chain phosphorylation is not significantly affected under conditions in which the enzyme is activated. Both observations would be expected from the *in vitro* studies on the system.

Cardiac myosin, however, does possess one feature that appears to distinguish it from the myosins of skeletal muscle. On isoelectric focusing, the P light chain migrates as two bands, designated light chain P1 and P2, which, on phosphorylation, either *in vitro* or in the intact muscle, were phosphorylated to give forms designated light chains P3 and P4, respectively (49). Similar forms had been observed in an earlier study (12), but light chain P2 (referred to as the satellite form) was considered to be a modification of light chain P1. Careful reexamination indicates that light chains P1 and P2 are probably different gene products, the relative amounts of which differ between species and during development (49).

Phosphorylation of the P Light Chain and ATPase Activity of Myosin from Striated Muscle

In view of the known effect of phosphorylation on the activity of a number of enzymes, a possible role for the phosphorylation of the P light chain is the regulation of the myosin ATPase activity. It would appear unlikely on general grounds that phosphorylation, in striated muscle at least, is a process that converts enzymically inactive myosin to an active form; it has been demonstrated that the light chains, and particularly the P light chain, are not essential for the ATPase activity of myosin

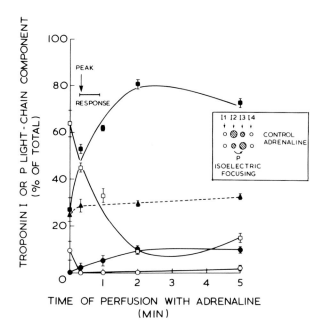

FIG. 2. Changes in the level of phosphorylation of the P light chain of myosin and troponin I in the perfused rabbit heart after treatment with adrenaline. Amounts of phosphorylated forms determined from densitometric determinations on two-dimensional electrophoretic gels. *Insert* is diagrammatic representation of electrophoretic pattern given by the troponin I. It is considered that spots I2 and I3 represent troponin I with serine 20 unphosphorylated and phosphorylated respectively (49). Spots I1 and I4 are minor unidentified forms. The *arrow* indicates the peak of the inotropic response. *Open circles,* troponin I1; *open squares,* troponin I2; *closed squares,* troponin I3; *closed circles,* troponin I4; *triangles,* phosphorylated myosin light chains P3 + P4. (From Westwood and Perry, ref. 49.)

(14,46,48). The Mg^{2+} ATPase of myosin isolated in the dephosphorylated state from skeletal and cardiac muscle can be activated by actin. In our laboratory, phosphorylation of skeletal myosin or heavy meromyosin produced only a very slight increase in the Ca- or Mg-activated ATPase of skeletal myosin and actomyosin (28). Similar, but possibly higher, activation was obtained with cardiac myosin (34). In all cases the activation did not exceed 20 to 30% and in view of the possibility of pseudo-ATPase activity (28,34), due to the presence of kinase and phosphatase in the preparation, these effects were not considered significant. On the other hand, under certain relative action concentrations and ionic conditions, Pemrick (30) has observed activation of the actin-activated ATPase of up to 180%.

Myosin Light Chain Kinase and Actin-Activation of Vertebrate Smooth Muscle Myosin

It might be presumed that the role of phosphorylation would be similar in all types of myosin; nevertheless, in contrast to the situation with myosin from striated muscle, there have been persistent reports of phosphorylation being essential for the actin-activation of myosin from vertebrate smooth and nonmuscle cells (3,15,41). This, however, has not been the experience of all investigators (5,9), and recently Persechini et al. (36) also query whether phosphorylation of the P light chain is the only requirement for actin-activation of the ATPase.

Myosin preparations isolated in our laboratory from chicken gizzard were found to possess different specific activities of the Mg^{2+} ATPase, depending on the procedure employed. For example, preparations made by a procedure similar to that of Ebashi (7) had low actin-activated ATPase, whereas those prepared by a procedure involving precipitation with ammonium sulphate and similar to that of Katoh and Kubo (22) had significantly higher actin-activated ATPase. The myosins prepared by either method were dephosphorylated. The rate of hydrolysis of ATP by these myosin preparations in the presence of actin and tropomyosin remained constant despite the fact that the P light chains were rapidly phosphorylated virtually to completion in the early stages of the incubation either with endogenous or added kinase (Fig. 3). Addition of excess of a partially purified gizzard muscle P light chain phosphatase to the actomyosin of high ATPase activity had no significant effect on the rate of ATPase hydrolysis, despite the fact that the extent of P light chain phosphorylation was reduced to about 15% of the fully phosphorylated control. These experiments suggested that the actin-activation of the ATPase of certain myosin preparations was not directly correlated to the extent of P light chain phosphorylation.

The results could be explained by the presence of varying amounts of an activator of the smooth muscle actomyosin ATPase in different myosin preparations. This was shown to be a possible explanation by the isolation of an activating fraction from chicken gizzard myosin preparations. After fractionation by ammonium sulphate precipitation, the crude activator preparation contained myosin light chain kinase but activated the Mg^{2+}-stimulated ATPase of actomyosin in a manner that

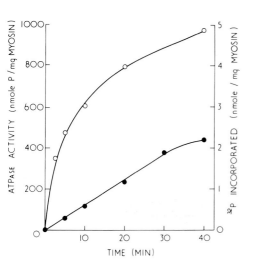

FIG. 3. Comparison of P light chain phosphorylation and the Mg^{2+}-stimulated ATPase of gizzard actomyosin. Incubation conditions, Ebashi-type chicken gizzard myosin (0.77 mg/ml), 50 mM Tris HC2(pH 7.5), 2.5 mM $MgC2_2$, 0.125 mM $CaC2_2$, 2.5 mM DTT, 2.5 mM Na ATP, or [γ-^{32}P]NaATP 0.6 Ci/mol, gizzard tropomyosin (0.1 mg/ml), skeletal actin (0.4 mg/ml), calmodulin (.01 mg/ml), light chain kinase (0.03 mg/ml), 25°C. *Closed circles*, ATPase; *open circles*, P light chain phosphorylation.

did not correlate with the extent of phosphorylation of the P light chain. Indeed, the activator was effective on fully phosphorylated myosin (Fig. 4). The activation produced by the fraction was Ca^{2+}-dependent and was increased by the addition of calmodulin to the system. The activator was specific for the actin-activated ATPase of smooth myosin, for it had no effect on the Mg^{2+}-, Ca^{2+}-, or EDTA/KCl-activated ATPases of gizzard myosin. The Mg^{2+}-activated ATPase of desensitized actomyosin prepared from skeletal muscle was not affected. We conclude, therefore, that a system consisting of an unidentified factor and calmodulin can activate the Mg^{2+}-stimulated ATPase of gizzard smooth muscle actomyosin up to rates of hydrolysis of 200 nmoles P/min and higher. At this stage, one cannot exclude the possibility of a phosphorylation step in the activation process, but the results indicate that activation does not depend on both heads of the myosin being fully phosphorylated. In some respects this system is similar to the leiotonin system of Mikawa et al. (26). These authors conclude, however, that although calmodulin will replace leitononin C, it is not identical with this protein. It is difficult to reconcile these findings with those of Gorecka et al. (15) and Chacko et al. (3) who report that P light chain phosphorylation is a requirement for the actin-activation of smooth muscle and nonmuscle myosin. The simplest explanation may be that there are two mechanisms for the activation of smooth actomyosin that are displayed by different types of myosin preparation. Certainly the *in vitro* results in which the activation of the ATPase smooth muscle actomyosin does not correlate with the extent of P light chain phosphorylation fits in well with the studies on the intact tissue in which

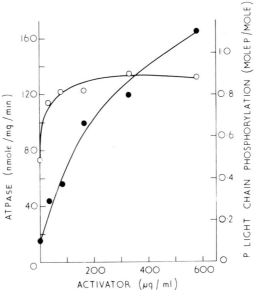

FIG. 4.   Effect of activator on the actin-activated ATPase of chicken gizzard myosin. Incubation conditions, 35 mM Tris-HCl, pH 7.5, 4 mM $MgCl_2$, 0.25 mM $CaCl_2$, 5 mM DTT, 4 mM ATP, skeletal actin (1 mg/ml), gizzard myosin (2 mg/ml), gizzard tropomyosin (0.4 mg/ml), 25°C. *Closed circles*, ATPase; *open circles*, P light chain phosphorylation.

the physiological properties suggest that P light chain has a modulating function in the regulation of contractile activity of smooth muscle (21). Such a role has the merit that it could be reconciled with a function for P light chain phosphorylation that is similar in all actomyosin-based contractile systems.

I FILAMENT REGULATION

Actin and tropomyosin are the most strongly conserved of the I filament proteins in striated muscle, suggesting that the characteristic contractile properties of the different types of muscle are more likely, in part at least, to arise from the differences in properties of the remaining component of the I filament, the troponin complex. Troponin I and troponin T are the most highly specialized proteins of the complex and may be restricted in distribution to striated muscle as comparable proteins have not yet been characterized in other tissues. Troponin C, on the other hand, is very similar in structure to the widely distributed calcium-binding protein, calmodulin.

Troponin I has a special significance in the troponin regulatory system in that it possesses well-defined properties, one of which is apparently to block the actin-myosin interaction and thereby produce inhibition of the Mg^{2+}-stimulated ATPase of actomyosin. This effect is neutralized by troponin C and much increased by tropomyosin (for a review, see 32). All these properties must be of special significance for the role of troponin I in the regulation of the interaction of actin and myosin. It follows, therefore, that detailed knowledge of the interaction of troponin I with the other proteins of the I filament is vital to our understanding of the mechanism of the process of Ca^{2+} regulation as a whole. The paradox about troponin I is that although its properties seem very appropriate for the regulation of acto-myosin interaction by a mechanism of interaction with actin that is neutralized by troponin I in the presence of Ca^{2+}, there is only one (8,38) or possibly two (43) molecule(s) of troponin I for every seven actin monomers in the I filament. One molecule of troponin I could be expected to block sterically the myosin site on one actin or possibly two. This fact, coupled with X-ray observations implying movement of the tropomyosin filaments in the grooves of the actin filament, have lead to the postulation of the "steric" hypothesis, which proposes that the interaction between actin and myosin is directly regulated by tropomyosin rather than troponin I (17,19,29). Although this hypothesis has many attractions, it has yet to be supported by compelling biochemical data (6). Indeed, recent kinetic studies on the actin-myosin interaction in regulated and nonregulated systems throw further doubt on the validity of the steric mechanism (4).

Different forms of troponin I and troponin T are present in fast and slow skeletal and in cardiac muscle. These three forms of the troponin components are different gene products and, in the case of troponin I, about 60% of the residues are identical when any two forms are compared (50). The differences in the troponin I are no doubt responsible, in part at least, for the characteristic features of the contraction-relaxation cycles in the different striated muscle types. Thus, studies of the structure-function relationship of all three types of troponin I enable conclusions to be made about both the structural features that are conserved and presumably essential for the basic mechanism, and those features that have evolved to account for the properties unique to the different muscle types.

As an approach to this problem, the interaction of troponin I from fast skeletal muscle with troponin C has been studied using proton magnetic resonance techniques (16). On digestion with cyanogen bromide, troponin I from rabbit fast skeletal muscle is cleaved into 10 peptides, one of which, residues 1-21 (peptide CN5), has been previously identified to interact with troponin C and another, consisting of residues 96-116 (peptide CN4), with troponin C and actin (45). These peptides were the only ones of the 7 examined, representing 91% of the total sequence, that showed evidence of interaction with troponin C. Perturbation of the side chains of isoleucine 10, threonine 11, alanine 12, and of the arginine present in the N terminal peptide CN5 were observed in interaction with troponin C in the presence of Ca^{2+}. In the case of the peptide CN4, Ca^{2+}-dependent interaction of troponin C with the side chains of phenylalanine, lysine, and leucine was observed. Perturbation of the arginine side chains, all of which are concentrated in the C terminal half of peptide CN4, was observed on interaction with actin whether Ca^{2+} was present or not. In resting muscle, the region represented by peptide CN4 in troponin I probably interacts with both troponin C and actin. It is, therefore, suggested that interaction with troponin C may be restricted to the region in the N terminal half of the peptide containing the three lysine residues (Fig. 5). These results confirm the general conclusions made from studies on the biochemical properties of the isolated peptides (45), but enable the interaction to be more precisely defined.

If the interaction with troponin C is restricted to the N-terminal region of peptide CN4, the C-terminal region of the peptide containing the arginine residues is still available for interaction with actin. Whether interaction in this region still exists between actin under these circumstances is not certain, but if it does, the enzymic data indicate that the nature of the interaction differs from that in the absence of troponin C. In some respects, the enzymic data obtained from *in vitro* systems are

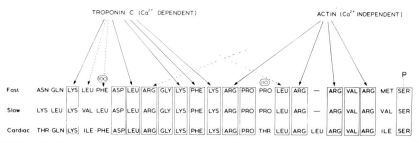

FIG. 5. Region of the primary structure of troponin I from rabbit fast skeletal muscle that interacts with troponin C and actin. The sequence (residues 96 to 117) represents the cyanogen bromide peptide CN4 and serine 117, the site of phosphorylation by cyclic AMP-dependent protein kinase. Residues perturbed on addition of troponin C and actin are indicated by *arrows*. The sequences of the slow skeletal and cardiac forms of troponin I are also shown for comparison. Residues identical on all three forms are enclosed in *boxes*. *Full arrows* to these residues indicate that they may be of particular significance in the interaction.

more simply explained by displacement of troponin I from interaction with actin on addition of troponin C. This effect is obtained in the absence or presence of Ca^{2+} (35). Clearly, with the intact myofibril, the interaction between the proteins may be slightly different from that which occurs in the reconstructed *in vitro* system.

The strongly conserved nature of the two regions of troponin I and their involvement in interaction with troponin C suggest that these interactions are fundamental to the general mechanism of function of troponin I. In troponin I from cardiac muscle, however, there is an additional 26 residue peptide at the N terminus. This results in the introduction into the molecule of an additional phosphorylation site, the serine residue at position 20, which is rapidly phosphorylated by cyclic AMP-dependent kinase. Unlike the phosphorylation sites on fast skeletal troponin I that lie close to the interaction sites and that are blocked by interaction with troponin C, serine 20 of cardiac troponin I is readily phosphorylated in the presence of troponin C complex or in the whole troponin complex in the intact muscle (27,33). Thus, in the intact heart, the extent of phosphorylation of serine 20 depends on the relative activities of the cAMP-dependent kinase and the phosphatase that acts on cardiac troponin I. Studies on the rabbit heart in the intact animal and perfused under the conditions of the Langendorff technique indicate that, in the normal beating heart, serine 20 is about 30% phosphorylated (27). On treatment of the perfused rabbit heart with adrenaline, the level of phosphorylation rises to 90 to 100%. This does not occur synchronously with the increase in force but usually rises to a maximum about 2 min after the peak in force is reached (49). Studies on the ATPase of regulated actomyosin and myofibrils indicate that phosphorylation of the troponin I leads to an increase in the Ca^{2+} concentration at which 50% activation of the ATPase occurs (39,40,42). Thus, when the troponin I is fully phosphorylated after intervention with adrenaline, the full activation of myofibrillar ATPase requires a higher Ca^{2+} concentration than in the normal beating heart. When the Ca^{2+} concentration falls as the heart enters diastole, the myofibrillar ATPase falls more rapidly than it would if the troponin I were at a lower level of phosphorylation. Phosphorylation of troponin I is one of the mechanisms for changing heart rate that, as is well known, is due mainly to shortening or prolongation of diastole.

Skeletal muscle does not have the same system as cardiac muscle for regulating response of the contractile system. Increase in force in skeletal muscle is obtained by recruitment of fibres and, to some extent, by posttetanic potentiation rather than manipulating the calcium flux in any given fibre. By the very nature of its role, the cardiac muscle cell must be much more adaptable in response. In the beating heart every cell contracts and increased force is obtained by increasing the force exerted by each cell. This is achieved by increasing the calcium flux that occurs when the heart comes under the influence of the catecholamines. To smooth out and balance this process, a negative feedback system, which leads to decreased sensitivity of the actomyosin ATPase to Ca^{2+}, has evolved as part of the adrenergic response.

GENERAL CONCLUSIONS

The binding of Ca^{2+} to specific target proteins of high affinity would be expected to be a more rapid process than phosphorylation and, hence, particularly suitable for initiating contraction in fast muscles. There is little doubt that this is the case with regulation through the actin filament involving the troponin system of striated muscle. It is a matter of controversy in some smooth muscles as to whether Ca^{2+} binding directly to myosin or actin filaments or phosphorylation of the myosin is the trigger for contraction. In adductor muscle of molluscs, direct Ca^{2+} binding to myosin initiates contraction and there is no evidence for the phosphorylation of the regulatory light chain of adductor myosin (11,23). The bulk of the evidence with vertebrate smooth muscle, which has high myosin light chain kinase activity, has so far favored phosphorylation of the P light chain as the triggering event. Nevertheless, not all experimental results support the view that both heads of smooth myosin have to be fully phosphorylated for hydrolysis of ATP by actomyosin. Several groups of workers have obtained preparations of myosin with which there is no direct correlation between the actin-activated ATPase activity and the extent of phosphorylation of the P light chains. The controversy regarding the mode of regulation of smooth myosin has persisted for some five or six years and the apparently contradictory results no doubt reflect the subtleties of the system. Although it need not be the case, one would expect the effect of phosphorylation to be similar in all myosin systems. Nevertheless, the myosin light chain kinase of smooth muscle possesses certain special features that distinguish it from the skeletal enzyme, implying that its function may be different in the former muscle. A role for P light chain phosphorylation in modulating the contractile response would have the virtue that it could be similar in all muscle types. If that were the case, the consequence of phosphorylation on the A filament would be somewhat analogous to that associated with the troponin type regulation in cardiac muscle. Whatever the explanation, one can say with confidence that there is still much to learn about the mechanism of regulation of the ATPase of smooth muscle actomyosin.

ACKNOWLEDGMENTS

The Medical Research Council has supported this work by the provision of a program grant (to S. V. Perry) and a research fellowship (to B. A. Levine).

REFERENCES

1. Barany, K., and Barany, M. (1977): Phosphorylation of the 18000 dalton light chain of myosin during a single tetanus of frog muscle. *J. Biol. Chem.*, 252:4752–4754.
2. Butler, T. M., and Siegman, M. J. (1981): Chemical energetics of contracting smooth muscle. *35th Annual Meeting of Soc. of Gen. Physiologists (Abstr.)*, Woods Hole, September 1981, p. 8a.
3. Chacko, S., Conti, M. A., and Adelstein, R. S. (1977): Effects of phosphorylation of smooth muscle myosin on actin activation and Ca^{2+} regulation. *Proc. Natl. Acad. Sci. USA*, 74:129–133.
4. Chalovitch, J. M., Chock, P. B., and Eisenberg, E. (1981): Mechanisms of action of troponin:tropomyosin. *J. Biol. Chem.*, 256:575–578.
5. Cole, H. A., Grand, R. J. A., and Perry, S. V. (1980): Phosphorylation of smooth muscle myosin and actin activation of the ATPase. *J. Muscle Res. Cell Motil.*, 1:468.

6. Eaton, B. L. (1976): Tropomyosin binding to F-actin induced by myosin heads. *Science*, 192:1337–1339.
7. Ebashi, S. (1976): A simple method of preparing actin-free myosin from smooth muscle. *J. Biochem. (Tokyo)*, 79:229–231.
8. Ebashi, S., Endo, M., and Ohtsuki, I. (1969): Control of muscle contraction. *Q. Rev. Biophys.*, 2:351–384.
9. Ebashi, S., Toyo-Oka, T., and Nonomura, Y. (1975): Gizzard troponin. *J. Biochem. (Tokyo)*, 78:859–861.
10. England, P. J. (1975): Correlation between contraction and phosphorylation of the inhibitory subunit of troponin in perfused rat heart. *FEBS Lett.*, 50:57–60.
11. Frearson, N., Focant, B. W. W., and Perry, S. V. (1976): Phosphorylation of a light chain component of myosin from smooth muscles. *FEBS Lett.*, 63:27–32.
12. Frearson, N., and Perry, S. V. (1975): Phosphorylation of the light chain components of myosin from cardiac and red skeletal muscles. *Biochem. J.*, 151:99–107.
13. Frearson, N., Salaro, R. J., and Perry, S. V. (1976): Changes in phosphorylation of the P light chain of myosin in the perfused rabbit heart. *Nature*, 264:801–802.
14. Gazith, J., Himmelfarb, S., and Harrington, W. F. (1970): Studies on the subunit structure of myosin. *J. Biol. Chem.*, 245:15–22.
15. Gorecka, A., Aksoy, M. O., and Hartshorne, D. J. (1976): The effect of phosphorylation of gizzard myosin on actin activation. *Biochem. Biophys. Res. Commun.*, 71:325–331.
16. Grand, R. J. A., Levine, B. A., and Perry, S. V. (1981): PMR studies on the interaction of troponin I with troponin C and actin. *Biochem. J.*, 203:61–68.
17. Haselgrove, J. C. (1972): X-ray evidence for a conformational change in the actin-containing filaments of vertebrate striated muscle. *Cold Spring Harbor Symp. Quant. Biol.*, 37:341–352.
18. Holroyde, M. J., Small, D. A. P., Howe, E., and Solaro, R. J. (1979): Isolation of cardiac myofibrils and myosin light chains with *in vivo* level of light chain phosphorylation. *Biochem. Biophys. Acta*, 587:628–637.
19. Huxley, H. E. (1972): Structural changes in the actin- and myosin-containing filaments during contraction. *Cold Spring Harbor Symp. Quant. Biol.*, 37:361–376.
20. Jeacocke, S. A., and England, P. J. (1980): Phosphorylation of myosin light chains in perfused rat heart. *Biochem. J.*, 763–768.
21. Kamm, K. E., Aksoy, M. O., Dillon, P. F., and Murphy, R. A. (1981): Maximum shortening velocity with no external load (V_0) is proportional to the fraction of phosphorylated 20000 dalton myosin light chains in arterial smooth muscle. *35th Annual Meeting of the Soc. of Gen. Physiologists (Abstr.)*, Woods Hole, September 1981, p. 20a.
22. Katoh, N., and Kubo, S. (1977): Purification and some properties of rabbit stomach myosin. *J. Biochem. (Tokyo)*, 81:1497–1503.
23. Kendrick-Jones, J., and Jakes, R. (1977): Myosin linked regulation: A chemical approach. In: *Myocardial Failure*, edited by G. Riecker, A. Weber, and J. Goodwin, pp. 28–40. Springer-Verlag, Berlin.
24. Kopps, S. J., and Barany, M. (1979): Phosphorylation of the 19000 dalton light chain of myosin in perfused rat heart under the influence of negative and positive inotropic agents. *J. Biol. Chem.*, 254:12007–12012.
25. Kushmerick, M. J., and Crow, M. T. (1981): Chemical energetics, mechanics and phosphorylation of regulatory light chains in mammalian fast- and slow-twitch muscles. *35th Annual Meeting of the Soc. of Gen. Physiologists (Abstr.)*, Woods Hole, September 1981, p. 7a.
26. Mikawa, T., Toyo-Oka, T., Nonomura, Y., and Ebashi, S. (1977): An essential factor of gizzard troponin fraction. A new type of regulatory protein. *J. Biochem. (Tokyo)*, 81:273–275.
27. Moir, A. J. G., Solaro, R. J., and Perry, S. V. (1980): The site of phosphorylation of troponin I in the perfused rabbit heart: the effect of adrenaline. *Biochem. J.*, 185:505–513.
28. Morgan, M., Perry, S. V., and Ottaway, J. (1976): Myosin light chain phosphatase. *Biochem. J.*, 157:687–697.
29. Parry, D. A. D., and Squire, J. M. (1973): Structural role of tropomyosin in muscle regulation: analysis of the X-ray diffraction patterns from relaxed and contracting muscles. *J. Mol. Biol.*, 75:33–55.
30. Pemrick, S. (1980): The phosphorylated L_2 light chain of skeletal myosin is a modifier of the actomyosin ATPase. *J. Biol. Chem.*, 255:8836–8841.
31. Perrie, W. T., Smillie, L. B., and Perry, S. V. (1973): A phosphorylated light chain component of myosin from skeletal muscle. *Biochem. J.*, 135:151–164.

32. Perry, S. V. (1979): The regulation of contractile activity in muscle. *Biochem. Soc. Trans.*, 7:593–617.
33. Perry, S. V., and Cole, H. A. (1974): Phoshorylation of troponin and the effects of interactions between the components of the complex. *Biochem. J.*, 141:733–743.
34. Perry, S. V., Cole, H. A., Frearson, N., Moir, A. J. G., Nairn, A. C., and Solaro, R. J. (1978): Phosphorylation of the myofibrillar proteins. *Proceedings of 12th FEBS Meeting, Dresden, Vol. 54*, edited by L-G. Krause, L. Pinna, and A. Wollenberger, pp. 147–159. Pergamon Press, Oxford.
35. Perry, S. V., Cole, H. A., Head, J. F., and Wilson, F. J. (1972): Localization and mode of action of the inhibitory protein component of the troponin complex. *Cold Spring Harbor Symp. Quant. Biol.*, 37:251–262.
36. Persechini, A., Mrwa, U., and Hartshorne, D. J. (1981): Effect of phosphorylation on the actin activated ATPase activity of myosin. *Biochem. Biophys. Res. Commun.*, 98:800–805.
37. Pires, E. M. V., and Perry, S. V. (1977): Purification and properties of myosin light chain kinase from fast skeletal muscle. *Biochem. J.*, 167:137–146.
38. Potter, J. D. (1971): The content of troponin, tropomyosin, actin and myosin in rabbit skeletal muscle myofibrils. *Arch. Biochem. Biophys.*, 162:436–441.
39. Ray, K. P., and England, P. J. (1976): Phosphorylation of the inhibitory subunit of troponin and its effect on the calcium dependence of cardiac myofibril adenosine triphosphatase. *FEBS Lett.*, 70:11–16.
40. Reddy, Y. S., and Wyborny, L. E. (1976): Phosphorylation of guinea pig cardiac natural acto-myosin and its effect on ATPase activity. *Biochem. Biophys. Res. Commun.*, 73:703–709.
41. Sobieszek, A., and Small, J. V. (1976): Myosin linked calcium regulation in vertebrate smooth muscle. *J. Mol. Biol.*, 102:75–92.
42. Solaro, R. J., Moir, A. J. G., and Perry, S. V. (1976): Phosphorylation of troponin I and the inotropic effect of adrenaline in the perfused rabbit heart. *Nature*, 262:615–617.
43. Sperling, J. E., Feldmann, K., Meyer, H., Jahnke, U., and Heilmeyer, L. M. G. (1979): Isolation of characterization and phosphorylation pattern of troponin complexes TI₂C and I₂C. *Eur. J. Biochem.*, 101:581–592.
44. Stull, J. T., Manning, D. R., High, C. W., and Blumenthal, D. K. (1980): Phosphorylation of contractile proteins in heart and skeletal muscle. *Fed. Proc.*, 39:1552–1557.
45. Syska, H., Wilkinson, J. M., Grand, R. J. A., and Perry, S. V. (1976): The relationship between biological activity and primary structure of troponin I from white skeletal muscle of the rabbit. *Biochem. J.*, 153:375–387.
46. Wagner, P. D., and Giningere, E. (1981): Hydrolysis of ATP and reversible binding of actin by myosin heavy chain free of all light chains. *Nature*, 292:560–562.
47. Walsh, M. P., Cavadore, J-C., Molla, A., and Demaille, J. G. (1980): Skeletal muscle calmodulin-dependent myosin light chain kinase: Purification of a 155000 dalton myofibrillar enzyme. *Proc. 4th Int. Conf. on Cyclic Nucleotides (Abstr.)*, Brussels, THC6.
48. Weeds, A. G. (1969): Light chains of myosin. *Nature*, 223:1362–1364.
49. Westwood, S. A., and Perry, S. V. (1981): The effect of adrenaline on the phosphorylation of the P light chain of myosin and troponin I in the perfused rabbit heart. *Biochem. J.*, 197:185–195.
50. Wilkinson, J. M., and Grand, R. J. A. (1978): Comparison of amino acid sequence of troponin I from different striated muscles. *Nature*, 271:31–35.

Basic Biology of Muscles: A Comparative
Approach, edited by B. M. Twarog,
R. J. C. Levine, and M. M. Dewey.
Raven Press, New York © 1982.

Role of the Myosin Light Chains in the Regulation of Contractile Activity

John Kendrick-Jones, Ross Jakes, Phillip Tooth, Roger Craig,
and Jonathan Scholey

MRC Laboratory of Molecular Biology, Cambridge CB2 2QH, United Kingdom

Muscular contraction and various types of nonmuscle cell motility are generated by the interaction between myosin and actin. The intracellular concentration of free calcium (Ca^{2+}) controls this interaction by mechanisms that differ in different muscles and nonmuscle cells. In striated muscles, actin and myosin are organised into a lattice of stable thin and thick filaments, which slide past one another as the muscle changes length (18). In these muscles, actin-linked and myosin-linked calcium regulatory mechanisms serve to regulate the interaction between the myosin cross-bridge and actin subunits and, thus, to control muscle contraction. In vertebrate striated muscles, the dominant calcium regulatory system consists of the actin-linked tropomyosin-troponin complex (12,48), whereas in molluscan muscles, although this actin-linked system may be present (14), the myosin and associated regulatory light chains constitute the major myosin-linked calcium regulatory system (9,21,22). In other muscles, actin-linked or myosin-linked or both regulatory systems operate (25). In nonmuscle cells, the regulation of contractile activity appears to be far more complex. In these cells, the contractile apparatus is often assembled transiently and, therefore, the cells must possess mechanisms for regulating the assembly of actin and myosin filaments, as well as for controlling the interaction between myosin and actin.

Our interest is in the role of the myosin light chains in regulating muscle contraction and cell motility. Regulatory light chains appear to play a crucial role not only in the control of myosin interaction with actin, but also in the regulation of myosin filament assembly in certain contractile systems.

All the myosins studied to date, with the exception of Acanthamoeba myosin I (32), have a similar structure, composed of two heavy chains wound around each other at the C terminal end to form a stable α helical rod, whereas at the N terminal end they diverge and fold into two globular heads, which contain the actin binding and ATPase activities and serve as the myosin cross-bridges (Fig. 1). Each myosin head (S_1) contains two different noncovalently bound light chains. One type of light chain (LC_2) has been called the essential light chain (22), although its precise

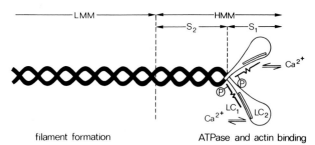

FIG. 1. Diagrammatic representation of a myosin molecule illustrating the positions of the proteolytic subfragments and the distribution of functional properties. Myosin is an asymmetric molecule comprised of two heavy chains (~200,000 molecular weight) and two pairs of light chains of two
types (MW ~20,000). The C terminal halves of the heavy chain are wound around each other to form
an α helical coiled coil region (LMM and S2) and the N terminal halves diverge to form two globular
heads (S1). Each S1 contains one of each type of light chain, called regulatory (LC1) and essential
(LC2) light chains. The precise structure and location of these light chains on the myosin head (S1)
is not yet known. The regulatory light chains control myosin interaction with actin mediated by Ca^{2+}
binding to S1 in molluscan myosins and by phosphorylation (ⓅP) in vertebrate smooth muscle and
nonmuscle cells.

function has not yet been established (47); the other, the regulatory light chain
(LC$_1$), in response to either phosphorylation or direct Ca^{2+} binding to the myosin
performs a regulatory function, at least in some contractile systems. Myosins may
be divided into three classes on the basis of the role of these regulatory light chains
in controlling myosin interaction with actin.

In Class 1—(molluscan and certain other invertebrate muscle myosins (25,45)—
the regulatory light chains inhibit myosin interaction with actin in the absence of
Ca^{2+} ($<10^{-7}$ M). When the muscle is activated, the intracellular Ca^{2+} concentration
increases to $>10^{-6}$ M, and calcium binds to specific sites on the myosin molecule,
relieving the inhibitory effect of the light chain (9,21,22). The myosin can now
bind to actin and the muscle contracts.

In Class 2—vertebrate smooth muscle and nonmuscle myosin—calcium triggers
myosin interaction with actin by binding to the calmodulin subunit of a specific
light chain kinase, activating it to phosphorylate the 20,000 molecular weight (MW)
regulatory light chains (2,16,39). Only when this light chain is phosphorylated can
the myosin interact with actin. There is evidence (33,44) that these light chains
also regulate the assembly of myosin into filaments. Although the light chains play
a vital role in regulation in these cells, it is possible that a number of other regulatory
mechanisms may also operate (12,27).

In Class 3—vertebrate striated muscle myosins—there are light chains homologous to those of molluscan and vertebrate smooth muscle myosin, but at present
there is no clear evidence that they play a role in the primary control of myosin-
actin interaction. It is possible, however, that they may have a modulatory effect
on myosin interaction with actin *in vivo* (7).

We are interested in elucidating how the myosin regulatory light chains, in
response to the appropriate regulatory signal, switch on and off muscle contraction

and cell motility. In this chapter we will discuss studies on the role of the regulatory light chains in molluscan myosins and the use of scallop myosin hybrids for analysing the functional capability of light chains. We will then compare the role of the homologous light chains in smooth muscle and nonmuscle myosins, and briefly discuss the role of those light chains in myosin filament assembly.

ROLE OF THE REGULATORY LIGHT CHAINS IN MOLLUSCAN MYOSINS

In molluscan muscles, it has been clearly demonstrated that Ca^{2+} regulates contraction by acting directly on the myosin molecule (21,45). In these muscles, the myosins require at least 10^{-6} M Ca^{2+} to interact with purified actin. Most of the studies on regulation in these muscles by ourselves and by Szent-Györgyi and his colleagues have been carried out on scallop myosin, because it was the first myosin shown to contain a regulatory light chain and it is the only myosin that has been found where the regulatory light chains can be reversibly removed without destroying the myosin (45). In isolated scallop myofibrils, myosin interaction with actin, as measured by the steady state actin-activated myosin MgATPase, requires the presence of calcium (Table 1). Each myosin molecule contains two regulatory light chains and two specific high affinity sites that bind Ca^{2+} in the presence of millimolar Mg^{2+} (K < 10^{-6} M) (45). Chantler and Szent-Györgyi (9) have shown that if scallop myosin is treated with EDTA at 25 to 30°C, both regulatory light chains are released and the myosin becomes desensitized. The specific calcium binding sites are lost and calcium is no longer required for myosin interaction with actin. When two moles of scallop regulatory light chains rebind to the desensitized myosin, the two Ca^{2+} binding sites and the Ca^{2+} requirement for actin-myosin interaction are restored (9). Confirmatory evidence that the regulatory light chains control contraction in molluscan muscles has been provided by the demonstration that, in chemically skinned scallop muscle fibre bundles, the Ca^{2+} sensitive development of tension is mediated by the regulatory light chains (38).

The exact location of specific calcium binding sites in scallop myosin, however, remains in doubt. Although there is a direct correlation between the regulatory light chain content of the myosin and the number of Ca^{2+} binding sites, the nonspecific calcium binding sites observed on the isolated regulatory light chains (8) do not appear to be competent to act as the specific calcium regulatory sites (4). Presumably, the heavy chain and possibly the essential light chain contribute to the formation of the specific calcium binding sites present on the intact myosin.

The principal features of the regulatory system in molluscan myosins may, therefore, be briefly summarised as follows. In relaxed muscles, the regulatory light chains inhibit the correct binding of the myosin heads to actin. On activation, when calcium binds to the specific sites on the myosin, this inhibition is removed so that the myosin can specifically bind to actin and the muscle contracts.

TABLE 1. *Calcium regulation in scallop myofibrils and desensitized scallop myofibril/regulatory light chain hybrids*[a]

	Actin-myosin interaction (AM Mg²⁺ ATPase) (μmole ATP split min^{-1}·mg^{-1})		Number specific (~) Ca²⁺ sites/myosin (Kd~10^6 M^{-1})	Percent relaxation[b]
	<10^{-7} M Ca²⁺	~10^{-5} M Ca²⁺		
Scallop myofibrils	0.014	0.378	1.7	96
Desensitized myofibrils (DMF)	0.267	0.184	< 0.1	nil
Class 1 DMF/scallop LC hybrid	0.016	0.402	1.6	96
Class 2 DMF/gizzard (or thymus) LC hybrid	0.019	0.286	1.3	92
Class 3 DMF/vert cardiac (or rabbit sk) LC hybrid	0.017	0.021	0.1	inhibition

[a]Scallop myofibrils were prepared as described previously (45) and desensitized by the procedure (9) outlined in Fig. 2. The hybrids were prepared by incubating desensitized scallop myofibrils with the regulatory light chains (LC) at a 2.5:1 light chain/myosin molar ratio, overnight with stirring at 4°C. The hybrids were extensively washed to remove unbound light chains and actin-activated Mg²⁺-dependent ATPase activities and calcium binding measurements were performed by the procedures outlined by Kendrick-Jones et al. (22). The regulatory light chain content of the hybrids was determined by densitometry of urea and SDS polyacrylamide gels (22). Similar results were obtained using purified desensitized scallop myosin. The approximate number of high affinity specific calcium binding sites, measure in the presence of 1 mM Mg²⁺, were determined by Scatchard plots.

[b]Percent relaxation = $100 - \dfrac{\text{ATPase} < 10^{-7} \text{M Ca}^{2+}}{\text{ATPase} \sim 10^{-5} \text{M Ca}^{2+}} \times 100$

SCALLOP MYOSIN/REGULATORY LIGHT CHAIN HYBRIDS

The ability to dissociate reversibly regulatory light chains from scallop myosin can be exploited further by using this myosin to test the regulatory capability of light chains isolated from a variety of myosins (Fig. 2) (22,36). When scallop myosin is desensitized by EDTA treatment at 25 to 30°C (9), two moles of scallop regulatory light chains are released, and two moles of regulatory light chains isolated from other myosins will bind to this desensitized myosin in the presence of Mg²⁺, to form scallop myosin/light chain hybrids (36). Sellers et al. (36) have divided the light chains into three classes based on their ability to restore calcium sensitivity to desensitized scallop myosin. We have confirmed many of their observations and have made a number of additional observations that may be relevant to the mechanism of regulation mediated by the myosin light chains (Table 1).

Class 1 Hybrids

Regulatory light chains from molluscan myosins and certain other invertebrate muscle myosins, which require calcium binding for myosin interaction with actin,

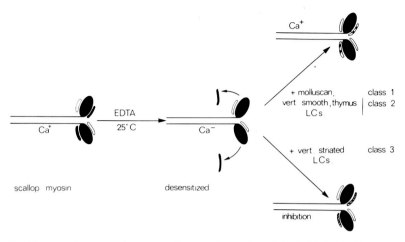

FIG. 2. The procedure used to prepare scallop myosin-regulatory light chain hybrids. Scallop myosin requires calcium for interaction with actin. When treated with 10 mM ethylenediaminetetraacetic acid (EDTA) at 25 to 30°C, it has been shown (9) that both regulatory light chains can be released and calcium regulation is abolished. In the presence of Mg^{2+}, regulatory light chains isolated from a variety of myosins will rebind to this desensitized myosin and, on the basis of their ability to restore calcium regulation to the desensitized scallop myosin preparation, they may be divided into three classes.

behave like the scallop regulatory light chains. They restore the high affinity calcium binding sites (K < 10^{-6} M) and form hybrids that possess Ca^{2+}-sensitive actin activated MgATPase activity.

Class 2 Hybrids

Regulatory light chains (MW 20,000) from gizzard, platelet, and thymus myosins, which require phosphorylation for their parent myosin to interact with actin, also form calcium regulated hybrids. However, Sellers et al. (36) observed that the restored calcium binding sites in these hybrids have lower affinities (K $\simeq$ 5 × 10^{-5} M) and, thus, higher levels of free Ca^{2+} concentrations are required for switching on the actin-activated MgATPase activity.

The observation that the 20,000 MW regulatory light chains of vertebrate non-muscle and smooth muscle myosins can substitute for the regulatory light chains of scallop myosin suggests that the role of the regulatory light chains in molluscan myosins and in vertebrate smooth muscle and nonmuscle myosins is similar, that is, at rest, both types of light chains inhibit myosin interaction with actin and this inhibition is removed by either calcium binding in molluscan myosins or by light chain phosphorylation in smooth muscle/nonmuscle myosins. In both regulatory systems, the basis of regulation exerted by the light chains is repression-depression.

Class 3 Hybrids

Regulatory light chains from vertebrate striated myosins form hybrids that contain no specific high affinity calcium binding sites. Sellers et al. (36) observed that these light chains have no effect on the scallop myosin actin activated MgATPase activity, that is, it remains high in the presence and absence of calcium. We observe that in these hybrids, actin-myosin interaction measured either by ATPase activity or by turbidity measurements is always inhibited in the presence and absence of Ca^{2+}. This discrepancy in results is difficult to explain; it could be due to slight differences in the binding of these light chains to scallop myosin. To exclude the possibility that our results might be due to the light chains irreversibly modifying the myosin, the hybrids were subjected to the EDTA/light chain dissociation procedure (9) and then incubated with scallop regulatory light chains in the presence of Mg^{2+} (Table 2). In the case of the rabbit light chain hybrid, about 40% of this light chain is released and a roughly corresponding amount of scallop light chain rebinds with a partial recovery of calcium sensitive ATPase activity. However, none of the cardiac light chain is released from its hybrid by this procedure, which again may be a reflection of slight differences in the binding of these light chains to the myosin. These results would tentatively suggest, at least in the rabbit light chain hybrid, that the inhibitory effect of these light chains is not due to irreversible modification of scallop myosin. Our results would support the proposal that Ca^{2+} acts in scallop myosin as a derepressor, that is, it overcomes the inhibited state imposed by the regulatory light chains. Thus, because no high affinity calcium-specific binding sites are present, the inhibition in the vertebrate striated light chain hybrid is not relieved in the presence of calcium.

Our results with this hybrid would also agree with the observation that the actin-activated activity of purified vertebrate striated myosin is insensitive to calcium

TABLE 2. *Is the inhibitory effect of the vertebrate striated muscle regulatory light chain in its hybrid due to irreversible modification of the scallop myosin?[a]*

	Actin-myosin interaction (AM Mg^{2+} ATPase) (μmole ATP $min^{-1} \cdot mg^{-1}$)		Percent LC bound[b]	
	$<10^{-7}$ M Ca^{2+}	$\sim 10^{-5}$ M Ca^{2+}	scallop LC	vert. LC
DMF/scallop LC hybrid				
after EDTA treatment	0.327	0.249	6	—
+ scallop LC	0.033	0.380	92	—
DMF/rabbit sk. LC hybrid				
after EDTA treatment	0.213	0.186	—	58
+ scallop LC	0.041	0.178	36	56
DMF/vert. cardiac LC hybrid				
after EDTA treatment	0.027	0.027	—	94
+ scallop LC	0.027	0.023	6	94

[a]Desensitized scallop myofibril/regulatory light chain hybrids prepared as described in Table 1 were treated with 10 mM EDTA at 25°C by the procedure described for the preparation of desensitized scallop myofibrils (9). Aliquots were mixed with scallop regulatory light chains (2.5:1 molar ratio light chain:myosin) and incubated overnight at 4°C. MgATPase activities and the regulatory light chain content of the samples were determined as described previously (22).
[b]100% light chain (LC) bound was taken as a 1:1 ratio of regulatory light chain:scallop essential light chain.

(31), demonstrating that a direct Ca^{2+} switch on the myosin molecule equivalent to that operating in molluscan myosins is absent in vertebrate striated myosins.

EFFECT OF PHOSPHORYLATION ON THE INTERACTION BETWEEN MYOSIN AND ACTIN IN SCALLOP MYOSIN/LIGHT CHAIN HYBRIDS

The observations (1,3,37,40) that, in vertebrate smooth muscle and nonmuscle myosins, phosphorylation of the regulatory light chains regulates myosin interaction with actin, prompted us to determine whether phosphorylation plays any role in regulation in molluscan myosins. We have been unable to detect any significant phosphorylation of the regulatory light chains of scallop myosin by endogenous kinases or after addition of purified light chain kinases isolated from rabbit skeletal and gizzard muscles (Table 3). We believe, therefore, that phosphorylation is not involved in regulation in molluscan myosins. When we tested the effect of phosphorylation on the regulatory light chain hybrids, however, we observed the following interesting results (Table 3). When the nonphosphorylated gizzard light chain hybrid was incubated with gizzard light chain kinase and ATPγS, the light

TABLE 3. *The effect of light chain phosphorylation on calcium regulation in scallop myosin hybrids[a]*

Class	Moles PO_4/ mole LC (n)	Actin-myosin interaction (AM Mg^{2+} ATPase) (μmole ATP split $min^{-1} \cdot mg^{-1}$)		Percent relaxation
		$<10^{-7}$ M Ca^{2+}	$\sim 10^{-5}$ M Ca^{2+}	
DMF/scallop LC hybrid + kinase + Ca^{2+}	0.1	0.014	0.376	96
DMF/gizzard LC hybrid + kinase + Ca^{2+} + kinase + EGTA	0.8 0.1	0.277 0.021	0.304 0.290	9 93
DMF/cardiac LC hybrid + kinase + Ca^{2+}	0.8	0.014	0.018	inhibition

[a]The hybrids (~5 mg·ml^{-1}) in 100 mM NaCl, 5 mM $MgCl_2$ 10 mM tris-HCl buffer pH 7.5, 0.2 mM $CaCl_2$ or 1 mM EGTA, 1 mM dithiothreitol (DTT), and 25 μg·ml^{-1} brain calmodulin were incubated at 25°C in the presence of ~50 μg·ml^{-1} gizzard muscle light chain kinase and 2 mM ATPγS for 10 min. Aliquots were taken for MgATPase activity measurements and for 8 M urea gel electrophoresis, and the remainder of the hybrid samples washed. No difference in MgATPase activities was observed in control hybrid samples incubated as above but in the absence of kinase or ATPγS. The approximate level of thiophosphorylation of the light chains was measured by the following indirect method. The washed thiophosphorylated hybrids and aliquots of the initial hybrid samples as controls, were suspended in the above assay solution in the presence of 0.2 mM $CaCl_2$ and gizzard kinase and incubated at 25°C with a final concentration of 5 mM $[^{32}P]\gamma$ATP (5μ Ci·μmole^{-1}). At suitable time intervals aliquots were taken, quenched in 5% (W/V) TCA containing 1 mM ATP and 20 mM pyrophosphate and the amount of $[^{32}P]$ incorporation determined by the millipore filtration assay (37). The specificity of phosphate incorporation was verified by urea and SDS gel electrophoresis and autoradiography. In the control gizzard and cardiac light chain hybrids, the level of $[^{32}P]$ phosphate incorporation was >80% whereas these hybrids thiophosphorylated in the presence of Ca^{2+} incorporated <10% $[^{32}P]$ phosphate. It was assumed that subtracting these values would give us the level of thiophosphorylation of the hybrids. The levels of thiophosphorylation determined by this procedure and estimated by 8 M urea gel electrophoresis of the thiophosphorylated light chain hybrids were in agreement. Percent relaxation, see Table 1.

chain became thiophosphorylated and calcium regulated was abolished, that is, the actin-activated MgATPase activity was high in the absence of calcium. In the case of the cardiac light chain hybrid, although this light chain is almost fully thio-phosphorylated, there was no effect on the inhibition of actin-myosin interaction observed with this hybrid. Similar results were obtained when the gizzard light chain hybrid was incubated with kinase and [^{32}P]γATP, although there was some variability between experiments (50 to 80% loss of calcium regulation). In all experiments, we observed a reasonable correlation between the degree of light chain phosphorylation and the loss of calcium regulation (Table 4). To test whether the effect of light chain phosphorylation was reversible, the gizzard light chain hybrid was phosphorylated using [^{32}P]γATP and kinase and then incubated with a gizzard muscle phosphatase preparation (Table 4). Phosphorylation of the hybrid caused a loss of calcium regulation, and dephosphorylation was accompanied by a restoration in calcium regulation. These results demonstrate that, in the gizzard light chain hybrid, the light chains are inhibitory and this inhibition can be overcome by either phosphorylation or calcium binding to the myosin.

MECHANISM OF ACTION OF THE REGULATORY LIGHT CHAINS

In view of the results with the scallop myosin hybrids, it is tempting to speculate that the regulatory systems in molluscan and vertebrate smooth muscles, mediated by the regulatory light chains, may operate by similar basic mechanisms, although the regulatory signals differ. Comparative studies on these systems may help to elucidate the mechanisms involved and establish whether there are significant dif-ferences. To understand how the regulatory light chains control the interaction between myosin and actin, we need to know their structure and their precise location on the myosin molecule.

On the basis of the sequence homology between the regulatory light chains and other members of the calcium binding protein family, for example, troponin C and

TABLE 4. *The effect of phosphorylation/dephosphorylation on calcium regulation in the gizzard light chain hybrid[a]*

DMF/gizzard LC hybrid	Moles ^{32}P/ mole LC	Actin-myosin interaction (AM Mg ATPase) (μmole ATP min$^{-1} \cdot$mg^{-1})		Percent relaxation
		$<10^{-7}$ M Ca^{2+}	$\sim 10^{-5}$ M Ca^{2+}	
Control	0.1	.024	.258	91
+ kinase + Ca^{2+}	0.6	.186	.284	35
then + phosphatase	0.1	.054	.248	80

The gizzard light chain hybrid was phosphorylated by gizzard kinase in the presence of 5 mM [^{32}P] γATP using the conditions described in Table 3. It was washed and suspended in 100 mM NaCl, 1 mM MgCl$_2$, 10 mM tris-HCl buffer pH 7.5, 1 mM DTT, and incubated at 25°C in the presence and ab-sence of ~100 μg·ml^{-1} gizzard muscle phosphatase for 30 min. The control sample was subjected to the same procedure but without addition of kinase and phosphatase. The level and specificity of [^{32}P] phosphate incorporation in the hybrid was determined as outlined in Table 3. Percent relaxation, see Table 1.

calmodulin, Kretsinger (24) predicted that they are structurally related. The common feature of this protein family is the presence of domains related to the EF hand of carp parvalbumin, the conformation of which has been determined by X-ray crystallography (24) and comprises an α helix E, a Ca^{2+} binding loop and an α helix F. Kretsinger suggested that the four domains (identified on the basis of sequence homology with parvalbumin) in the light chains might be arranged as predicted for troponin C, with the domains arranged in pairs and these pairs arranged so as to generate an overall compact molecule with a hydrophobic core. Hydrodynamic studies (42), however, indicate that the isolated regulatory light chains are rather long, asymmetric molecules. In an attempt to reconcile these apparently contradictory observations, a model similar to that shown in Fig. 4 was proposed by Bagshaw and Kendrick-Jones (5). Selective proteolytic cleavage studies on the regulatory light chains provide preliminary evidence to support this model. All the regulatory light chains tested—for example, molluscan (scallop, merceneria), vertebrate striated (skeletal, cardiac), and vertebrate smooth muscle (gizzard)—are selectively cleaved by carboxypeptidase Y preparations (Sigma) to generate two fragments (Fig. 3). Cleavage appears to be due to a contaminating proteinase A (at a level of about 1%) and involves a single cleavage in the centre of these light chains. The fragments remain intact even after prolonged exposure to carboxypeptidase, indicating that they must contain a certain degree of organised structure. The C terminal fragment binds to desensitized scallop myosin in the presence and absence of divalent cations but does not restore calcium regulation, whereas the N terminal fragment does not bind. Other members of the calcium binding protein family, such as troponin C, calmodulin, and the essential light chains, are not cleaved under these conditions, which suggests that these proteins have slightly different structures. In Fig. 4, these observations are summarised in the form of a speculative model for the structure of the regulatory light chains. The four domains in the regulatory light chains are arranged in pairs, with the N terminal fragment containing domains 1 and 2 and the C terminal fragment domains 3 and 4; these pairs are arranged so as to maximise the axial ratio. The region between the pairs of domains contains the site of proteolytic cleavage and may serve as a flexible hinge. The C terminal fragment (domains 3 and 4) may be involved in permanent attachment to the myosin head; the N terminal fragment (domains 1 and 2), which contains the nonspecific divalent cation binding site (4) and the phosphorylation site in the gizzard light chain (19), may be involved in switching on and off myosin head interaction with actin. Obviously, further work investigating the structure of the regulatory light chains when attached to the myosin head is required to test whether this model is correct.

WHERE ARE THE LIGHT CHAINS LOCATED ON THE MYOSIN MOLECULE?

Detailed studies by the groups of Andrew Szent-Györgyi and Carolyn Cohen have been carried out on the location of the regulatory light chains on scallop

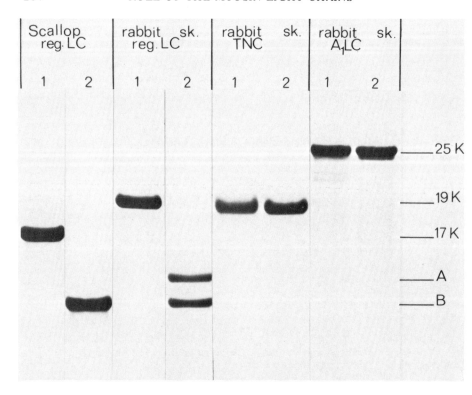

FIG. 3.   Selective proteolytic cleavage of the regulatory light chains. The proteins were digested with carboxypeptidase Y (Sigma) at a 125:1 protein:enzyme ratio (W/W) at 37°C for 5 hr (samples labeled *2*). The control samples (labeled *1*) were also incubated at 37°C for the same length of time. Digestion was terminated by addition of a final concentration of 1% sodium dodecyl sulphate (SDS) and the samples run on 20% polyacrylamide slab gels in the presence of SDS and 0.1 M tris-bicine pH 8.1. Digestion of all the vertebrate regulatory light chains results in two bands: *A* is the N terminal fragment; *B* is the C terminal fragment. Cleavage of the molluscan light chains yields only one band, that is, the two fragments have the same mobility in this gel system. The fragments can be separated by DEAE-cellulose chromatography using the buffer system described by Jakes et al. (19).

myosin. Their initial studies showed that the S_1 proteolytic fragment of myosin is not calcium sensitive whereas HMM and single-headed myosin, both of which contain the S_2 region of the myosin, are calcium sensitive (43). These results suggest that the S_2 region of the myosin molecule is essential for calcium regulation. Electron microscopic studies on scallop myosin and its subfragments by these groups (11,13) have further shown that the regulatory light chains are located in the neck region of the myosin head and may extend across the S_1-S_2 hinge region of the myosin. Although at present there is no detailed information on the position of the regulatory light chains in vertebrate smooth muscle and nonmuscle myosins, the results of proteolytic digestion studies would suggest the light chains in these myosins have a similar location to those in molluscan myosins, i.e. in the neck region of the myosin head near the S_1-S_2 junction. Gizzard HMM requires phosphorylation for interaction with actin (30,35), whereas gizzard S_1 can interact with actin regardless of whether the light chain is phosphorylated (P. Tooth and J. Kendrick-Jones,

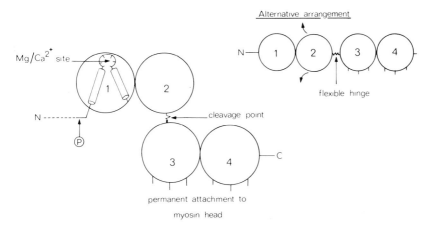

FIG. 4. A speculative model illustrating a possible "arrangement" of the myosin regulatory light chains, based on a model originally proposed by Bagshaw and Kendrick-Jones (5). The regulatory light chains on the basis of sequence homology are believed to contain four domains related to the E-F hand structure of parvalbumin. It is predicted (24) that these domains are arranged in pairs, i.e., 1 + 2 and 3 + 4. In between the pairs of domains, there may be a short flexible region, containing the site of proteolytic cleavage, which may allow the domain pairs to assume a number of alternative arrangements relative to each other so as to maximize their axial ratio. This hinge region may therefore play a role in the function of the light chains. Ⓟ denotes the position of the phosphoserine residue in the vertebrate regulatory light chains (19) and the *dotted line* indicates the extra residues at the N-terminal of these light chains compared with the scallop light chain.

unpublished results). This location is also supported by results from electron microscopy. When actin is decorated with gizzard S1 and HMM, containing intact regulatory light chains, the arrowheads formed have a "barbed" appearance (Fig. 5). They are similar to those obtained using scallop S1 or HMM containing regulatory light chains (11). Three-dimensional reconstructions of the scallop arrowheads indicate that the barbs are a result of the presence of the regulatory light chains in the neck region of the head (46). The barbs obtained with the gizzard proteins suggest a similar location in smooth muscle myosin.

This rather indirect evidence suggests that, in molluscan, smooth muscle, and nonmuscle myosins, the regulatory light chains may be located near the hinge region, linking the S_1 to the S_2 portion of the myosin molecule. In this position, one could speculate that the light chains could function by controlling the structure or mobility of the myosin head. However, to understand precisely how these light chains control myosin interaction with actin and decide whether they operate by similar mechanisms, we need further information on the structure of these light chains, their exact position on the myosin, and the changes that may occur in response to Ca^{2+} binding or phosphorylation.

ROLE OF LIGHT CHAIN PHOSPHORYLATION IN FILAMENT ASSEMBLY

The initial demonstration that in vertebrate smooth muscle myosin, light chain phosphorylation not only controls actin-myosin interaction but also appears to play

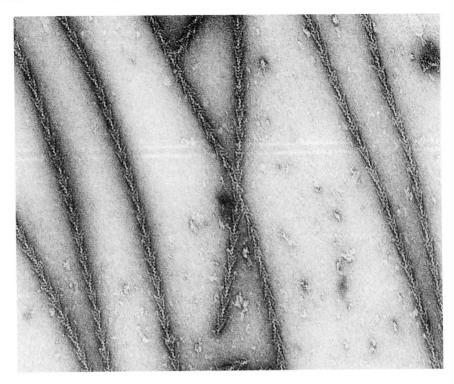

FIG. 5. Electron micrograph of "barbed" arrowheads formed by decorating F-actin with gizzard S1, containing intact, unphosphorylated regulatory light chain. S1 was prepared by papain digestion (500:1 protein to enzyme ratio W/W) of gizzard myosin ($\sim$ 10 mg.ml^{-1}) in 100 mM NaCl, 25 mM tris-HCl pH 7.5, 0.2 mM DTT at 25°C for 5 min. After removal of the undigested myosin and rod subfragments, the S1 containing supernatant in 40 mM NaCl, 5 mM Imidazole buffer pH 7.0, 0.2 mM DTT was loaded on to an ATP affinity column equilibrated in the same buffer plus 1 mM EDTA. S1 containing a degraded 20,000 MW light chain was eluted at 80 mM NaCl, whereas S1 containing an intact 20,000 MW light chain was eluted at 160 mM NaCl. Decoration was carried out according to the method of Craig et al. (11) and the grid was stained with 1% uranyl acetate. The filaments lie in a sheet of stain suspended over a hole in the carbon support film. Breakage and consequent anisotropic shrinkage of the stain in the electron beam leads to the somewhat variable appearance of the arrowheads seen. × 180,000.

a role in controlling myosin filament assembly was made by Suzuki et al. (44). We have confirmed these observations and have extended them to include vertebrate nonmuscle myosins (thymus and platelet) (23,33,34). Similar results have recently been obtained with the myosin isolated from thyroid cells (28).

The effect of light chain phosphorylation on the stability of vertebrate smooth muscle (gizzard) and nonmuscle myosins (thymus) was initially monitored by turbidity measurements using the procedure originally described by Suzuki et al. (44) (Fig. 6). Nonphosphorylated gizzard or thymus myosin filaments were incubated with calmodulin and light chain kinase in the absence of Ca^{2+}, and the turbidity recorded. The turbidity remained constant until ATP was added, whereupon a rapid drop in turbidity was observed, which remained low until Ca^{2+} was added to activate

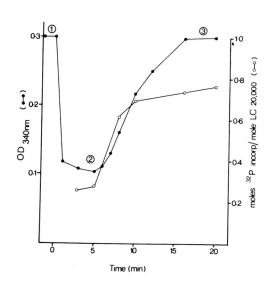

FIG. 6. Phosphorylation of the regulatory light chains of thymus myosin induces filament assembly. Nonphosphorylated myosin filaments (0.4 mg/ml) in 0.15 M NaCl, 10 mM MgCl₂, 0.5 mM EGTA, 0.2 mM DTT, 25 mM Imidazole pH 7.0 were incubated with 25 μg.ml⁻¹ brain calmodulin and ~ 50 μg.ml⁻¹ kinase at 20°C. Turbidity at λ = 340 nm *(closed circles)* was recorded in a Perkin Elmer double beam spectrophotometer. At zero time, 2.5 mM [³²P] γATP (2.5 μCi/μmole) was added and after 5 min calcium was added (Ca²⁺ ‹ 10⁻⁴ M). Aliquots were taken at the times indicated for determining [³²P] phosphate incorporation *(open circles)* into the light chain by the procedures outlined in Table 3. The specificity of phosphate incorporation into the 20,000 MW light chain was checked by SDS and urea gel electrophoresis and autoradiography (33).

the calmodulin/kinase complex. The resulting steady increase in turbidity was accompanied by phosphorylation of the regulatory light chains of the myosin. In the absence of kinase, there was no light chain phosphorylation or turbidity increase on addition of Ca²⁺. Filaments prepared from phosphorylated thymus and gizzard myosins were stable in mM ATP.

To verify that the turbidity measurements were monitoring the stability of the myosin filaments, aliquots of the myosin solution were taken for electron microscope examination (Fig. 7). The initial myosin solution contained numerous filaments, but on addition of ATP, these filaments disappeared. When Ca²⁺ was added, the steady increase in turbidity and in the level of light chain phosphorylation correlated with the reappearance of myosin filaments. These observations demonstrate that, under conditions of physiological ionic strength and MgATP concentrations, purified vertebrate smooth muscle and nonmuscle myosins are only assembled into filaments when the regulatory light chains are phosphorylated.

At present it is difficult to visualise how light chain phosphorylation associated with the myosin heads controls the interactions in the tail regions of the myosins believed to be involved in filament formation. The observation that filaments prepared from scallop and rabbit myosins and from nonphosphorylated gizzard light chain/scallop myosin hybrids are stable in mM MgATP indicates that this feature is a property of both the heavy chain and light chain of vertebrate smooth muscle and nonmuscle myosins. The demonstration that stoichiometric levels of MgATP cause nonphosphorylated gizzard and thymus myosin filaments to disassemble, whereas filaments formed from their respective rod subfragments were stable in mM MgATP (33), suggests that the MgATP binding site resides in the S₁ region of the molecule. An obvious candidate with the required affinity is the MgATPase site. The recent results (23) that indicate that ADP-vanadate (but not ADP or

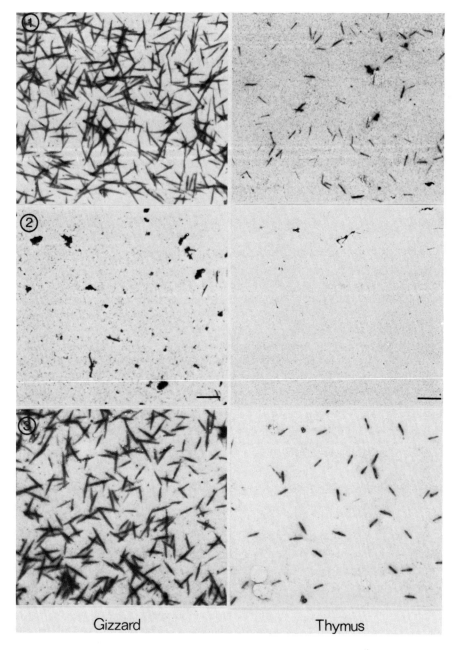

FIG. 7. Electron micrographs of thymus and gizzard myosin taken at the time points one, two, and three in the filament disassembly-assembly assay described in Fig. 5; 10 μl aliquots of the myosin filament assay solution described in Fig. 5 were taken and prepared for electron microscopy by the procedure described previously (33). The time points were: one, before ATP addition; two, 5 min after ATP addition; three, after calcium addition to 10^{-4} M Ca^{2+}, time 20 min. Bar = 1 μm.

vanadate alone or the ATP analogue AMP.PNP) disassemble nonphosphorylated gizzard and thymus myosin filaments would further suggest that the formation of a stable ternary complex (15), the myosin steady state complex M**.ADP.Pi, on the nonphosphorylated myosin head may be the event that leads to filament disassembly. Preliminary hydrodynamic studies indicate that the ATP-disassembled myosin species is a dimer (23,44). The polarity of this dimer has not yet been established but it is tempting to speculate that it has bipolar symmetry and may represent the building unit for the assembly of side polar or mixed polarity filaments observed in previous studies (10,17) on vertebrate smooth muscle and nonmuscle myosins. Further work is required to establish how phosphorylation/dephosphorylation of the light chain affects these myosin units and their packing into filaments.

Do these *in vitro* studies have any relevance to the state of myosin filament assembly in living vertebrate smooth muscle and nonmuscle cells? A plausible scheme for myosin assembly *in vivo*, based on these results, could be proposed. In relaxed smooth muscle or resting nonmuscle cells, the myosin is inhibited from assembling into filaments by the nonphosphorylated regulatory light chains. Following activation, the calmodulin-dependent kinase phosphorylates the light chains and removes their inhibitory effect. The myosin then assembles into filaments that interact with actin to generate the force for movement. We believe that such a scheme may be relevant to the situation in vertebrate nonmuscle cells, where motile events occur at different times and in different places and, hence, a transient assembly of the contractile filaments may be an essential requirement. In vertebrate smooth muscles, however, it is rather more difficult to envisage the physiological advantage of a regulated myosin filament assembly/disassembly process associated with the contraction/relaxation cycle. In these muscles, there are also a number of ultrastructural studies (39,41) that show that myosin filaments are present in the relaxed muscle. The discrepancy between these observations and the results obtained with purified smooth muscle myosin could be explained by postulating the existence of a nonmyosin component that stabilises the myosin filaments when the muscle relaxes. Thus, to determine whether myosin filament assembly mediated by light chain phosphorylation is involved in excitation-contraction coupling in vertebrate smooth muscle and nonmuscle cells, detailed ultrastructural studies are required to correlate myosin assembly with varying levels of light chain phosphorylation during *in vivo* contractile activity.

CONCLUSIONS

Regulatory light chains play a vital role in regulating muscle contraction and cell motility. In resting molluscan muscles, vertebrate smooth muscles, and vertebrate nonmuscle cells, the regulatory light chains inhibit the interaction between myosin and actin. When the cell is activated, this inhibition is released by calcium acting directly on the myosin in molluscan myosins or by calcium promoting the phosphorylation of the light chains in vertebrate nonmuscle and smooth muscle cells. Light chain phosphorylation also promotes the assembly of vertebrate nonmuscle

and smooth muscle myosins into filaments *in vitro*. At present, the *in vivo* significance of this finding is unclear, but it may well be involved in the transient assembly
of the contractile apparatus in nonmuscle cells. In vertebrate smooth muscles,
although the ultrastructural studies (39,41) suggest that myosin filament assembly
is not involved in excitation-contracting coupling, that is, filaments are present in
relaxed and contracting muscles, we believe that light chain phosphory-
lation/dephosphorylation may have an effect on the structure of the myosin filaments
in these muscles *in vivo*. There is no clear evidence that the homologous "regulatory"
light chains in vertebrate striated muscle myosins play a role in the primary control
of muscle contraction, either in response to direct calcium binding to myosin (6)
or in response to phosphorylation (29). It is possible that these light chains have
lost their regulatory function during the differentiation process or during isolation
and purification of the myosin. Alternatively, it is possible that these light chains
are involved in the modulation of the primary calcium response (7).

Although the regulatory systems mediated by direct calcium action in molluscan
myosins and by light chain phosphorylation in vertebrate smooth muscle and nonmuscle initially appear to be distinct, they do possess a number of common features.
The regulatory light chains from both systems are functionally interchangeable, at
least in scallop myosin. Sequence data (20,26) and divalent cation binding studies
(4) indicate that they probably have similar structures. Structural studies suggest
that they are probably located in the same region of their parent myosin molecules,
i.e. in the neck region of S_1, extending into the S_1-S_2 hinge region where they could
control the mobility and/or conformation of the myosin head. Thus, in the relaxed
state, the regulatory light chains may inhibit the interaction between myosin and
actin by preventing the myosin head from adopting the appropriate structure or
orientation necessary for its specific binding to actin which is crucial for contraction.

ACKNOWLEDGMENTS

We wish to thank Drs. Hugh Huxley, Ken Taylor, and Kate Shoenberg for
continued interest and support. R. Craig holds a fellowship from the MRC.

REFERENCES

1. Adelstein, R. S., and Conti, M. A. (1975): Phosphorylation of platelet myosin increases actin-
   activated myosin ATPase activity. *Nature*, 256:597–598.
2. Adelstein, R. S., and Eisenberg, E. (1980): Regulation and kinetics of the actin-myosin-ATP
   interaction. *Annu. Rev. Biochem.*, 49:921–956.
3. Aksoy, M. O., Williams, D., Sharkey, E. M., and Hartshorne, D. J. (1976): A relationship between
   Ca^{2+} sensitivity and phosphorylation of gizzard actomyosin. *Biochem. Biophys. Res. Commun.*,
   69:35–41.
4. Bagshaw, C. R., and Kendrick-Jones, J. (1979): Characterisation of homologous divalent metal
   ion binding sites of vertebrate and molluscan myosins using electron paramagnetic resonance
   spectroscopy. *J. Mol. Biol.*, 130:317–336.
5. Bagshaw, C. R., and Kendrick-Jones, J. (1980): Identification of the divalent metal ion binding domain of myosin regulatory light chains using spin-labelling techniques. *J. Mol. Biol.*, 140:411–433.
6. Bagshaw, C. R., and Reed, G. H. (1977): The significance of the slow dissociation of divalent metal
   ions from myosin 'regulatory' light chains. *FEBS Lett.*, 81:386–390.

7. Barany, K., Barany, M., Gillis, J. M., and Kushmerick, M. J. (1980): Myosin light chain phosphorylation during the contraction cycle of frog muscle. *Fed. Proc.*, 39:1547–1551.
8. Chantler, P. D., and Szent-Györgyi, A. G. (1978): Spectroscopic studies on invertebrate myosins and light chains. *Biochemistry*, 17:5440–5448.
9. Chantler, P. D., and Szent-Györgyi, A. G. (1980): Regulatory light chains and scallop myosin: Full dissociation, reversibility and cooperative effects. *J. Mol. Biol.*, 138:473–492.
10. Craig, R., and Megerman, J. (1977): Assembly of smooth muscle myosin into side-polar filaments. *J. Cell Biol.*, 75:990–996.
11. Craig, R., Szent-Györgyi, A. G., Beese, L., Flicker, P., Vibert, P., and Cohen, C. (1980): Electron microscopy of thin filaments decorated with a Ca^{2+}-regulated myosin. *J. Mol. Biol.*, 140:35–55.
12. Ebashi, S. (1980): Regulation of muscle contraction. *Proc. R. Soc. Lond. [Biol.]*, 207:259–286.
13. Flicker, P., Walliman, T., and Vibert, P. (1981): Location of regulatory light chains in scallop myosin. *Biophys. J.*, 33:279a.
14. Goldberg, A., and Lehman, W. (1978): Troponin-like proteins from muscles of the scallop, *Aequipecten irradians. Biochem. J.*, 171:413–418.
15. Goodno, C. C. (1979): Inhibition of myosin ATPase by vandate ion. *Proc. Natl. Acad. Sci. USA*, 76:2620–2624.
16. Hartshorne, D. J., and Gorecka, A. (1980): The biochemistry of the contractile proteins of smooth muscle. In: *Handbook of Physiology, Section 2, The Cardiovascular System. Vol. II. Vascular Smooth Muscle*, edited by D. F. Bohr, A. P. Somlyo, and H. V. Sparks, pp. 93–120. American Physiology Society, Bethesda, Maryland.
17. Hinssen, H., D'Haese, J., Small, J. V., and Sobieszek, A. (1978): Mode of filament assembly of myosins from muscle and nonmuscle cells. *J. Ultrastruct. Res.*, 64:282–302.
18. Huxley, H. E. (1969): The mechanism of muscular contraction. *Science*, 164:1356–1366.
19. Jakes, R., Northrop, F., and Kendrick-Jones, J. (1976): Calcium binding regions of myosin regulatory light chains. *FEBS Lett.*, 70:229–234.
20. Kendrick-Jones, J., and Jakes, R. (1976): Myosin-linked regulation: A chemical approach. In: *Internal Symposium on Myocardial Failure*, edited by G. Riecker, A. Weber, and J. Goodwin, pp. 28–40. Springer-Verlag, Berlin.
21. Kendrick-Jones, J., Lehman, W., and Szent-Györgyi, A. G. (1970): Regulation in molluscan muscles. *J. Mol. Biol.*, 54:313–326.
22. Kendrick-Jones, J., Szentkiralyi, E. M., and Szent-Györgyi, A. G. (1976): Regulatory light chains in myosins. *J. Mol. Biol.*, 104:747–775.
23. Kendrick-Jones, J., Tooth, P., Taylor, K. A., and Scholey, J. M. (1981): Regulation of myosin filament assembly by light chain phosphorylation. *Cold Spring Harbor Symp. Quant. Biol. (in press)*.
24. Kretsinger, R. H. (1980): Structure and evolution of calcium modulated proteins. *CRC Crit. Rev. Biochem.*, 8:119–174.
25. Lehman, W., and Szent-Györgyi, A. G. (1975): Regulation of muscular contraction. *J. Gen. Physiol.*, 66:1–30.
26. Maita, T., Chen, J.-I, and Matsuda, G. (1981): Amino acid sequence of the 20,000 molecular weight light chain of chicken gizzard muscle myosin. *Eur. J. Biochem.*, 117:417–424.
27. Marston, S. B., Trevett, R. M., and Walters, M. (1980): Calcium ion regulated thin filaments from vascular smooth muscle. *Biochem. J.*, 185:355.
28. Martin, F., Gabrion, J., and Cavadore, J-C. (1981): Thyroid myosin filament assembly-disassembly is controlled by myosin light chain phosphorylation-dephosphorylation. *FEBS Lett.*, 131:235–238.
29. Morgan, M., Perry, S. V., and Ottaway, J. (1976): A myosin light chain phosphatase from skeletal muscle. *Biochem. J.*, 157:687–697.
30. Onishi, H., and Watanabe, S. (1979): Chicken gizzard heavy meromyosin that retains the two light chain components, including a phosphorylatable one. *J. Biochem.*, 85:457–472.
31. Perry, S. V. (1979): The regulation of contractile activity in muscle. *Biochem. Soc. Trans.*, 7:593–617.
32. Pollard, T. D., and Korn, E. D. (1973): Acanthamoeba Myosin: Isolation from Acanthamoeba Castellanii of an enzyme similar to muscle myosin. *J. Biol. Chem.*, 248:4682–4690.
33. Scholey, J. M., Taylor, K. A., and Kendrick-Jones, J. (1980): Regulation of nonmuscle myosin assembly by calmodulin-dependent light chain kinase. *Nature*, 287:233–235.
34. Scholey, J. M., Taylor, K. A., and Kendrick-Jones, J. (1981): The role of myosin light chains in regulating actin-myosin interaction. *Biochimie*, 63:255–271.
35. Seidel, J. C. (1978): Chymotryptic HMM from gizzard myosin. A proteolytic fragment with regulatory properties of the intact myosin. *Biochem. Biophys. Res. Commun.*, 85:107–113.

36. Sellers, J. R., Chantler, P. D., and Szent-Györgyi, A. G. (1980): Hybrid formation between scallop myofibrils and foreign regulatory light chains. *J. Mol. Biol.*, 144:223–245.
37. Sherry, J. M. F., Gorecka, A., Aksoy, M. O., Dabrowska, R., and Hartshorne, D. J. (1978): Roles of calcium and phosphorylation in the regulation of the activity of gizzard myosin. *Biochemistry*, 17:4411–4418.
38. Simmons, R. M., and Szent-Györgyi, A. G. (1978): Reversible loss of calcium control of tension in scallop striated muscle associated with the removal of regulatory light chains. *Nature*, 273:62–64.
39. Small, J. V., and Sobieszek, A. (1980): The contractile apparatus of smooth muscle. *Int. Rev. Cytol.*, 64:241–306.
40. Sobieszek, A., and Small, J. V. (1976): Myosin-linked calcium regulation in vertebrate smooth muscle. *J. Mol. Biol.*, 102:75–92.
41. Somlyo, A. V. (1980): Ultrastructure of vascular smooth muscle. In: *Handbook of Physiology, Section 2, The Cardiovascular System. Vol. II. Vascular Smooth Muscle*, edited by D. F. Bohr, A. P. Somlyo, and H. V. Sparks, pp. 33–67. American Physiology Society, Washington, D.C.
42. Stafford, W. F., III, and Szent-Györgyi, A. G. (1978): Physical characterisation of myosin light chains. *Biochemistry*, 17:607–614.
43. Stafford, W. F., III, Szentkiralyi, E. M., and Szent-Györgyi, A. G. (1979): Regulatory properties of single-headed fragments of scallop myosin. *Biochemistry*, 18:5273–5280.
44. Suzuki, H., Onishi, H., Takahashi, K., and Watanabe, S. (1978): Structure and function of chicken gizzard myosin. *J. Biochem.*, 84:1529–1542.
45. Szent-Györgyi, A. G., Szentkiralyi, E. M., and Kendrick-Jones, J. (1973): The light chains of scallop myosin as regulatory subunits. *J. Mol. Biol.*, 74:179–203.
46. Vibert, P., and Craig, R. (1982): Three dimensional reconstruction of thin filaments decorated with a Ca^{2+} regulated myosin. *J. Mol. Biol.*, 157:299–319.
47. Wagner, P. D., and Giniger, E. (1981): Hydrolysis of ATP and reversible binding to F actin by myosin heavy chains free of all light chains. *Nature*, 292:560–562.
48. Weber, A., and Murray, J. M. (1973): Molecular control mechanisms in muscle construction. *Physiol. Rev.*, 53:612–673.

Basic Biology of Muscles: A Comparative Approach, edited by B. M. Twarog, R. J. C. Levine, and M. M. Dewey. Raven Press, New York © 1982.

Regulation of Contractile Proteins by Reversible Phosphorylation of Myosin and Myosin Kinase

Robert S. Adelstein, Mary D. Pato, James R. Sellers, Primal de Lanerolle, and Mary Anne Conti

Laboratory of Molecular Cardiology, National Heart, Lung and Blood Institute, Bethesda, Maryland 20205

Myosins isolated from vertebrate smooth muscle and nonmuscle cells (e.g., platelets, macrophages, fibroblasts) share important structural and enzymatic properties (1). These myosins are hexamers, being composed of a pair of heavy chains (molecular weight = 200,000) and two pairs of light chains (20,000 and 16,000 daltons) (Fig. 1). Smooth muscle myosin and myosins isolated from nonmuscle cells differ from vertebrate cardiac and skeletal muscle myosins with regard to their ability to hydrolyze MgATP in the presence of actin. This biochemical property is of importance since the actin-activated MgATPase activity of myosin is regarded to be the *in vitro* correlate of muscle contraction. Unlike striated muscle myosins, smooth muscle and nonmuscle myosins cannot be activated by actin unless they are covalently modified by the phosphorylation of the 20,000 dalton myosin light chain (1,14).

This chapter deals with the regulation of smooth muscle and nonmuscle myosins by kinases and phosphatases. It focuses on two different phosphorylation reactions.

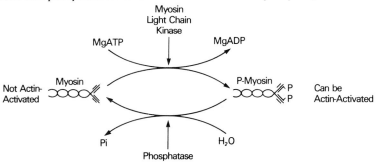

FIG. 1. Stick diagram of the myosin molecule illustrating the reversible phosphorylation of the regulatory light chains of myosin and the effect on the actin-activated MgATPase activity of myosin isolated from smooth muscle and nonmuscle cells.

First, the reversible phosphorylation of myosin, in which case phosphorylation is associated with an increase in contractile activity. Second, the reversible phosphorylation of the enzyme myosin kinase, in which case phosphorylation is associated with a decrease in contractile activity. Central to both these reactions are Ca^{2+} and the Ca^{2+}-binding protein calmodulin.

THE REVERSIBLE PHOSPHORYLATION OF MYOSIN

Recently we have presented evidence that the actin-activated MgATPase activity of turkey gizzard myosin as well as turkey gizzard heavy meromyosin (HMM), a chymotryptic fragment of myosin that retains its enzymatic activity, is dependent on the phosphorylation of the 20,000 dalton light chain of myosin (26). This experiment was carried out by reconstituting the purified components of a reversible phosphorylating system: myosin or HMM, myosin kinase, calmodulin (necessary for the activation of myosin kinase by Ca^{2+}), a smooth muscle phosphatase, and actin (Fig. 2). The results, which are presented in Table 1, show that unphosphorylated and dephosphorylated myosin and HMM cannot be activated to any significant extent by actin, but the fully phosphorylated (i.e., 2 moles of P_i incorporated/mol of myosin) and rephosphorylated species can be activated. Similar results were found when human platelet myosin was substituted for turkey gizzard smooth muscle myosin (Table 1).

Based on similar *in vitro* studies from a number of laboratories, the following hypothesis has been constructed. The actin-activated MgATPase activity of vertebrate smooth muscle and nonmuscle myosins is suppressed by the unphosphorylated 20,000 dalton light chain of myosin. This inhibition can be relieved when myosin kinase transfers the γ-phosphate from ATP to a particular serine residue located on this light chain. Suppression of the actin-activated MgATPase activity is restored when the myosin light chain is dephosphorylated (Fig. 1).

This hypothesis is consistent with the finding that proteolytic fragments of smooth muscle myosin, which lack the 20,000 dalton light chain, lose their ability to be regulated by phosphorylation. These fragments, which retain their ability to hydrolyze MgATP, are activated by actin in the absence of any covalent modification or added proteins (20,24). In contrast, proteolytic fragments, such as the HMM described above, which retain an intact 20,000 dalton light chain, continue to have their actin-activated MgATPase activity regulated by phosphorylation (26).

It is worth noting that the following difficulties have been encountered by investigators using these *in vitro* preparations. Denatured myosin can still undergo reversible phosphorylation. Hence it is possible to prepare a myosin that can be reversibly phosphorylated, but that cannot undergo actin-activation of its MgATPase activity. Smooth muscle myosin can undergo irreversible alterations such as oxidation of -SH groups (25) and, as outlined above, proteolysis of the 20,000 dalton myosin light chain (20,24), which result in a MgATPase activity that can be activated by actin, whether or not the 20,000 dalton light chain is phosphorylated. Thus, an important first step in reconstituting a reversible phosphorylating system is the preparation of a nondenatured, nonoxidized, and nondegraded myosin.

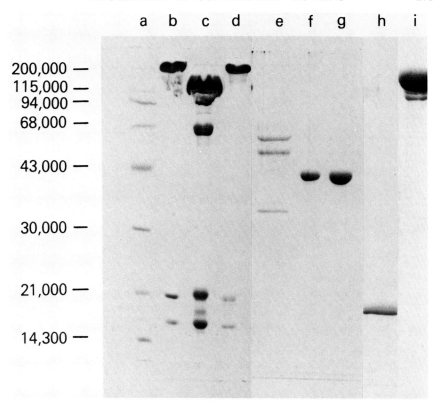

FIG. 2. 1% SDS - 12.5% polyacrylamide slab gel electrophoresis of the purified proteins used in studying the effects of reversibly phosphorylating myosin and HMM. **a**, molecular weight standards (BIORAD): 200,000, myosin heavy chains; 115,000, β-galactosidase; 94,000, phosphorylase b; 68,000, bovine serum albumin; 43,000, ovalbumin; 30,000, carbonic anhydrase; 21,000, soybean trypsin inhibitor; 14,300, lysozyme. **b**, turkey gizzard smooth muscle myosin; **c**, turkey gizzard heavy meromyosin; **d**, human platelet myosin; **e**, smooth muscle phosphatase I; **f**, rabbit skeletal muscle actin; **g**, turkey gizzard smooth muscle actin; **h**, porcine brain calmodulin; **i**, turkey gizzard myosin kinase. (From Sellers et al., ref. 26.)

The Role of Ca^{2+}

In the phosphorylating system outlined above, the role of Ca^{2+} is to activate the substrate-specific enzyme myosin light chain kinase. The part played by calmodulin in this reaction has recently been documented for the smooth muscle and platelet myosin kinase (2,9,10,15). Myosin kinase, unlike most calmodulin dependent enzymes, is *completely* inactive in the absence of the calcium-binding protein calmodulin. Calmodulin bears a close structural resemblance to the regulatory protein troponin C and, similar to it and other calcium binding proteins, calmodulin contains multiple sites for the binding of Ca^{2+} (18). In the relaxed muscle where $[Ca^{2+}] = 10^{-7}$ M, calmodulin is not capable of activating myosin kinase, but at Ca^{2+} levels of 10^{-5} M or greater, Ca^{2+} occupies three or four of the binding sites in calmodulin and the Ca^{2+}-calmodulin complex activates myosin kinase (4). When

TABLE 1. *Dependence of the actin-activated MgATPase activity of myosin on phosphorylation*

	Phosphate incorporation (mol P_i/mol myosin)	MgATPase (nmol P_i released/min/mg)	
		+ Actin	− Actin
Smooth muscle myosin			
Unphosphorylated	0	4	< 2
Phosphorylated	1.9	51	< 2
Dephosphorylated	0.1	5	
Rephosphorylated	2.0	46	
Smooth muscle HMM			
Unphosphorylated	0	10	< 2
Phosphorylated	1.9	357	< 2
Dephosphorylated	0.10	20	
Rephosphorylated	2.1	371	
Platelet myosin			
Unphosphorylated	0	5	
Phosphorylated	1.9	89	
Dephosphorylated	0.1	5	

From Sellers et al. (26).

TABLE 2. *Dependence of smooth muscle myosin light chain kinase on calcium and calmodulin*

Myosin kinase assay[a]	Incorporation into L.C./min[b] (dpm)	Kinase specific activity (μmol/mg per min)
+ Ca^{2+} + calmodulin	278,000	3.0
+ EGTA + calmodulin	3,000	—
+ Ca^{2+}	3,900	—
+ EGTA	2,900	—

[a]Myosin light chain kinase activity was assayed in 0.1 ml, 20 mM Tris-HCl, pH 7.3, 10 mM $MgCl_2$, 0.1 mM AT^{32}P (0.5 Ci/mmole), 10^{-8} M myosin kinase, 0.2 mg/ml smooth muscle myosin light chain at 24°C. Either 0.2 mM $CaCl_2$ (in excess over EGTA) or 2 mM EGTA was present in the assay and 10^{-6} M calmodulin was added to the assay as indicated above. Background level of dpm for the Millipore assay performed under the above conditions is 3,000 to 4,000. Assays in the absence of calmodulin or calcium showed no increase in dpm with time over the background (7).
[b]L.C. = 20,000 dalton myosin light chains.

the concentration of free Ca^{2+} decreases below 10^{-6} M, calmodulin dissociates from myosin kinase and the enzyme is inactivated (2). Table 2 shows the dependence of myosin kinase activity on Ca^{2+} and calmodulin.

Does Ca^{2+} also bind to and activate smooth muscle myosin directly, as in the case of scallop myosin (6)? This possibility has been suggested by work with myosin isolated from guinea pig vas deferens (5) and arterial smooth muscle (23). However, even in these muscles, where Ca^{2+} binding to myosin appears to play a role, this effect appears to be a secondary one because prior phosphorylation of the myosin is required.

Intact Preparations of Smooth Muscle

Work with intact smooth muscle preparations has, in general, supported the *in vitro* findings, but it also has raised at least one unanswered question. The state of myosin phosphorylation appears to correlate with tension so that high levels of phosphorylation (i.e., 0.7 moles of P_i/mole of 20,000 dalton light chain) appear to be associated with the greatest tension (3,11,12). Moreover, the time course of phosphorylation appears to precede or parallel that for the development of tension (11).

A major difficulty relates to the dephosphorylation of myosin and the decrease in smooth muscle tension. Although some investigators have found that complete relaxation of a previously contracted muscle correlates with complete dephosphorylation of myosin (3,11), an exact relationship between dephosphorylation and relaxation has not been established. Recently, Dillon et al. (12) have shown that a decrease in myosin phosphorylation to about 50% of maximum does not result in a decrease in tension. Until further experiments are carried out, it is difficult to know the exact meaning of these findings. One possibility might involve a site for Ca^{2+}-binding on myosin (see above) that remains occupied, so that despite a decrease in myosin phosphorylation, tension is maintained. A second possibility relates to the nature of the dephosphorylation reaction. A decrease in tension might require dephosphorylation of both heads of myosin. If there is an ordered dephosphorylation, so that one head on a myosin molecule is dephosphorylated in preference to both heads on the molecule, then a decrease in phosphate content would occur prior to decrease in tension. There are, of course, other possibilities, including a separate regulatory mechanism, not associated with myosin phosphorylation. The description of an alternative regulatory system to myosin phosphorylation appears in an accompanying chapter by Ebashi *(this volume)* (also see 13). Walters and Marston have recently suggested still another form of regulation in smooth muscle (29).

In summary, the *in vitro* data relating the actin-activated MgATPase activity to the state of myosin phosphorylation is quite reasonable. Although studies with intact smooth muscle preparations tend to agree with the *in vitro* studies, there appears at certain times to be a lack of correlation between dephosphorylation and relaxation. The cause for this apparent dissociation is still to be elucidated.

REVERSIBLE PHOSPHORYLATION OF MYOSIN KINASE

The enzyme that catalyzes the phosphorylation of myosin is itself a substrate for another kinase (8). This latter kinase is cAMP-dependent protein kinase, and phosphorylation of myosin kinase results in a decrease in myosin kinase activity at a given concentration of calmodulin. The decrease in myosin kinase activity is observed only if myosin kinase is phosphorylated when calmodulin is not bound. Under these circumstances, phosphate is introduced into two different serine residues and there is a major (15- to 20-fold) decrease in the affinity of myosin kinase for calmodulin (Fig. 3) (8). This decrease in the affinity of myosin kinase for calmodulin is reversible. Pato, of this laboratory, has purified a phosphatase (21) from smooth muscle that rapidly dephosphorylates myosin kinase and restores the tight binding between myosin kinase and calmodulin (8). Figure 3 is a double reciprocal plot showing the effect of phosphorylation and dephosphorylation on the apparent binding of calmodulin to myosin kinase.

Myosin Kinase Phosphatase

The enzyme catalyzing dephosphorylation of phosphorylated myosin kinase has been purified to homogeneity. Since we have purified a number of different phosphatases from smooth muscle, we have designated this enzyme as smooth muscle phosphatase I (21). It was purified using an affinity column of thiophosphorylated myosin light chain-Sepharose 4B. Thiophosphorylated light chains were chosen because they are hydrolyzed much more slowly than the ordinary phosphate derivative. Figure 4 shows the elution profile from this affinity column. The purified

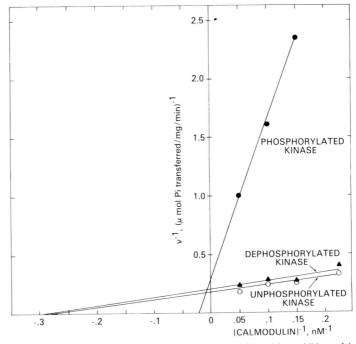

FIG. 3. Double reciprocal plot of the activation curves obtained by addition of increasing concentrations of calmodulin to aliquots of unphosphorylated *(open circles)*, phosphorylated *(closed circles)*, and dephosphorylated *(closed triangles)* smooth muscle myosin kinase. Myosin kinase was phosphorylated by incubation with the catalytic subunit of cAMP-dependent protein kinase and [^{32}P]ATP. It was dephosphorylated by incubation with smooth muscle phosphatase I. Note the marked effect of phosphorylation on the K_{app} for calmodulin (for details, see ref. 8).

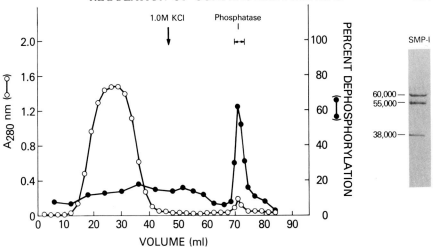

FIG. 4. Affinity chromatography of smooth muscle phosphatase I on thiophosphorylated myosin light chains-Sepharose. The column (1.5 × 11.5 cm) was equilibrated with 20 mm KCl, 20 mm Tris·HCl (pH 7.4), 2 mm EGTA, 5 mm EDTA, 1 mm dithiothreitol, at 4°C at 35 ml/hr. After loading the sample, the column was washed with about 1.5 column volumes of the equilibrating buffer. Elution of the phosphatase was carried out with a buffer containing 1 m KCl. Fractions of 1.2 ml were collected and assayed for phosphatase activity *(closed circles)*. The absorbance at 280 nm was measured *(open circles)*. An SDS-polyacrylamide gel of the purified phosphatase is shown on the *right*. SMP-I = smooth muscle phosphatase I. (From Pato and Adelstein, ref. 21.)

phosphatase is a trimer composed of different polypeptide chains (M_r = 68,000, 55,000 and 38,000) in a molar ratio of 1:1:1. It is very similar in its structure to a phosphatase isolated from other sources (19,28).

Despite the fact that this phosphatase actively dephosphorylates the isolated 20,000 dalton light chain of myosin (as well as myosin kinase), it has no significant activity when incubated with intact myosin. On the other hand, freezing the phosphatase in the presence of 0.2 M 2-mercaptoethanol results in the dissociation of the three subunits. Following this dissociation, which reveals that the 38,000 dalton peptide is the catalytic subunit, the phosphatase actively dephosphorylates intact myosin, in addition to the isolated light chain of myosin (22). Recently, we have identified a separate phosphatase in smooth muscle cells that appears to be specific for myosin (M. D. Pato and R. S. Adelstein, *unpublished observation*).

Significance of Myosin Kinase Phosphorylation

To date there is no direct evidence from preparations of intact smooth muscle concerning the role of myosin kinase phosphorylation in decreasing contractile activity. Silver and DiSalvo (27) showed that increasing cAMP-dependent protein kinase activity in a crude preparation of smooth muscle actomyosin resulted in a decrease in the phosphorylation of the 20,000 dalton myosin light chain. Although it is possible that myosin kinase was phosphorylated by the protein kinase, the enzyme was not identified in this preparation. Working with skinned smooth muscle fibers, Kerrick et al. (17) demonstrated the following: Diffusing the catalytic subunit of cAMP-dependent protein kinase into muscle fibers that had been previously contracted resulted in their immediate relaxation, with a concomitant decrease in myosin light chain phosphorylation. One of the proteins that was phosphorylated in this preparation was found to comigrate with smooth muscle myosin kinase. If, prior to introducing protein kinase, a high concentration of calmodulin was diffused into the preparation, relaxation did not occur. This latter finding is consistent with the *in vitro* observation that there is no change in the affinity of myosin kinase for calmodulin if the kinase is phosphorylated while calmodulin is bound.

A definitive physiological experiment establishing the role of myosin kinase phosphorylation in muscle relaxation has still to be carried out. The importance of this putative mechanism for muscle relaxation is that it might offer one explanation for why an increase in cAMP is associated with a decrease in contractile activity in both nonmuscle and smooth muscle cells.

SUMMARY

In vitro experiments support the idea that the actin-activated MgATPase activity of smooth muscle myosin and myosin from nonmuscle cells is regulated by the phosphorylation of the 20,000 dalton light chain of myosin. Experiments with intact smooth muscles support this mechanism but also raise the possibility that tension may be maintained in the presence of partial dephosphorylation (12). The possibility that smooth muscle contraction may also be modulated by additional regulatory systems (13,29) is to be expected based on experience with other types of muscle.

The enzyme myosin light chain kinase catalyzes the phosphorylation of the 20,000 dalton light chain of myosin. This enzyme requires Ca^{2+}-calmodulin for activity. The activity of myosin kinases that have been isolated from avian smooth muscle cells (8) or human platelets (16) can be decreased by phosphorylation. This phosphorylation is catalyzed by cAMP-dependent protein kinase and decreases myosin kinase activity by interfering with the binding of Ca^{2+}-calmodulin.

A number of different phosphatases have been purified from smooth muscle (22). These phosphatases play an important role in determining the state of phosphorylation of myosin and myosin kinase.

Two areas of particular interest at present are the regulation of phosphatase activity and the physiological significance of myosin kinase phosphorylation.

ACKNOWLEDGMENTS

The authors wish to acknowledge the expert editorial assistance of Mrs. Sharla Goldstein.

REFERENCES

1. Adelstein, R. S., and Eisenberg, E. (1980): Regulation and kinetics of the actin-myosin-ATP interaction. *Annu. Rev. Biochem.*, 49:921–956.
2. Adelstein, R. S., and Klee, C. B. (1981): Purification and characterization of smooth muscle myosin light chain kinase. *J. Biol. Chem.*, 256:7501–7509.

3. Barron, J. T., Barany, M., and Barany, K. (1979): Phosphorylation of the 20,000-dalton light chain of myosin of intact arterial smooth muscle in rest and contraction. *J. Biol. Chem.*, 254:4954–4956.
4. Blumenthal, D. K., and Stull, J. T. (1981): Activation of skeletal muscle myosin light chain kinase by calcium (2+) and calmodulin. *Biochemistry*, 19:5608–5614.
5. Chacko, S., Conti, M. A., and Adelstein, R. S. (1977): Effect of phosphorylation of smooth muscle myosin on actin-activation and on Ca^{++} regulation. *Proc. Natl. Acad. Sci. USA*, 74:129–133.
6. Chantler, P. D., and Szent-Györgyi, A. G. (1980): Regulatory light-chains and scallop myosin. Full dissociation, reversibility, and cooperative effects. *J. Mol. Biol.*, 138:473–492.
7. Conti, M. A., and Adelstein, R. S. (1980): Phosphorylation by cyclic 3'-5'-monophosphate-dependent protein kinase regulates myosin light chain kinase. *Fed. Proc.*, 39:1569–1573.
8. Conti, M. A., and Adelstein, R. S. (1981): The relationship between calmodulin binding and phosphorylation of smooth muscle myosin kinase by the catalytic subunit of 3' : 5' cAMP-dependent protein kinase. *J. Biol. Chem.*, 256:3178–3181.
9. Dabrowska, R., Aromatorio, D., Sherry, J. M. F., and Hartshorne, D. J. (1977): Composition of the myosin light chain kinase from chicken gizzard. *Biochem. Biophys. Res. Commun.*, 78:1263–1272.
10. Dabrowska, R., and Hartshorne, D. J. (1978): A Ca^{2+} and modulator-dependent myosin light chain kinase from non-muscle cells. *Biochem. Biophys. Res. Commun.*, 85:1352–1359.
11. de Lanerolle, P., and Stull, J. T. (1980): Myosin phosphorylation during contraction and relaxation of tracheal smooth muscle. *J. Biol. Chem.*, 255:9993–10000.
12. Dillon, P. F., Aksoy, M. O., Driska, S. P., and Murphy, R. A. (1981): Myosin phosphorylation and the cross-bridge cycle in arterial smooth muscle. *Science*, 211:495–497.
13. Ebashi, S. (1980): Regulation of muscle contraction. *Proc. R. Soc. Lond. [Biol.]*, 207:259–286.
14. Hartshorne, D. J., and Gorecka, A. (1980): The biochemistry of the contractile proteins of smooth muscle. In: *Handbook of Physiology Vol II*, edited by R. M. Berne, A. P. Somlyo, and H. V. Sparks, pp. 93–120. American Physiological Society, Bethesda, Maryland.
15. Hathaway, D. R., and Adelstein, R. S. (1979): Human platelet myosin light chain kinase requires the calcium-binding protein calmodulin for activity. *Proc. Natl. Acad. Sci. USA*, 76:1653–1657.
16. Hathaway, D. R., Eaton, C. R., and Adelstein, R. S. (1981): Regulation of human platelet myosin light chain kinase by the catalytic subunit of cyclic AMP-dependent protein kinase. *Nature*, 291:252–254.
17. Kerrick, W. G. L., Hoar, P. E., Cassidy, P. S., and Bridenbaugh, R. L. (1981): Skinned muscle fibers: The functional significance of phosphorylation and calcium-activated tension. *Cold Spring Harbor Conf. Cell Prolif.*, 8:887–900.
18. Klee, C. B., Crouch, T. H., and Richman, P. G. (1980): Calmodulin. *Annu. Rev. Biochem.*, 49:489–515.
19. Lee, E. Y. C., Mellgren, R. L., Killilea, S. D., and Aylward, J. H. (1978): Regulatory mechanisms of carbohydrate metabolism. In: *FEBS Symposium. Vol 42*, edited by W. Esmann, pp. 327–346. Pergamon Press, New York.
20. Okamoto, Y., and Sekine, T. (1978): Effects of tryptic digestion on the enzymatic activities of chicken gizzard myosin. *J. Biochem.*, 83:1375–1379.
21. Pato, M. D., and Adelstein, R. S. (1980): Dephosphorylation of the 20,000 dalton light chain of myosin by two different phosphatases from smooth muscle. *J. Biol. Chem.*, 255:6535–6538.
22. Pato, M. D., and Adelstein, R. S. (1982): Smooth muscle phosphatases from turkey gizzard. *Biophys. J.*, 37:263a.
23. Rosenfeld, A., and Chacko, S. (1981): Phosphorylation and calcium binding by myosin isolated from arterial smooth muscle. *Fed. Proc.*, 40:1786a.
24. Seidel, J. C. (1980): Fragmentation of gizzard myosin by α-chymotrypsin and papain, the effects on ATPase activity and the interaction with actin. *J. Biol. Chem.*, 255:4355–4361.
25. Seidel, J. C., and Nath, N. (1980): Sulfhydryl groups and enzymatic activity of gizzard myosin: chemical modification of the 17,000 dalton light chain. *Fed. Proc.*, 39:1934a.
26. Sellers, J. R., Pato, M. D., and Adelstein, R. S. (1981): Reversible phosphorylation of smooth muscle myosin, heavy meromyosin and platelet myosin. *J. Biol. Chem.*, 256:13137–13142.
27. Silver, P. J., and DiSalvo, J. (1979): Adenosine 3'-5'-monophosphate-mediated inhibition of myosin light chain phosphorylation in bovine aortic actomyosin. *J. Biol. Chem.*, 254:9951–9954.
28. Tamura, S., and Tsuiki, S. (1980): Purification and subunit structure of rat liver phosphoprotein phosphatase, whose molecular weight is 260,000 by gel filtration (Phosphatase IB). *Eur. J. Biochem.*, 111:217–224.
29. Walters, M., and Marston, S. B. (1981): Phosphorylation of the calcium ion-regulated thin filaments from vascular smooth muscle. *Biochem. J.*, 197:127–139.

Basic Biology of Muscles: A Comparative
Approach, edited by B. M. Twarog,
R. J. C. Levine, and M. M. Dewey.
Raven Press, New York © 1982.

Diversity of Regulatory Mechanisms and Regulation in Smooth Muscle

S. Ebashi

Department of Pharmacology, Faculty of Medicine, University of Tokyo, Hongo,
Tokyo 113, Japan

From the era of Galenos, the skeletal muscle had been considered as the most representative muscle, as if it were only muscle that could deserve scientific research. There can be found some reason for this, but it had often distorted studies on other muscles.

The scientists in the field of cardiac and smooth muscle were aware of the importance of Ca^{2+} in contractility in early days. This was particularly so in the case of cardiac muscle, as exemplified by the historical work of S. Ringer in the nineteenth century. This was not so clearly demonstrated in smooth muscle, where the whole situation is more complicated than cardiac muscle, but already at the beginning of this century we could see several papers that emphasized the importance of Ca^{2+} to the contractility of this muscle.

The apparent antagonizing effects of Ca^{2+} of skeletal muscle contractility, however, had prevented most scientists from thinking of the essential nature of Ca^{2+} in contractility. Because no conceptual distinction between excitation and contraction had existed, they tacitly ascribed differences in the response to Ca^{2+} in various muscles to differences in the intrinsic contractile mechanisms of these muscles.

As is well known, the first person who suggested the crucial role of Ca^{2+} in contractility was Heilbrunn (16), who showed in 1940 that if the frog sartorius muscle bundle, of which both ends had been cut, was immersed in isotonic $CaCl_2$, it shortened extensively until its length became less than one-fifth the original. From a current point of view, this phenomenon was of a complicated nature and did not necessarily indicate the direct role of Ca^{2+} in contractile processes. Even so, his intuition had sensed the hidden truth.

More direct evidence for the importance of Ca^{2+} was presented by Kamada and Kinosita (19). Kamada, who developed the technique to inject chemicals into cells using micropipette, and even measured the membrane potential of paramecium in 1934 (18) using the micropipette or "microelectrode," applied his technique to muscle and found that injected Ca^{2+} caused a local slow contraction:

The fibre can restore the original state once more so as to *respond to a second injection* as before. Hence the reaction induced by the micro-injection may be considered.as fundamentally of *a reversible nature* (italics added).[1]

Because this work was performed in the midst of World War II and published in a Japanese journal, it could not influence the following muscle research. Heilbrunn and Wiercinski (17) have made a similar but more extensive study in 1947 that is widely accepted as the pioneering work, having presented the evidence for the important role of Ca^{2+} in contractility.

Even the latter work, however, did not arouse due interest in Ca^{2+} among muscle scientists. The impact of the actomyosin-ATP system explored by A. Szent-Györgyi and his school was so great that virtually no attention had been paid to this "minor" molecule for many years until it was biochemically reexamined (cf. 8).

TROPONIN AND SMOOTH MUSCLE REGULATION

The recognition of the role of Ca^{2+} in skeletal contractile system has led us to find out "native tropomyosin" (4), namely, the troponin-tropomyosin system (10). This led us to believe that the same system should have been operating in smooth muscle, too, because we were also convinced of the traditional view that skeletal muscle should represent all kinds of muscle; everything existing in skeletal muscle would be somehow found in other muscles. We isolated native tropomyosin from gizzard (9) and rashly believed that it should have been composed of tropomyosin and troponin; no further investigation into the nature of gizzard native tropomyosin was then carried out.

Even the work of Kendrick-Jones et al. (20), which has shown the absence of troponin but the presence of myosin-linked regulation in some kinds of animals, did not change our minds. It took a few more years for us to become aware of our fault until the publication of Bremel (1), indicating the myosin-linked nature of gizzard actomyosin system.

We then restarted the work on smooth muscle and found that desensitized natural actomyosin, which was obtained by washing natural actomyosin, or myosin B, with a low ionic strength solution, was not activated by MgATP unless it was supplemented by native tropomyosin (14,15), in sharp contrast with the desensitized skeletal natural actomyosin, which is activated by MgATP without the aid of anything else. We wondered if this might not be due to denaturation incurred by desensitization procedures, but the success of developing a new method for preparing myosin in a pure state (5) has enabled us to conclude that the myosin-actin-ATP interaction in smooth muscle is fundamentally different from that in vertebrate skeletal and cardiac muscle, as illustrated in Fig. 1. The difference is so great that molluscan striated muscle is rather akin to vertebrate skeletal muscle.

---

[1]In work prior to this, Chambers and Hale (3) used micropipette for applying a strong $CaCl_2$ solution to an almost frozen muscle fiber, but this was made outside the cell, not into cytoplasm. It is interesting and strange, however, that the fiber responded to the solution with a vigorous contraction.

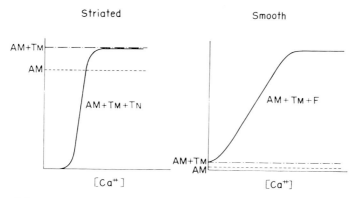

FIG. 1. Schematic illustration of the myosin-actin-ATP interaction of vertebrate smooth muscle as a function of free Ca^{2+} concentration in comparison with that of vertebrate striated muscle. AM: actomyosin. TM: tropomyosin. TN: troponin. F: regulatory factor of smooth muscle. (From Ebashi et al., ref. 14.)

In addition to this, we have noticed that gizzard myosin B was insensitive to the change in the MgATP concentration; the substrate-inhibition type of response to MgATP, quite pronounced in vertebrate skeletal muscle, cannot be seen with this muscle (gizzard myosin B is not sensitive also to the change in ionic strength, but this is not as marked as the above).

Because the competition test presumes that the fundamental reaction of the smooth muscle actomyosin system is essentially the same as that of the skeletal actomyosin system, that is, gizzard myosin and actin by themselves could actively respond to MgATP, the above result obtained by Bremel (1) did not necessarily indicate that the gizzard actomyosin system would be myosin-linked. However, it clearly pointed out our early misunderstanding (9) and has forced us to confront a new problem.

LEIOTONIN SYSTEM

Our subsequent effort to find out the regulatory factor, namely, the activating factor, led us to find an actin-linked factor, termed leiotonin, composed of leiotonin A, the regulatory moiety, and leiotonin C, the Ca-binding moiety. As for the nature of leiotonin, we have recently written several reviews (6,7,12,13,24), and, to avoid overlapping with them, will refer briefly to a problem concerning the preparation of leiotonin.

Previous preparation of leiotonin (23) was made when the involvement of acidic components, leiotonin C (22), and calmodulin (25), in the leiotonin and light chain kinase activities was not noticed. Furthermore, in examining the purity of preparation with sodium dodecyl sulfate polyacrylamide gel electrophoresis, we used a phosphate-buffered system, which is now shown to have poor resolution. In addition to the above, the activity of leiotonin is intensely lowered by DEAE column chromatography (myosin light chain kinase is also affected by this procedure). As a result, the previous preparation was not only low in its activity and yield, but also contaminated by light chain kinase, more or less, if examined in the presence of calmodulin. So we have to develop a new method of preparing leiotonin, which is now under progress. Figures 2 and 3 show our preliminary result along this line (13).

PHOSPHORYLATION OF MYOSIN LIGHT CHAIN

Even if native tropomyosin (9) or a similar preparation, which contains light chain kinase as well as leiotonin, is deprived of calmodulin, it usually retains a considerable part, sometimes almost full of activating effect on the actin-myosin-ATP interaction without phosphorylation of light chain (Fig. 4). This confirms the presence of an activating factor other than light chain kinase, but cannot exclude the role of the kinase in contractile processes (26).

Chacko et al. (2) and Onishi and Watanabe (25) reported the method to isolate a mysoin preparation of which the light chain had completely been phosphorylated, and that myosin thus prepared could positively interact with actin, irrespective of the presence or absence of Ca^{2+}. Using pure myosin as the starting material and following the method of Chacko et al. (2), we confirmed their results (S. Nakamura, *personal communication*).

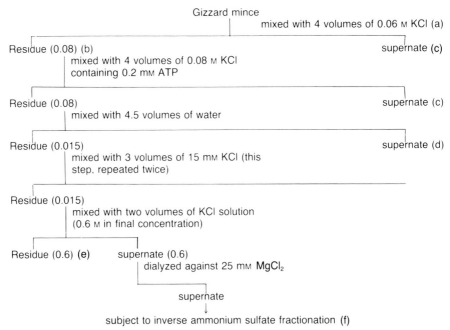

FIG. 2.   Preliminary steps for preparation of leiotonin. At each step, the mixture is incubated at 0°C for 1 hr and then centrifuged at 11,000 xg for 10 min. (a) All solutions contain 1 mM $NaHCO_3$, (b) figures in parentheses indicate ionic strengths, (c) starting material for myosin, (d) starting material for tropomyosin, (e) starting material for actin, and (f) see Fig. 3.

CONTRACTION   KINASE

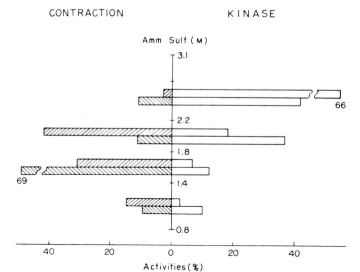

FIG. 3.   Contractile and myosin light chain kinase activities of each fraction obtained by inverse ammonium sulfate fractionation. Two cases are illustrated. Inverse ammonium sulfate fractionation means that the pellet sedimented by high concentrations of ammonium sulfate, e.g., 55 g per 100 ml of ammonium sulfate, was sequentially extracted by lower concentrations of ammonium sulfate (for details, see ref. 11). The fraction obtained between 1.4 and 1.8 M ammonium sulfate is used for further purification.

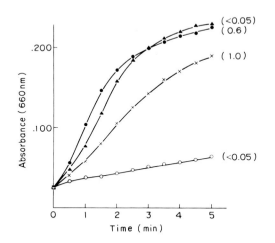

FIG. 4.   Superprecipitation induced by crude fractions containing leiotonin and light chain kinase. The reaction mixtures contain: 0.03 M KCl, 8 mM $MgCl_2$ 0.02 M Tris-maleate (pH 6.8), 0.5 mM ATP, 1.3×10^{-5} M Ca^{2+} given by Ca-GEDTA (glycoletherdiaminetetra-acetic acid, EGTA) buffer, 0.2 mg/ml myosin, 0.2 mg/ml actin, 0.05 mg/ml tropomyosin, and a fraction. The reaction, carried out at 21°C, was started by adding actin. *Open circles*, no addition; *x*, native tropomyosin (9), 35 μg/ml; *triangles*, crude leiotonin fraction (final fraction in Fig. 2) treated with Ultrogel (AC44), 3 μg/ml; *closed circles*, triangles plus 0.5 μg/ml calmodulin. *Figures in parentheses* indicate the kinase activities contained in each system expressed in relative values to that with native tropomyosin.

On the other hand, we can demonstrate such an experiment, as schematically illustrated in Fig. 5, without fail, which does not favor the active role of the kinase in activating the contractile process.

Our urgent problem is to find a solution that will reconcile these apparently contradicting results.

Ca²⁺ IS NOT NECESSARILY THE ACTIVATOR

The mechanism of troponin regulation is complicated, as shown in Fig. 1. As a whole, however, the role of Ca^{2+} is the activation in this case, too. Thus, it was our common belief that Ca^{2+} would be an activator in principle.

However, Kohama (21) has isolated from slime mold a factor that, in the presence of Ca^{2+}, acts as a true repressor on the actomyosin system, as shown in Fig. 6. At present we do not know whether or not the inhibitory factor is the real regulator of cell motility of slime mold, but this experimental fact indicates that the mode of action of Ca^{2+} is very diverse.

We have already been acquainted with the unique action of Ca^{2+} on the contractile element of vortex (27), Ca^{2+} being the energizer itself of the contractile system. Thus, the mechanisms explaining how Ca^{2+} controls biological systems are quite diverse. We should not have any foresight or prejudice about the modes of action of Ca^{2+}.

ACKNOWLEDGMENT

The author is indebted to the grant-in-aid from the Muscular Dystrophy Association, the Ministry of Education, Science and Culture, Japan; the Ministry of Health and Welfare, Japan; and the Iatrochemical Foundation.

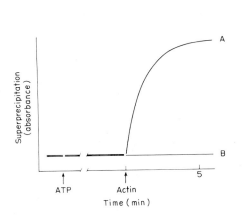

FIG. 5. Design of the experiment to show the effect of the light chain kinase activity on superprecipitation. Experimental conditions are similar to those described in the legend to Fig. 3, but [Ca²⁺] = 1 ~ 2 × 10⁻⁵ M is given by adding CaCl₂, not by Ca-GEDTA buffer. As the activating factor, native tropomyosin or crude leiotonin fraction, which contain a definite light chain kinase activity, is used. At the *first arrow*, ATP is added and after 5 to 10 min, when myosin is fully phosphorylated, actin is added together with GEDTA *(B)*, 0.5 ~ 1 mM in final concentration, or without GEDTA *(A)*. Since crude fractions contain a phosphatase activity more or less, myosin in *B* will eventually be dephosphorylated, but for the first 1 or 2 min, within which superprecipitation reaches nearly maximum, phosphorylated state is retained. Using the fraction which is completely devoid of phosphatase activity, the same result as above has been obtained. (S. Nakamura, *personal communication*.)

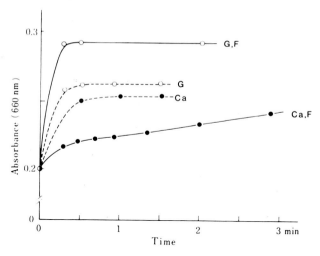

FIG. 6. Effect of Ca^{2+}-dependent inhibitory factor of slime mold on superprecipitation. Ca, G, and F indicate the presence of Ca^{2+}, GEDTA, and the inhibitory factor. Reaction mixture contains: 0.31 mg/ml slime mold myosin, 0.06 mg/ml slime mold actin, 8 mM $MgCl_2$, 10 mM KCl, 20 mM Tris-maleate buffer (pH 6.0), 10 μM $CaCl_2$ or 0.1 mM GEDTA, 0.2 mM ATP, and 0.03 mg/ml inhibitory factor. The reaction was started by adding ATP. (From Kohama, ref. 21.)

REFERENCES

1. Bremel, R. D. (1974): Myosin-linked calcium regulation in vertebrate smooth muscle. *Nature*, 252:405–407.
2. Chacko, S., Conti, M. A., and Adelstein, R. S. (1977): Effect of phosphorylation of smooth muscle myosin on actin activation and Ca^{2+} regulation. *Proc. Natl. Acad. Sci. USA*, 74:129–133.
3. Chambers, R., and Hale, H. P. (1932): The formation of ice in protoplasm. *Proc. R. Soc. Lond. [Biol.]*, 110:336–352.
4. Ebashi, S. (1963): Third component participating in the superprecipitation of 'natural actomyosin'. *Nature*, 22:1010–1012.
5. Ebashi, S. (1976): A simple method of preparing actin-free myosin from smooth muscle. *J. Biochem.*, 79:229–231.
6. Ebashi, S. (1980): Regulation of muscle contraction. *Proc. R. Soc. Lond. [Biol.]*, 207:259–286.
7. Ebashi, S. (1982): Actin-linked mechanism: leiotonin system. In: *Smooth Muscle Contraction*, edited by N. L. Stephens. Marcel Dekker, New York *(in press)*.
8. Ebashi, S., and Endo, M. (1968): Calcium ion and muscle contraction. *Prog. Biophys. Mol. Biol.*, 18:123–183.
9. Ebashi, S., Iwakura, H., Nakajima, H., Nakamura, R., and Ooi, Y. (1966): New structural proteins from dog heart and chicken gizzard. *Biochem. Z.*, 345:201–211.
10. Ebashi, S., and Kodama, A. (1965): A new protein factor promoting aggregation of tropomyosin. *J. Biochem.*, 58:107–108.
11. Ebashi, S., and Nakasone, H. (1981): Notes on preparation of leiotonin. *Proc. Japan Acad.*, 57B:217–221.
12. Ebashi, S., Nonomura, Y., and Hirata, M. (1982): Mode of calcium binding to smooth muscle contractile system. In: *Ca^{2+}-binding Proteins in Smooth Muscle*, edited by S. Kakiuchi and H. Hidaka, pp. 189–197. Plenum, New York.
13. Ebashi, S., Nonomura, Y., Nakamura, S., Nakasone, H., and Kohama, K. (1982): Regulatory mechanism in smooth muscle actin-linked regulation. *Fed. Proc.* (in press).
14. Ebashi, S., Nonomura, Y., Toyo-oka, T., and Katayama, E. (1976): Regulation of muscle contraction by the calcium-troponin-tropomyosin system. In Calcium in biological systems, *Symp. Soc. Exp. Biol.*, no. 30, pp. 349–360.

15. Ebashi, S., Toyo-oka, T., and Nonomura, Y. (1975): Gizzard troponin. *J. Biochem.*, 78:859–861.
16. Heilbrunn, L. V. (1940): The action of calcium on muscle protoplasm. *Physiol. Zool.*, 13:88–94.
17. Heilbrunn, L. V., and Wiercinski, F. J. (1947): The action of various cations on muscle protoplasm. *J. Cell. Comp. Physiol.*, 29:15–32.
18. Kamada, T. (1934): Some observations on potential differences across the ectoplasm membrane of Paramecium. *J. Exp. Biol.*, 11:94–102.
19. Kamada, T., and Kinosita, H. (1943): Disturbances initiated from naked surface of muscle protoplasm. *Japan J. Zool.*, 10:469–493.
20. Kendrick-Jones, J., Lehman, W., and Szent-Györgyi, A. G. (1970): Regulation in molluscan muscles. *J. Molc. Biol.*, 54:313–326.
21. Kohama, K. (1981): Ca-dependent inhibitory factor for the myosin-actin-ATP interaction of Physarum polycephalum. *J. Biochem.*, 90:1829–1832.
22. Mikawa, T., Nonomura, Y., Hirata, M., Ebashi, S., and Kakiuchi, S. (1978): Involvement of an acidic protein in regulation of smooth muscle contraction by the tropomyosin-leiotonin system. *J. Biochem. (Tokyo)*, 84:1633–1636.
23. Mikawa, T., Toyo-oka, T., Nonomura, Y., and Ebashi, S. (1977): Essential factor of gizzard 'troponin' fraction (A new type of regulatory protein). *J. Biochem. (Tokyo)*, 81:273–275.
24. Nonomura, Y., and Ebashi, S. (1980): Calcium regulatory mechanism in vertebrate smooth muscle. *Biomed. Res.*, 1:1–14.
25. Onishi, T., and Watanabe, S. (1979): Calcium regulation in chicken gizzard muscle and inosine triphosphate-induced superprecipitation of skeletal acto-gizzard myosin. *J. Biochem. (Tokyo)*, 86:569–573.
26. Sherry, J. M. F., Gòrecka, A., Aksoy, M. O., Kabrowska, R., and Hartshorne, D. J. (1978): Roles of calcium and phosphorylation of the activity of gizzard myosin. *Biochemistry*, 17:4411–4418.
27. Weis-Fogh, T., and Amos, W. B. (1972): Evidence for a new mechanism of cell motility. *Nature*, 236:301–304.

Basic Biology of Muscles: A Comparative Approach, edited by B. M. Twarog,
R. J. C. Levine, and M. M. Dewey.
Raven Press, New York © 1982.

Catch Muscle

Rudolf K. Achazi

*Free University Berlin, Institute of Animal Physiology and Applied Zoology,
D-1000 Berlin 41, Germany*

Catch refers to a phenomenon characteristic of certain muscles, which enter into a prolonged state of stretch resistance once they are contracted and maintain this state for hours, days, and even weeks. Catch muscles are usually smooth muscles. They are characterized by: (a) myofilaments of exceptional length; (b) slow shortening speed and high tension development; (c) the ability to perform both phasic and tonic contractions; during phasic contraction there is a direct relationship between the duration of stimulation and contraction, and during tonic contraction an active phase, in which tension and stretch resistance develop, is followed by a passive phase, in which the muscle shows stretch resistance only; (d) a low metabolic rate during the passive state; and (e) the serotonin-sensitivity of stretch resistance (77,98).

Since the second half of the last century, catch muscles were of scientific interest because of these peculiarities. In 1904, Frank (33) and Grützner (39) already debated the main topics, for example, the neural regulation of catch and the molecular basis of contraction, stretch resistance, and relaxation. These questions are still in discussion. Publications before 1975, have already been subject to extensive reviews by Hanson and Lowy (42), Needham (72), Rüegg (77), and Twarog (98,99). This chapter mainly deals with later publications and tries to put into perspective the importance of paramyosin phosphorylation and serotonin (5-HT) induced accumulation of cAMP (adenosine 3′, 5′-monophosphate) for stretch resistance and relaxation.

OCCURRENCE OF CATCH MUSCLES

According to the definition above, the following muscles show catch: (a) slow adductors of bivalves with smooth fibers, often referred to as white or opaque; (b) byssus retractor muscles of *Mytilus*, *Modiolus*, and *Pinna*, which are derived from the intrinsic foot muscles; (c) the intrinsic foot muscles of bivalves and the extrinsic foot muscles of some bivalves; and (d) the siphon muscles and the mantle-edge muscles of bivalves (15). In other systematic groups, catch is demonstrated for the pharynx and the penis retractor muscles of snails (16,106), and the body-wall muscles of nematomorpha (88). There are reports on catch-like tension in muscles of holothurioidea (33,104), sipunculidea (42), and even arthropoda (111). Particularly suitable for studying catch are the adductor muscles of bivalves and, above all, the anterior byssus retractor muscle (ABRM) of mytilidae.

STRUCTURE OF CATCH MUSCLES

Structure and Fine Structure of the Muscle Fibers

On the whole, catch muscles are organized like vertebrate smooth muscles. They are invested by an epimysium and separated into fiber bundles by a perimysium, which transsects the muscle radially (15). In the ABRM, the fiber bundles are 100 to 200 μm in diameter and run the length of the entire muscle (101). The uninucleate cells measure 1.2 to 1.8 mm in length and about 5 μm in diameter. They are encased in collagen sheets (endomysium) that bind the cells. In the endomysium, only glio-interstitial cells are observed, not connective tissue cells. The fibers are interconnected by numerous gap junctions. These probably facilitate the fast propagation of the cholinergic excitatory depolarization through the entire fiber bundle, which induces tonic contraction. Dense bodies (hemidesmosomes) are anchored in the cell membranes in regular intervals. They are particularly abundant in the cell apexes and often found opposed to dense bodies of neighboring cells. They are connected by fibrillar material with the intercellular collagen and may serve as cell junctions for the propagation of tension along the whole fiber bundle. They give rise to thin filaments (35,82,97a,102).

The sarcoplasmic reticulum is more elaborated than in most vertebrate smooth muscles. Arising in the superficial volume of the fiber, there is a net of tubular and cisternal structures that penetrates the depth of the fiber. Membrane junctions constitute dyad- and triad-like structures and are located close to the cell membrane.

Mitochondria are situated close to the sarcoplasmic reticulum. The fiber membrane, the vesicles of the sarcoplasmic reticulum, and the mitochondrial membranes are claimed to be involved in the excitation-contraction coupling (9,34,44,84,97a,102).

The Structure of the Contractile Apparatus

The Thin Filament Fraction

The contractile apparatus of catch muscles consists of thin and thick filaments running parallel to the main fiber axis. The thin filaments are considerably longer than those of vertebrate striated muscles (50). In the ABRM, they measure about 11 μm in length and 4 to 6 nm in diameter (46,82). The double helical filaments consist of bead-like globular subunits and are polar (24). Actin (MW 46,000) and tropomyosin (MW 35,000) can easily be isolated from the thin filament fraction (90). Troponin-like proteins with a molecular weight of 18,000, 23,000, and 60,000 could not be demonstrated until 1981. The thin filament preparation activates rabbit myosin 2 to 3 times in presence of Ca^{2+} (57).

In the ABRM, 60 to 80 thin filaments are attached to both ends of one dense body. These structures are 1.0 to 1.8 μm in length and 120 to 200 nm wide. They are distributed uniformly throughout the whole fiber. About 7.5% are attached to the cell membrane. In these dense bodies thin filaments are connected only with one side (46,82,97a,102).

The Thick Filament Fraction

Thick filaments of catch muscles are 5 to 20 times larger than those of noncatch muscles of bivalves and of vertebrate or insect striated muscles (50,59). In the smooth adductors of *Mercenaria* and *Anodonta*, they are 30 to 40 μm in length and up to about 140 nm in diameter (42,115). In the ABRM, they measure 20 to 45 μm in length. Their diameter has been said to change with the functional state of the muscle from 45 to 70 nm, being greatest in catch (36,82).

The extraordinary size of the thick filaments is connected with the amount of paramyosin present. As known from paramyosin free mutants of *Caenorhabditis elegans*, a nematode, the length of the thick filaments may be determined by paramyosin, whereas the diameter is to be attributed to myosin as well as paramyosin (63). Comparison of the size of paramyosin-containing thick filaments from differently organized muscles reveals that the length as well as the diameter increases with the paramyosin content (59).

Thick filaments of catch muscles are bipolar and taper gradually at each end. Their cross-sections are round to oval but somewhat irregular. Their core consists of paramyosin. Myosin forms an external layer on this core. The central zone is probably myosin free, as in the thick filaments of striated mollusc muscles. They are of rough appearance with globular portions projecting from the surface, due to myosin heads (19,27,28,73,91).

The main protein of the thick filaments is paramyosin (59). It is a rod-like molecule with an average length of 127.5 nm (19). It has a coiled-coil structure consisting of two α-helices wound around each other with both N-terminal groups probably on one end, and both C-terminal groups at the other end as in earthworm muscle (108). In *Mercenaria*, the subchains are probably identical (109), whereas in *Pecten*, the population of subchains seems to be heterogeneous (107). In *Mercenaria*, each molecule contains up to five covalently bound phosphate groups (21). As phosphate is incorporated into paramyosin of the ABRM and the adductor of *Mytilus* as well as of the adductors and the foot muscles of *Anodonta*, this protein can be classified as a phosphoprotein (2,4,38). The molecular weight of the native subchain is in the order of 100,000 to 110,000 dalton (4,83,91,112).

The spatial arrangement of the paramyosin rods in the core of the thick filaments is still debated, although there is agreement on the fact that the net pattern and the band pattern are images of the same structure from different visual angles. Heumann (45) proposed a helical arrangement of the planar sheets of assembled parallel paramyosin molecules based on micrographs of freeze-substituted ABRM. Elliott (29), however, who developed this helical model in colaboration with Lowy (30), argues for a planar arrangement of the paramyosin sheets.

The myosin molecule of smooth as well as of striated adductor muscles consists of one pair of heavy chains of about 195,000 to 200,000 dalton and two pairs of light chains of about 18,000 dalton (54,90). In clam foot muscle, the light chains have a molecular weight of 16,000 and 17,000 daton (8). With 1 to 10 mM EDTA (ethylene-diaminetetraacetic acid), one of the light chains can be stripped off the myosin head, as in case of the foot muscle the 17,000 dalton light chain. The EDTA-treated myosin preparation is desensitized; its activity is no longer suppressed in the absence of Ca^{2+}. This EDTA-light chain serves as a regulatory protein in molluscs. A very simple model assumes that, in the absence of Ca^{2+}, the actin binding sites of the myosins are blocked by this chain (94). The muscles of bivalves, therefore, contain a myosin-linked as well as an actin-linked regulatory system.

Tension-Length Behavior of a Catch Muscle

The contraction of catch muscles proceeds according to the sliding filament mechanism (41). The investigation of the tension-length behavior of the ABRM under defined experimental conditions results in a tension-length curve, hitherto only found in single fiber cell preparations of frog skeletal muscles (37). The curves obtained show three distinct phases: (a) an ascending limb, where the active tension production increases with decreasing length until the abrupt start of the plateau; (b) a well-defined plateau, on which tension remains constant despite decreasing length; and (c) a descending limb, which also starts abruptly (23). The simplest explanation of this behavior is to assume that the whole muscle consists of defined contractile units. Although sarcomeres have never been identified in electron micrographs, they have been deduced on the basis of quantitative analysis of micrographs (82).

The length-tension relationship, especially the abrupt changes on either side of the plateau, is consistent with a model having: (a) a highly ordered myofilament arrangement; (b) a constant length of thick and thin filaments; (c) a myosin-free central zone of the thick filaments; (d) the same amount of filament overlap in the "sarcomeres" of each cell at any particular muscle length; and (e) the behavior of the whole ABRM as a single cell. The implications of this model are not consistent in some points with our present knowledge, and need to be further investigated.

FUNCTIONAL STATES OF THE CATCH MUSCLES

Phasic Contraction

Catch muscles display both phasic and tonic contraction. In the ABRM, phasic contraction is elicited by stimulation of the cerebropedal connective or by brief repetitive pulses of current directly applied to the muscle (98). The muscle develops tension, which ceases when the stimulation is terminated (51,113). This behavior of the muscle is attributed to the simultaneous stimulation of excitatory and relaxing nerve fibers, the former releasing acetylcholine (ACh) and the latter serotonin (5-HT) and/or dopamine. The presence of these transmitters in the ganglia as well as in the muscle has been proved by biochemical and histochemical methods (18,47,

48a,65,78,89,94,103,106,110,114). Phasic contraction is elicited, too, by simultaneous application of 10^{-6} M 5-HT and 10^{-4} M ACh, 10^{-9} M 5-HT and 10^{-4} M ACh at temperatures above 30°C (95,96), and K^+ in concentrations higher than 200 mM in presence of external Ca^{2+} (43,94).

Changes in the intracellular free Ca^{2+} concentration during phasic contraction were studied with murexide as a Ca^{2+} indicator. The Ca^{2+} transient reaches its maximum during the rising phase of the isometric contraction and has almost returned to its initial level when the tension reaches its peak. There is a linear relation between the peak height of the isometric contraction and the total area of the Ca^{2+} transient (56). Thus, the muscle relaxes spontaneously after reduction of the level of free Ca^{2+} in the myoplasm. Therefore, excitation-contraction coupling should proceed as in the vertebrate cross-striated muscle (Fig. 4, SI).

In catch muscles, both the actin-linked and the myosin-linked regulatory systems are present (53,57,58). It is not yet known which one regulates the phasic contraction.

During phasic contraction, the maximal active tension recorded from the ABRM was 10.7 kg·cm^{-2} (51).

Tonic Contraction

Active Phase

In catch muscles, tonic contraction is induced by cathodal direct current pulses or short-time application of 10^{-6} to 10^{-3} M ACh (48,94,96). More prolonged exposure to concentrations higher than $5.5 \cdot 10^{-5}$ M ACh have been reported to cause irreversible damage to muscles (23). Tonic contraction is elicited, too, by 25 to 160 mM K^+ or caffeine in the external medium (43,86,87,97).

During ACh-induced contraction, Na^+ seems to be responsible for depolarization of the fiber membrane (48). This causes the release of Ca^{2+} from intracellular stores and its accumulation in the myoplasm (9). The Ca^{2+} transient during tonic contraction shows the same time course as during phasic contraction, that is, it has almost returned to its initial level when the tension reaches its peak (56). During ACh-induced contraction, Ca^{2+}-influx across the fiber membrane (102) is probably of minor importance because of the following: (a) ACh does not significantly alter the $^{45}Ca^{2+}$-influx (93), (b) extracellular Ca^{2+} has almost no effect on the contraction elicited by ACh, and (c) procaine inhibits contraction induced by ACh, which is thought to prevent the release of Ca^{2+} from vesicles of the sarcoplasmic reticulum (86). There is general agreement that Ca^{2+} in the K^+-induced contraction enters from the extracellular medium whereas in caffeine-induced contraction Ca^{2+} is released from internal storage sites (43,66,67,85,87).

In the ABRM, metabolic rate (107), directly related tension development (77), and recovery after quick release (51) are directly related during phasic contraction and the active phase of tonic contraction. In addition, the muscle may exhibit a sevenfold increase in stretch resistance (77). During tonic contraction tension, development is dependent on the increase in free Ca^{2+} concentration in the region of the myofilaments. The processes leading to tension development appear identical to those during phasic contraction. Simultaneously with tension development, the muscle acquires stretch resistance, may be Ca^{2+} independent (9,22,64). This has suggested the existence of a system in addition to the tension-developing system, which is in control of stretch resistance as already discussed by Rüegg (77).

Catch

After removal of ACh, the fiber membrane repolarizes gradually, but tension persists for about 1 hr or even longer (48,94,96). In this functional state, quick release is followed only by a partial recovery of tension (51). After vibration-induced inhibition of the contractile tension, the recovery of tension can only be recorded at the onset of the tonic contraction and not later on (60). Therefore, the muscle is in a state in which detached cross-bridges are no longer reattached to the same extent as during the active phase. Despite this "inactivity," the muscle displays a sevenfold increase in stretch resistance compared to a muscle in phasic contraction (77). In this functional state, the metabolic rate is reduced to about 9% of that during tension development (11). The muscle is in catch.

The molecular basis of catch is still unknown. Two hypothesises are discussed: the linkage hypothesis and the paramyosin hypothesis. The linkage hypothesis implies that actin myosin cross-bridges are the basis of stretch resistance. According to Lowy and Millman (61,62) and Nauss and Davies (71), the cross-bridges are formed during the active phase. In catch they are either locked in the attached state (71) or detach very slowly (61,62). Based on the energy requirement of muscles in catch, however, Baguet and Gillis (11) postulate a continuous cycling of the cross-bridges at low level during catch. The paramyosin hypothesis supposes that two different kinds of linkages are active in catch muscles. The tension development is caused by linkages between thick and thin filaments, whereas stretch resistance is brought about by interaction of thick filaments only. The interactions in catch may involve either paramyosin cross-bridges or merely frictional or shearing forces between adjacent thick filaments (52,75,76). Arguments against or in favor of one of these hypotheses can be based on biochemical or ultrastructural evidences.

Biochemical Investigations of Catch

A reasonable starting-point for biochemical investigations of catch is the analysis of the actomyosin-ATPase activity, as the low energy expenditure during catch implies that the state of activity of the ATPase has altered at the transition from the active phase of tonic contraction to catch. In vitro, the ATPase of catch muscles shows a maximal activity in the same order of magnitude as in vertebrate and insect striated muscles (31,42,77). Under experimental conditions the ATPase activity is directly regulated by Ca^{2+}; the concentration necessary for half maximal activation being only slightly higher than in vertebrate and insect muscles (80). The ATPase of molluscan actomyosin, however, exhibits a time-dependent reduction in activity (Fig. 1), especially at higher Ca^{2+} concentrations (31,81).

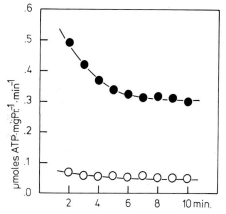

FIG. 1. ATPase activity of actomyosin of *Anodonta cygnea* adductor muscles at different times after the start of the reaction. Conditions: 25 mM K-phosphate (pH 7.0), 20 mM KCl, 1 mM MgCl₂, 0.1 mM DTE, 0.4 mM NADH, 0.4 mM DTE, 2.0 mM ATP, 1 unit LDH-PK. *Closed circles*, 0.1 mM CaCl₂; *Open circles*, 1 mM EGTA.

During phasic contraction and the active phase of tonic contraction, the ATPase activity *in vivo* seems also to be regulated by Ca^{2+}, whose concentration in the myoplasm is elevated compared to resting muscles (9,56). During catch, the Ca^{2+} concentration in the myoplasm seems to be as low as in resting muscles. If stretch resistance is maintained by cross-bridges, as assumed by the linkage hypothesis, the ATPase activity has to be controlled by a regulatory system independent of Ca^{2+}. This system might affect the actomyosin ATPase directly at low as well as at high Ca^{2+} concentrations because stretch resistance is already present during the active phase of tonic contraction. A clue to such a system furnishes the demonstration that the activity of actin-activated ATPase is regulated by the concentration of paramyosin (91). This inhibition, however, requires a special interaction of myosin and paramyosin. In co-filaments produced by slow precipitation, paramyosin activates the ATPase (31,73). In co-filaments prepared by rapid precipitation, paramyosin competes with F-actin in its effect on myosin. This result is not consistent with the suggestion that paramyosin stabilizes actomyosin interaction during catch, but is consistent with a system that regulates the affinity of myosin and F-actin (31).

Evidence in favor of a regulatory system based on paramyosin can be adduced as follows: Phosphorylation and dephosphorylation of enzymes and regulatory proteins is an important mechanism affecting many biological processes. The contractility of vertebrate smooth muscles, for example, is assumed to be controlled by the phosphorylation of myosin light chains (7). In the contractile apparatus of catch muscles, there are two phosphorylated proteins: an unidentified one with a molecular weight of about 300,000 dalton, and paramyosin (2,4,21,38). The extent of phosphorylation determines the solubility of paramyosin (21) and the degree of regularity of the 14.5 nm band pattern of paramyosin paracrystals (3). The extent of phosphorylation of paramyosin changes according to the state of contraction; for that reason, protein kinases incorporate 1.7 to 4.2 times as much phosphate into paramyosin of 5-HT-treated muscles as into paramyosin of ACh-treated or untreated muscles (4).

The degree of phosphorylation also affects the activity of actomyosin ATPase. This can be demonstrated with natural actomyosin as well as with actomyosin of elevated paramyosin content. Natural actomyosin of catch muscles can be separated into a pellet (P) with high myosin content and a supernatant (S) with low myosin content by reducing the KCl concentration from 600 to 350 mM. Both fractions contain about equal amounts of paramyosin. Into the proteins of the supernatant, 3.8 times as much phosphate was incorporated as into the proteins of the pellet. After phosphorylation of the supernatant, both fractions were mixed in the original ratio and coprecipitated. As shown in Fig. 2, the ATPase activity of the sedimented proteins is much more affected by the phosphorylated supernatant than by the unphosphorylated. The inhibition at 10^{-6} M Ca^{2+} is nearly two times higher than at 10^{-3} M Ca^{2+}. This result was confirmed using actomyosin of elevated paramyosin content extracted from 5-HT- or ACh-treated muscles (Fig. 3). These results indicate that paramyosin may adopt the function of a regulatory protein, controlling the ATPase activity of actomyosin according to the extent of its phosphorylation, independently of the amount of free Ca^{2+} present. As long as the molecular mechanism of paramyosin-phosphate actomyosin interaction is not known, there can only be speculations about its participation in catch. Considering the energy consumption during catch, an effect of paramyosin-phosphate on the velocity of the cross-bridge cycling seems reasonable, and not the stabilization of the cross-bridges (91). Paramyosin-phosphate may not, however, be involved in the catch if the mechanism involves merely a competition between paramyosin-phosphate and F-actin for myosin, as in rapidly precipitated co-filaments of myosin and paramyosin (31).

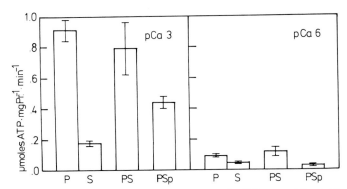

FIG. 2. The effect of phosphorylation on the actomyosin-ATPase activity at 10^{-3} M (pCa 3) and 10^{-6} M (pCa 6) Ca^{2+}. Actomyosin of *Mytilus eduls* catch muscles was separated into a pellet (P) and a supernatant (S) fraction, as described. The activity of the pellet proteins is compared with the activity of the coprecipitated proteins of unphosphorylated supernatant and pellet (PS) and of phosphorylated supernatant and pellet (PSp), at two different Ca^{2+} concentrations. Conditions: (a) phosphorylation: 50 mM di-Na-glycerol-1-phosphate (pH 6.5), 2 mM theophylline, 50 μM EGTA, 10 mM Mg-acetate, 20 mM NaF, 25 μg partially purified protein kinase, incubation temperature 25°C; after phosphorylation the proteins were dialyzed against ATPase buffer in the cold, to remove interfering salts. (b) ATPase: 5 mM imidazole (pH 7.0), 25 mM KCl, 2.5 mM $MgCl_2$, 5 mM ATP, 1 mM $CaCl_2$ (pCa 3) or 1 mM Ca-EGTA (pCa 6), incubation temperature 25°C; P_i liberated was measured with ammonium molybdate in a colorimetric test.

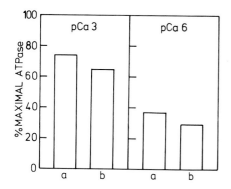

FIG. 3. Relative ATPase activity at 10^{-3} M (pCa 3) and 10^{-6} M (pCa 6) Ca^{2+} of phosphorylated actomyosin preparations extracted from ACh-treated (a) and from 5-HT-treated (b) catch muscles of *Mytilus edulis*. After phosphorylation, the actomyosin preparations were dialyzed against ATPase buffer to remove interfering salts. Controls were treated likewise, although ATP was omitted from the phosphorylation mixture. Conditions: see Fig. 2.

Cooley et al. (21) have proved that at pH 6.8 phosphorylated paramyosin is less soluble than unphosphorylated. Therefore, phosphorylation can be regarded as a mechanism to induce the crystallization of paramyosin and thus to increase the stiffness of thick filaments at the transition from active contraction to catch, as postulated by the paramyosin hypothesis (76,77). The consequence of increased stiffness of the individual thick filament is obviously not an elevated stiffness of the whole fiber, unless there are also linkages between the myofilaments. These linkages should, however, be actomyosin cross-bridges, because the thick filaments consist of a core of paramyosin covered by a layer of myosin molecules, with the possible exception of their central zone (27).

Both of these consequences of paramyosin phosphorylation could cooperate to elicit stretch resistance according to the following hypothesis. ACh activates the Ca^{2+}-dependent regulatory system, which induces tension development and the paramyosin-dependent regulatory system, which induces stretch resistance. The Ca^{2+}-dependent system is identical with that one active during phasic contraction. In the paramyosin-dependent system, ACh activates a paramyosin kinase, possibly a Ca^{2+}-calmodulin regulated enzyme. Thus, the phosphoryation of paramyosin is enhanced. The stiffness of the thick filaments increases and the velocity of the cross-bridge cycles slows down. The muscle is then in catch (Fig. 4).

Ultrastructural Changes of the Muscle in Catch

Reported changes in the organization of the contractile apparatus during induction of active contraction and catch in the adductor muscles of *Anodonta* (115), the penis retractor of snails (105), and the ABRM (36,46) include a multiplication of the number of thick filaments per cross-section, a decrease in the filament distances, and an increase in the diameter of thick filaments. In EDTA-treated fiber bundles of the ABRM, the number of thick filaments per cross-section increased up to threefold although the contraction of the muscle as a whole was isometric. This was interpreted as evidence that the thick filaments aggregate. There was no direct contact between individual thick filaments, nor did they fuse under these experimental conditions. During active contraction, cross-bridges between thick and thin

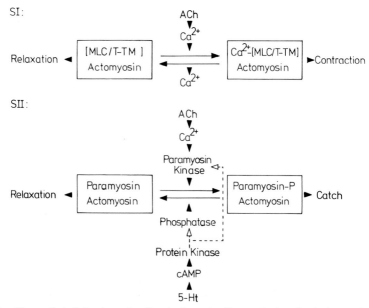

FIG. 4. The control of phasic contraction, tonic contraction, and relaxation in the catch muscle. **SI:** Ca²⁺-dependent regulatory system responsible for tension developement during phasic contraction and active phase of tonic contraction. **SII:** Paramyosin-dependent regulatory system responsible for stretch resistance during tonic contraction and for relaxation. ACh, acetylcholine; 5-HT, serotonin; cAMP, adenosine-3′,5′-monophosphate; MLC, myosin light chain; T-TM, troponin-tropomyosin. *Dashed lines* indicate alternative routes of the 5-HT-induced relaxation.

filaments were said to be clearly visible and abundant. After entering catch, a second kind of linkage became visible. These additional connections were said to be established between thick filaments only (36). The nature of the proteins participating in these linkages is unknown. These experiments were performed under conditions favoring the release of myosin molecules so that the observed linkages may be artefacts from the paramyosin core (73). Fusion of thick filaments was observed in glycerinated fibers, which did not exhibit 5-HT-sensitive stretch resistance (36).

Relaxation of Catch

Relaxation of muscles in catch can be elicted by brief repetitive current pulses (98), by 5-HT (88,94,106), and by dopamine (47,48a,74). Epinephrine, norepinephrine, indole-amines other than 5-HT, and catecholamine-related compounds also serve as relaxing compounds (20,67–69,94,100). Furthermore, two substances that do not interact with membrane-bound receptors induce relaxation: dibutyryl-cAMP, a membrane-soluble derivative of the second messenger cAMP, and theophylline, an inhibitor of cAMP degradation (20,25).

In the ABRM, 5-HT decreases stretch resistance gradually, the threshold concentration being 10^{-11} M. The relaxation is maximal at 10^{-7} to 10^{-6} M 5-HT (94).

Muneoka et al. (69) report on a seasonal variation in the sensitivity of catch muscles. Muscles of fresh winter animals seem to be less 5-HT-sensitive than those of fresh summer animals. The threshold concentration for dopamine is 10^{-8} M, and muscles of animals caught in winter seem to be more sensitive.

In *Mytilus californicus*, stimulation of the pedal ganglion causes release of 5-HT and dopamine in the muscle which is dependent on the stimulation frequency (78,79). Both compounds are probably released from different nerve endings (35,65). At a stimulation frequency of 10 Hz, the amount of dopamine released is about nine times greater than 5-HT (79). Nevertheless, most arguments speak in favor of 5-HT being the transmitter released from relaxing nerves (69,103).

At concentrations up to 10^{-6} M, 5-HT does not affect the membrane potential and probably does not increase the conductance for Na^+, K^+, or Cl^-. At concentrations of 10^{-5} M, 5-HT hyperpolarizes. Dopamine hyperpolarizes at 10^{-8} and greater (47,48a,66,94).

Intracellular Regulation of the Relaxation

Both 5-HT and dopamine induce relaxation by interaction with membrane-bound receptors (69,70,92). Thus, the relaxing mechanism should involve intracellular messengers, for example, a change in the concentration of free Ca^{2+} or cAMP. According to Nauss and Davies (71) and Bloomquist and Curtis, (12–14), 5-HT increases the Ca^{2+}-efflux of ABRM fiber bundles. Due to the time constant of this efflux, they assumed that Ca^{2+} is released by the proteins of the contractile apparatus. This explanation seems unlikely, as the regulatory protein-actomyosin complex should exhibit a high ATPase activity in presence of high Ca^{2+}-concentrations, even if the inhibitor, paramyosin-phosphate, is present. This is not the case in catch. The origin of this Ca^{2+}-efflux has not yet been explained satisfactorily.

In the ABRM, 5-HT is an extraordinarily potent activator of the adenylate cyclase (1,5,26,55). During phasic contraction and the active phase of the tonic contraction, the cAMP content of the muscle increases only slightly. After application of 5-HT to a muscle in catch, the cAMP content increases 5.2-fold prior to relaxation. Five minutes later the cAMP content has increased about 13-fold (1,5). In the ABRM, the quantity of cAMP accumulated in presence of 5-HT amounts to 1.2 nmoles·mg protein^{-1} in 10 min (25) and exceeds the amount accumulated in the heart-muscle of bivalves (49), the heart-muscle of frogs (17), the brain tissue of rats (32), and the integument of insects (6) by 50- to 300-times.

In the ABRM, the membrane-bound adenylate cyclase forms a complex with the 5-HT receptor protein and can be concentrated by differential centrifugation. In animals gathered in winter, the threshold of the receptor-enzyme complex is 10^{-7} M 5-HT in presence of GTP and an inhibitor of the monoamine-oxidase. At 10^{-8} M, 5-HT cAMP synthesis is already increased by about 30% however, not significantly. At 10^{-7} M 5-HT the activation is 254% and increases linearly up to 10^{-4} M 5-HT (A. Gies, *unpublished data*). In intact muscle fibers, the threshold is 10^{-6} M 5-HT (55). The difference in the threshold concentrations may be explained by the activity of the monoamine-oxidase present in the intact muscles.

Dopamine also activates the adenylate cyclase, but the accumulation amounts only to about 10% of that induced by 5-HT (55). The threshold concentration of the receptor-enzyme complex is 10^{-5} M dopamine (A. Gies, *unpublished data*). The 5-HT- and dopamine-induced cAMP accumulation in whole muscles is inhibited by methysergide, a weak blocker of 5-HT-receptors (55). The cAMP accumulation of muscle fibers and membrane preparations is inhibited by Ca^{2+} (5).

These results imply that cAMP may be the second messenger of 5-HT and dopamine during relaxation of catch. This assumption is supported by experiments with intact and EDTA- or Triton X-100-treated muscle fibers. As already pointed out, the catch in intact ABRM can be released by dibutyryl-cAMP and theophylline, and dibutyryl-cAMP potentiates the action of theophylline (20,26). In skinned muscle fibers, cAMP itself induces relaxation (10,64). In glycerinated muscle fibers neither cAMP nor 5-HT are effective. This may be explained by the fact that gylcerination removes nearly all soluble proteins and leaves the proteins of the contractile apparatus.

There are two possible ways for cAMP to release catch: by activation of the translocation of Ca^{2+} from the myofilaments into the intracellular stores and across the fiber membrane, or by activation of the phosphorylation and dephosphorylation of proteins of the contractile apparatus or proteins that affect the activity of these proteins.

The 5-HT-induced Ca^{2+}-efflux of the ABRM already mentioned speaks in favor of the translocation hypothesis (12–14,71). This is, however, not consistent with: (a) the Ca^{2+} transient during tonic contraction (56), (b) the location of Ca^{2+} during catch (9), (c) the maintenance of catch at Ca^{2+} concentrations as low as 10^{-9} M in EDTA- and Triton X-100-treated fibers (64), (d) the demonstration that catch can only be elicited at Ca^{2+} concentrations as low as 10^{-7} M in saponin-treated fibers (22), and (e) the lack of dependence of the Ca^{2+}-flux of membrane vesicles on cAMP, even when there is concurrent phosphorylation of sarcoplasmic proteins by cAMP dependent protein kinases (5,40).

The cAMP-system of catch muscles was investigated to check the possibility of cAMP-induced relaxation. In full homogenates of catch muscles of *Mytilus* and *Anodonta*, the presence of cAMP-dependent protein kinases could be demonstrated. A fraction of these enzymes is closely associated with the proteins of the contractile apparatus, especially with paramyosin. Soluble, as well as paramyosin-associated, protein kinases can be separated into cAMP-dependent and independent enzymes by chromatography (3,4,40). With regard to phosphorylated proteins, only a fraction of the contractile proteins have been analyzed (2,4,38). Proteins of the cytosol and the membranes, however, are also phosphorylated by cAMP-dependent kinases. Amongst the soluble proteins, a heat-stable modulator protein was isolated. It competitively inhibits the phosphorylation (40). These preliminary investigations prove that catch muscles contain cAMP-dependent and independent systems likely to belong to different regulatory pathways, including the ACh-induced stretch resistance and the 5-HT-induced relaxation.

The molecular mechanism of the 5-HT-induced cAMP regulated relaxation process of muscles in catch is not known. Assuming that stretch resistance is caused by phosphorylation of paramyosin, 5-HT-induced relaxation should involve the dephosphorylation of paramyosin. The dephosphorylation may be brought about either by activation of a paramyosin-specific phosphatase or, more likely, by inactivation of a paramyosin-specific protein kinase. Activation as well as inactivation may be due to cAMP-dependent protein kinases. In the second case, paramyosin phosphoryation may be continually counteracted by dephosphoryation (Fig. 4, SII).

CONCLUSION

Based on the results presented, the following mechanism controlling the functional states of the catch muscle is proposed. Phasic contraction is controlled by the Ca^{2+}-dependent regulatory system (Fig. 4, SI). Thereby, ACh activates the translocation of Ca^{2+} from the intracellular stores to the myofilaments. Ca^{2+} counteracts the inhibitory effect of the regulatory proteins (EDTA-light chains and/or troponin-tropomyosin) and the muscle develops tension. As soon as stimulation ceases, the Ca^{2+} flux reverses and relaxation occurs spontaneously; the presence of 5-HT inhibits the activation of the paramyosin-dependent system.

During tonic contraction, tension development is also controlled by the Ca^{2+}-dependent system. Simultaneously, the paramyosin-dependent system (Fig. 4, SII) is activated and induces stretch resistance. In this process ACh activates a paramyosin kinase by means of Ca^{2+}-calmodulin, and paramyosin is phosphorylated. The property of the actomyosin-ATPase changes, resulting in a decreased velocity of cross-bridge cycling, and increased stiffness of the thick filaments. The muscle becomes stretch resistant (catch).

Spontaneous relaxation of catch involves the slow dephosphorylation of paramyosin by intrinsic phosphatase; 5-HT accelerates the relaxation process by activating the dephosphorylation of paramysoin by a cAMP-dependent system. Dephosphoryation entails the reduction of the stiffness of the thick filaments and removes the inhibition of the actomyosin-ATPase.

REFERENCES

1. Achazi, R. K. (1973): Zyklisches AMP (cAMP) im vorderen Byssus-Retraktor-Muskel (ABRM) von *Mytilus edulis*. *Arzneim Forsch.*, 23:624.
2. Achazi, R. K. (1976): Das 3'-5'-Adenosinmonophosphatsystem im glatten Molluskenmuskel. *Verh. Dtsch. Zool. Ges.*, 1976:241.
3. Achazi, R. K. (1979): 5-HT induced accumulation of 3',5'-AMP and the phosphorylation of paramyosin in the ABRM of *Mytilus edulis*. *Malacologia*, 18:465–468.
4. Achazi, R. K. (1979): Phosphorylation of molluscan Paramyosin. *Pfluegers Arch.*, 379:197–201.
5. Achazi, R. K., Dölling, B., and Haakshorst, R. (1974): 5-HT-induzierte Erschlaffung und zyklisches AMP bei einem glatten Molluskenmuskel. *Pfluegers Arch.*, 349:1–9.
6. Achazi, R. K., Haakshorst, R., Volmer, H., and Surholt, B. (1977): The 3',5'-Adenosin-monophosphate system during the larval-adult moulting cycle in the migratory locust, *Locusta migratoria*. *Insect Biochem.*, 7:21–26.
7. Adelstein, R. S. (1979): Phosphorylation of muscle contractile proteins. *Fed. Proc.*, 39:1544–1546.

8. Ashiba, G., Asada, T., and Watanabe, S. (1980): Calcium regulation in clam foot muscle. *J. Biochem. (Tokyo)*, 88:837–846.

9. Atsumi, S., and Sugi, H. (1976): Localization of Calcium-accumulating structures in the anterior byssal retractor muscle of *Mytilus edulis* and their role in the regulation of active and catch contractions. *J. Physiol. (Lond.)*, 257:549–560.

10. Baguet, F., and Dumont, G. (1973): The muscular membrane and the relaxation effect of serotonin in a Lamellibrach smooth muscle (ABRM). *Arch. Int. Physiol. Biochim.*, 81:769–770.

11. Baguet, F., and Gillis, J. M. (1968): Energy cost of tonic contraction in a lamellibranch catch muscle. *J. Physiol. (Lond.)*, 198:127–143.

12. Bloomquist, E., and Curtis, B. A. (1972): The action of serotonin on calcium-45 efflux from the anterior byssal retractor muscle of *Mytilus edulis*. *J. Gen. Physiol.*, 59:476–485.

13. Bloomquist, E., and Curtis, B. A. (1975): Ca efflux from the anterior byssus retractor muscle in phasic and catch contraction. *Am. J. Physiol.*, 229:1237–1243.

14. Bloomquist, E., and Curtis, B. A. (1975): Net calcium fluxes in the anterior byssus retractor muscle with phasic and catch contraction. *Am. J. Physiol.*, 229:1244–1248.

15. Bowden, J. (1958): The structure and innervation of lamellibranch muscle. *Int. Rev. Cytol.*, 7:295–335.

16. Bozler, E. (1930): Untersuchungen zur Physiologie der Tonusmuskeln. *Z. Vergl. Physiol.*, 12:579–602.

17. Brooker, G. (1973): Oscillation of cyclic adenosine monophosphate concentration during the myocardial contraction cycle. *Science*, 182:933–934.

18. Cambridge, G. W., Holgate, J. A., and Sharp, J. A. (1959): A pharmacological analysis of the contractile mechanism of *Mytilus edulis*. *J. Physiol. (Lond.)*, 148:451–464.

18a. Cohen, C., and Holmes, K. C. (1963): X-ray diffraction evidence for α-helical coiled-coils in native muscle. *J. Mol. Biol.*, 6:423–132.

19. Cohen, C., Szent-Györgyi, A. G., and Kendrick-Jones, J. (1971): Paramyosin and the filaments of molluscan "catch" muscles. I. Paramyosin: Structure and assembly. *J. Mol. Biol.*, 56:223–237.

20. Cole, R. A., and Twarog, B. M. (1972): Relaxation of catch in a molluscan smooth muscle. I. Effects of drugs which act on the adenyl cyclase system. *Comp. Biochem. Physiol.*, 43A:321–330.

21. Cooley, L. B., Johnson, W. H., and Krause, S. (1979): Phosphorylation of paramyosin and its possible role in the catch mechanism. *J. Biol. Chem.*, 254:2195–2198.

22. Cornelius, F. (1980): The regulation of tension in a chemically skinned molluscan smooth muscle. *J. Gen. Physiol.*, 75:709–725.

23. Cornelius, F., and Lowy, J. (1978): Tension length behavior of a molluscan smooth muscle related to filament organization. *Acta Physiol. Scand.*, 102:167–180.

24. Craig, R., Szent-Györgyi, A. G., Beese, L., Flicker, P., Vibert, P., and Cohen, C. (1980): Electron microscopy of thin filaments decorated with a Ca^{2+}-regulated myosin. *J. Mol. Biol.*, 140:35–55.

25. Dölling, B. (1972): *Zur Erschlaffung von Muskeln nach dem Sperrtonus*. Staatsexamensarbeit, Münster, NRW.

26. Dölling, B., Achazi, R. K., Zebe, E., and Ahlert, U. (1972): 3', 5'-Adenosinmonophoshat im Byssusretraktor der Miesmuschel *Mytilus edulis*. *Naturwissenschaften*, 59:313.

27. Elfvin, M., Levine, R. J. C., and Dewey, M. M. (1976): Paramyosin in invertebrate muscles. I. Identification and localization. *J. Cell Biol.*, 71:261–272.

28. Elliott, A. (1974): The arrangement of myosin on the surface of paramyosin filaments in the white adductor muscle of *Crassostrea angulata*. *Proc. R. Soc. Lond. [Biol.]*, 186:53–66.

29. Elliott, A. (1979): Structure of molluscan thick filaments: A common origin for diverse appearances. *J. Mol. Biol.*, 132:323–341.

30. Elliott, A., and Lowy, J. (1970): A model for the coarse structure of paramyosin filmaments, *J. Mol. Biol.*, 53:181–203.

31. Epstein, A. F., Aronow, B. J. and Harris, H. E. (1976): Myosin-paramyosin cofilaments: Enzymatic interactions with F-actin. *Proc. Natl. Acad. Sci. USA*, 73:3015–3019.

32. Forn, J., Krueger, B. K., and Greengard, P. (1974): Adenosine 3',5'-monophosphate content in rat caudate nucleus: Demonstration of dopaminergic and adrenergic receptors. *Science*, 186:1118–1120.

33. Frank, O. (1904): Thermodynamik des Muskels. *Ergeb. Physiol.*, 3.II:348–514.

34. Gilloteaux, J. (1976): Les spécialisations membranaires des fibres musculaires lisses d'un mollusque: Ultrastructure du muscle rétracteur antérieur du byssus (ABRM) de *Mytilus edulis*, L. (Mollusca Pelecypoda). *Cytobiologie*, 12:440–456.

35. Gilloteaux, J. (1976): Les connexions intracellulaires d'un muscle lisse: Ultrastructure du muscle rétracteur du byssus (ABRM) de *Mytilus edulis*, L. (Mollusca Pelecypoda). *Cytobiologie*, 12:457–472.

36. Gilloteaux, J., and Baguet, F. (1977): Contractile filaments organization in functional states of the anterior byssus retractor muscle (ABRM) of *Mytilus edulis*, L. *Eur. J. Cell Biol.*, 15:192–220.

37. Gordon, A. M., Huxley, A. F., and Julian, F. J. (1966): The variation in isometric tension with sarcomer length in vertebrate muscle fibers. *J. Physiol. (Lond.)*, 184:170–191.

38. Gottschalk, E. (1979): *Paramyosin von Süßwassermuscheln und seine Phosphorylierung durch Kinase*. Staatsexamensarbeit, Münster, NRW.

39. Grützner, P. (1904): Die glatten Muskeln. *Ergeb. Physiol.*, 3.II:12–88.

40. Haakshorst, R. (1977): *Untersuchung über Vorkommen und Eigenschaften von Proteinkinasen in den Muskeln von* Mytilus edulis. Dissertation. Münster, NRW.

41. Hanson, J., and Lowy, J. (1959): Evidence for a sliding filament contractile mechanism in tonic smooth muscles of lamellibranch molluscs. *Nature*, 184:286–287.

42. Hanson, J., and Lowy, J. (1960): Structure and function of the contractile apparatus in the muscles in invertebrate animals. In: *Structure and Function of Muscles, Vol. I*, edited by G. H. Bourne, pp. 265–335. Academic Press, New York.

43. Hasumi, T., Kosaka, J., and Nagai, T. (1981): Potassium contracture of anterior byssus retractor muscle of Mytilus edulis. *Comp. Biochem. Physiol.*, 68A:9–16.

44. Heumann, H.-G. (1969): Calciumakkumulierende Strukturen in einem glatten Wirbellosenmuskel. *Protoplasma*, 67:111–115.

45. Heumann, H.-G. (1980): Paramyosin structures in the thick filaments of the anterior byssus retractor muscle of *Mytillus edulis*. *Eur. J. Cell Biol.*, 22:780–788.

46. Heumann, H. G., and Zebe, E. (1968): Über die Funktionsweise glatter Muskelfasern (Elektronenmikroskopische Untersuchungen am Byssusretraktor (ABRM) von *Mytilus edulis*). *Z. Zellforsch.*, 85:534–55.

47. Hidaka, T. (1969): Dopamine hyperpolarizes and relaxes *Mytilus* muscle. *Am. Zool.*, 9:251.

48. Hidaka, T., and Goto, M. (1973): On the relationship between membrane potential and tension in *Mytilus* smooth muscle. *J. Comp. Physiol.*, 82:357–364.

48a. Hidaka, T., Yamaguchi, H., Twarog, B. M., and Muneoka, Y. (1977): Neurotransmitter action on the membrane of *Mytilus* smooth muscle. II. Dopamine. *Gen. Pharmacol.*, 8:87–91.

49. Higgins, W. (1974): Intracellular actions of 5-hydroxytryptamine on the bivalve myocardium. *J. Exp. Zool.*, 190:99–110.

50. Huxley, H. E., and Hanson J. (1960): The molecular basis of contraction in cross-striated muscles. In: *The Structure and Function of Muscle, Vol. I*, edited by G. H. Bourne, pp. 183–227. Academic Press. New York.

51. Jewell, B. R. (1959): The nature of the phasic and tonic responses of the anterior byssal retractor muscle of *Mytilus*. *J. Physiol. (Lond.)*, 149:154–177.

52. Johnson, W. H., Kahn, J. S.,and Szent-Györgyi, A. G. (1959): Paramyosin and contraction of "catch muscles." *Science*, 130:160–161.

53. Kendrick-Jones, J., Lehman, W., and Szent-Györgyi, A. G. (1970): Regulation of molluscan muscles. *J. Mol. Biol.*, 54:313–326.

54. Kendrick-Jones, J., Szentkirályi, E. M., and Szent-Györgyi, A. G. (1972): Myosin-linked regulatory system: The role of the light chains. *Cold Spring Harbor Symp. Quant. Biol.*, 37:47–53.

55. Köhler, G., and Lindl, T. (1980): Effects of 5-hydroxytryptamine, dopamine, and acetylcholine on accumulation of cyclic AMP and cyclic GMP in the anterior byssus retractor muscle of *Mytilus edulis*, L. (Mollusca). *Pfluegers Arch.*, 383:257–262.

56. Kometani, K., and Sugi, H. (1978): Calicum transients in a molluscan smooth muscle. *Experientia*, 34:1469–1470.

57. Lehman, W. (1981): Thin filament-linked regulation in molluscan muscles. *Biochem. Biophys. Acta*, 668:349–356.

58. Lehman, W., and Szent-Györgyi, A. G. (1975): Regulation of muscular contraction. *J. Gen. Physiol.*, 66:1–30.

59. Levine, R. J. C., Elfvin, M., Dewey, M. M., and Walcott, B. (1976): Paramyosin in invertebrate muscles. II. Content in relation to structure and function. *J. Cell. Biol.*, 71:273–279.

60. Ljung, B., and Hallgren, P. (1975): On the mechanism of inhibitory action of vibrations as studied in a molluscan catch muscle and in vertebrate vascular smooth muscle. *Acta Physiol. Scand.*, 95:424–430.

61. Lowy, J., and Millman, B. M. (1959): Contraction and relaxation in smooth muscles of lamellibranch molluscs. *Nature*, 183:1730–1731.
62. Lowy, J., Millman, B. M., and Hanson, J. (1964): Structure and function in smooth tonic muscles of lamellibranch molluscs. *Proc. R. Soc. Lond. [Biol.]*, 160:525–536.
63. Mackenzie, J. M., and Epstein, H. F. (1980): Paramyosin is necessary for determination of Nematode thick filament length in vivo. *Cell*, 22:747–755.
64. Marchand-Dumont, G., and Baguet, F. (1975): The control mechanism of relaxation in molluscan catch-muscle (ABRM). *Pfluegers Arch.*, 354:87–100.
65. McLean, J. R., and Robinson, J. E. (1978): Histochemical identification of monoaminergic nerve cell bodies in the pedal ganglion which innervates the anterior byssus retractor muscle of the lamellibranch mollusc *Mytilus edulis*. *J. Anat.*, 126:640.
66. Muneoka, Y., Cottrell, G. A., and Twarog, B. M. (1977): Neurotransmitter action on the membrane of *Mytilus* smooth muscle. III. Serotonin. *Gen. Pharmacol.*, 8:93–96.
67. Muneoka, Y., Shiba, Y., and Kanno, Y. (1978): Effect of propanolol on the relaxation of molluscan smooth muscle; possible inhibition of serotonin release. *Hiroshima J. Med. Sci.*, 27:155–161.
68. Muneoka, Y., Shiba, Y., and Kanno, Y. (1978): Effects of neuroleptic drugs on the relaxing action of various monoamines in molluscan smooth muscle. *Hiroshima J. Med. Sci.*, 27:163–171.
69. Muneoka, Y., Shiba, Y., Maetani, T., and Kanno, Y. (1978): Further study on the effect of mersalyl, an organic mercurial, on releasing response of a molluscan smooth muscle to monoamines. *J. Toxicol. Sci.*, 3:117–126.
70. Muneoka, Y., and Twarog, B. M. (1977): Lanthanum block of contraction and of relaxation in response to serotonin and dopamine in molluscan catch muscle. *J. Pharmacol. Exp. Ther.*, 202:601–609.
71. Nauss, K. M., and Davies, R. E. (1966): Changes in inorganic phosphate and arginine during the development, maintenance and loss of tension in the anterior byssus retractor muscle of *Mytilus edulis*. *Biochem. Z.*, 345:173–187.
72. Needham, D. M. (1971): *Machina Carnis*. Cambridge University Press, London.
73. Nonomura, Y. (1974): Fine structure of the thick filament in molluscan catch muscle. *J. Mol. Biol.*, 88:445–455.
74. Northrop, R. B. (1964): Pharmacological responses of the anterior byssus retractor muscle of *Mytilus* to dopamine, serotonin and methylsergide. *Am. Zool.*, 4:423.
75. Rüegg, J. C. (1961): On the tropomyosin-paramyosin system in relation to the viscous tone of lamellibranch 'catch' muscle. *Proc. R. Soc. Lond. [Biol.]*, 154:224–249.
76. Rüegg, J. C. (1964): Tropomyosin paramyosin system and prolonged contraction in a molluscan smooth muscle. *Proc. R. Soc. Lond. [Biol.]*, 160:536–542.
77. Rüegg, J. C. (1971): Smooth muscle tone. *Physiol. Rev.*, 51:201–248.
78. Satchell, D. G., and Twarog, B. M. (1978): Identification of 5-hydroxy-tryptamine (serotonin) released from the anterior byssus retractor muscle of *Mytilus californicus* in response to nerve stimulation. *Comp. Biochem. Physiol.*, 59C:81–85.
79. Satchell, D. G., and Twarog, B. M. (1979): Identification and estimation of dopamine release from the anterior byssus retractor muscle of *Mytilus californicus* in response to nerve stimulation. *Comp. Biochem. Physiol.*, 64C:231–235.
80. Schädler, M. (1967): Proportionale Aktivierung von ATPase-Aktivität und Kontraktionsspannung durch Calciumionen in isolierten contractilen Strukturen verschiedener Muskelarten. *Pfluegers Arch.*, 296:70–90.
81. Schumacher, T. (1972): Zum Mechanismus der ökonomischen Halteleistung eines glatten Muskels (Byssus retractor anterior, *Mytilus edulis*). *Pfluegers Arch.*, 331:77–89.
82. Sobieszek, A. (1973): The fine structure of the contractile apparatus of the anterior byssus retractor muscle of *Mytilus edulis*. *J. Utrastruct. Res.*, 43:313–343.
83. Stafford, W. F., and Yphantis, D. A. (1972): Existence and inhibition of hydrolytic enzymes attacking paramyosin in myofibrillar extracts of *Mercenaria mercenaria*. *Biochem. Biophys. Res. Commun.*, 49:848–855.
84. Stössel, W., and Zebe, E. (1968): Zur intrazellulären Regulation der der Kontraktilität. *Arch. Ges. Physiol.*, 302:38–56.
85. Sugi, H., and Suzuki, S. (1978): The nature of potassium- and acetyl-choline-induced contractures in the anterior byssal retractor muscle of *Mytilus edulis*. *Comp. Biochem. Physiol.*, 61C:275–279.
86. Sugi, H., and Yamaguchi, T. (1976): Activation of the contractile mechanism in the anterior byssal retractor muscle of *Mytilus edulis*. *J. Physiol. (Lond.)*, 257:531–547.

87. Sugi, H., Yamaguchi, T., and Tanaka, H. (1977): The effect of hypertonic solutions on contracture, tension and volume in *Mytilus* smooth muscle. *Comp. Biochem. Physiol.*, 58A:405–407.
88. Swanson, C. J. (1971): Isometric response of the paramyosin smooth muscle of Paragordius varius (Leidy) (Aschelminthes, Nematomorpha). *Z. Vergl. Physiol.*, 74:403–410.
89. Sweeney, D. (1963): Dopamine: Its occurence in molluscan ganglia. *Science*, 139:1051.
90. Szent-Györgyi, A. G. (1975): Chemical and conformational events in regions of the myosin. *J. Supramol. Struct.*, 3:348–353.
91. Szent-Györgyi, A. G., Cohen, C., and Kendrick-Jones, J. (1971): Paramyosin and the filaments of molluscan "catch" muscles. *J. Mol. Biol.*, 56:239–258.
92. Takayanagi, J., Murakami, H., Iwayama, Y., Yoshida, Y., and Miki, S. (1981): Dopamine receptor in anterior byssus retractor muscle of *Mytilus edulis*. *J. Pharmacol.*, 31:249–252.
93. Tameyasu, T., and Sugi, H. (1976): Effect of acetylchloine and high external potassium ions on Ca movements in molluscan smooth muscle. *Comp. Biochem. Physiol.*, 53C:101–103.
94. Twarog, B. M. (1954): Responses of a molluscan smooth muscle to acetylcholine and 5-hydroxytryptomine. *J. Cell Comp. Physiol.*, 44:141–163.
95. Twarog, B. M. (1960): Effects of acetycholine and 5-hydroxytryptamine on the contraction of a molluscan smooth muscle. *J. Physiol. (Lond.)*, 152:236–242.
96. Twarog, B. M. (1967): Factors influencing contraction and catch in *Mytilus* smooth muscle. *J. Physiol. (Lond.)*, 192:847–856.
97. Twarog, B. M. (1967): Excitation of Mytilus smooth muscle. *J. Physiol. (Lond.)*, 192:857–868.
97a. Twarog, B. M. (1967): The regulation of catch in molluscan muscle. *J. Gen. Physiol.*, 50:157–169.
98. Twarog, B. M. (1976): Aspects of smooth muscle function in molluscan catch muscle. *Physiol. Rev.*, 56:829–838.
99. Twarog, B. M. (1977): Dissociation of calcium dependent reactions at different sites: Lanthanum block of contraction and relaxation in a molluscan smooth muscle. In: *Excitation-Contraction Coupling in smooth muscle.*, edited by R. Casteels, T. Godfraind, and J. C. Rüegg, pp. 261–271. Elsevier/North-Holland Biochemical Press, Amsterdam.
100. Twarog, B. M., and Cole, R. A. (1972): Relaxation of catch in a molluscan smooth muscle -II. Effects of serotonin, dopamine and related compounds. *Comp. Biochem. Physiol.*, 43A:331–335.
101. Twarog, B. M., Dewey, M. M., and Hidaka, T. (1973): The structure of Mytilus smooth muscle and the electrical constants of the resting muscle. *J. Gen. Physiol.*, 61:207–221.
102. Twarog, B. M., and Muneoka, Y. (1972): Calcium and the control of contraction and relaxation in a molluscan catch muscle. *Cold Spring Harbor Symp. Quant. Biol.*, 37:489–504.
103. Twarog, B. M., Muneoka, Y., and Ledgere, M. (1977): Serotinon and dopamine as neurotransmitters in *Mytilus:* Block of serotinon receptors by an organic mercurial. *J. Pharmacol. Exp. Ther.*, 201:350–356.
104. von Uexküll, J. (1926): Die Sperrmuskulatur von Holothurien. *Arch. Ges. Physiol.*, 212:1–14.
105. Wabnitz, R. W. (1975): Functional states and fine structure of the contractile apparatus of the penis retractor muscle (PRM) of *Helix pomatia*, L. *Cell Tissue Res.*, 156:253–265.
106. Wabnitz, R. W., and von Wachtendonk, D. (1976): Evidence for serotonin (5-hydroxytryptamine) as transmitter in the penis retractor muscle of *Helix pomatia*, L. *Experientia*, 32:707–709.
107. Walker, J. D., and Stewart, M. (1975): Paramyosin: Chemical evidence for chain heterogenity. *FEBS Lett.*, 58:16–17.
108. Weisel, J. W. (1975): Paramyosin segments: Molecular orientation and interactions in invertebrate muscle thick filaments. *J. Mol. Biol.*, 98:675–681.
109. Weisel, J. W., and Szent-Györgyi, A. G. (1975): The coiled-coil structure: Identity of the two chains of *Mercenaria* paramyosin. *J. Mol. Biol.*, 98:665–673.
110. Welsh, J. H., and Moorhead, M. (1960): The quantitative distribution of 5-hydroxytryptamine in the invertebrates, especially their nervous systems. *J. Neurochem.*, 6:146–169.
111. Wilson, D. M., and Larimer, J. L. (1968): The catch property of ordinary muscle. *Proc. Natl. Acad. Sci. USA*, 61:909–916.
112. Winkelman, L. (1976): Comparative studies of paramyosins. *Comp. Biochem. Physiol.*, 55B:391–397.
113. Winton, F. R. (1937): The changes in viscosity of an unstriated muscle (Mytilus edulis) during and after stimulation with alternating, interrupted and uninterrupted direct currents. *J. Physiol. (Lond.)*, 88:492–511.

114. York, B., and Twarog, B. M. (1973): Evidence for the release of serotonin by relaxing nerves in molluscan muscle. *Comp. Biochem. Physiol.*, 44A:423–430.
115. Zs.-Nagy, I., Salánki, J., and Garamvölgyi, N. (1971): The contractile apparatus of the adductor muscles in *Anodonta cygnea*, L. (Mollusca, Pelecypoda). *J. Ultrastruct. Res.*, 37:1–16.

Basic Biology of Muscles: A Comparative
Approach, edited by B. M. Twarog,
R. J. C. Levine, and M. M. Dewey.
Raven Press, New York © 1982.

Striated Myoepithelial Cells

Margaret Anderson

Department of Biological Sciences, Smith College, Northampton, Massachusetts 01063

The myoepithelial cells of the proventriculus of the marine polychaete worm
Syllis spongiphila are so named because they are tightly joined to each other to
form the covering of the proventriculus, thereby exhibiting epithelial characteristics,
and because they contain contractile elements organized into sarcomeres typical of
striated muscle cells. These cells exhibit three characteristics that are of interest to
physiologists. First, they are composed of only one or two sarcomeres, which may
reach 40 μm in length; second, each columnar myoepithelial cell is divided into a
peripheral myofibrillar ring and a central nonfibrillar core containing abundant
vesicles of crystalline material; third, the cells generate calcium action potentials,
which are accompanied by contractions, and the calcium channels appear to permit
the passage not only of Ca^{2+}, Sr^{2+}, and Ba^{2+} ions but also of Mn^{2+} ions.

MORPHOLOGY

The proventriculus of *S. spongiphila* is the section of the alimentary canal between
the proboscis and the intestine. The proventriculus measures 1 to 2 mm in length;
its function is to draw food into the alimentary canal by opening its lumen, and to
direct food into the intestine by closing the lumen. The proventriculus consists of
a single layer of myoepithelial cells (each 50 to 100 μm in length and about 25
μm in diameter), arranged radially in a regular array around the central lumen.

The proventriculi of several species of syllids were carefully described by Haswell
in 1890 (25). Figure 1 is a reproduction of a plate from his paper, which includes
drawings of a transverse section of the proventriculus (number 21) and myoepithelial
cells (number 22) of *Syllis nigropunctata*. The cells of *S. nigropunctata* closely
resemble those of *S. spongiphila*. The myoepithelial cells of different species of
syllids consist of varying numbers of sarcomeres (25,45); the cells of *Syllis kinbergiana*, illustrated by Haswell (Fig. 1, number 24) have three Z-lines. Haswell
(25,26) correctly concluded that the syllid myoepithelial cells resemble striated
muscle cells of arthropods and vertebrates. Smith et al. (41) subsequently showed
that, in *S. spongiphila*, the dark transverse striations are Z-lines dividing each cell

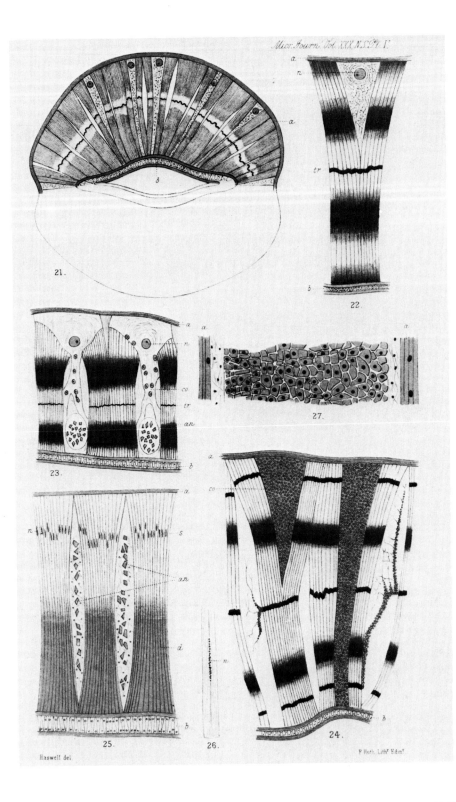

21.

22.

23.

27.

25.

26.

24.

Haswell del.

F Huth, Lith. Edin.

into two complete sarcomeres. The scanning electron micrograph of the proventriculus of *S. spongiphila* shown in Fig. 2 was produced by Smith et al. (41). These investigators (41) found that thin filaments (5 nm in diameter, presumably actin) insert on the Z-line and extend into the flanking sarcomeres; the second sets of thin filaments of the two sarcomeres insert on the ends of the cells. Orbitals of about 20 thin filaments interdigitate with the thick filaments, which are 25 μm in length and 90 nm in diameter at their centers, tapering to 20 nm at the ends. The thick filaments exhibit transverse striations with a periodicity of about 14 nm, which is characteristic of paramyosin-containing filaments (24). Cells along the dorsal and ventral margins of the proventriculus are particularly unusual: they do not have a Z-line, and they consist of a single sarcomere that is 40 to 50 μm in length. Smith et al. (41) also showed that highly branched transverse tubules invade the cells at all levels of the sarcomeres and that there is present only an extremely reduced sarcoplasmic reticulum (SR). The cells closely interdigitate with their neighbors; although no gap junctions have yet been described, the cells of the proventriculus are strongly electrically coupled to each other (3).

As Haswell (25) noted, each cell includes a distinct central core that extends the entire length of the cell. Smith et al. (41) showed that this core contains the nucleus, most of the cell's mitochondria, and unusual membrane-bound vesicles of crystalline material, which preliminary experiments suggested was largely Mg. More recent experiments performed in my laboratory in collaboration with Richard T. Briggs strongly suggest that Ca rather than Mg ions are present in high concentrations in the vesicles, and that other divalent cations could be present, but in lower concentrations. In these experiments, tissues were exposed overnight to 10 to 50 mM EGTA (which, at pH 7, exhibits log10 apparent association constants of 6.68 for Ca and 1.61 for Mg (1); the EGTA was applied both alone and in the presence of $CaCl_2$ or $MgCl_2$, after which the tissues were prepared for electron microscopy. Three micrographs are shown in Fig. 3. In the presence of EGTA alone (a), the crystalline material was extracted, leaving relatively empty vesicles. When $CaCl_2$ was added with EGTA (b), little or no crystalline material was extracted; this tissue resembles control tissues that were not treated with EGTA. When $MgCl_2$ was added with EGTA (c) some extraction of the crystalline material occurred. Because EGTA binds Ca more than four orders of magnitude more strongly than it binds Mg, the addition of Mg should have had relatively little effect on the calcium-binding

---

FIG. 1. Plate V reproduced from the work of Haswell (25). **21:** Drawing of a transverse section of the proventriculus of *Syllis nigropunctata*. The cells are oriented radially around the central lumen. A central Z-line is present in all cells, except those on the dorsal and ventral margins of the organ, and the central nonfibrillar cores are evident in several cells. **22:** A single myoepithelial cell of *S. nigropunctata* showing a single transverse Z-line. The partially visible core includes the nucleus of the cell, which Haswell originally interpreted as the nucleus of a "ganglion cell," which he believed overlay the exterior face of the myoepithelial cell. **24:** Cells from the proventriculus of *Syllis kinbergiana*. These cells are organized into fibrillar and nonfibrillar regions, but they are divided by three Z-lines into more sarcomeres than seen in *S. nigropunctata* or *S. spongiphila*.

FIG. 2. Scanning electron micrograph of a transversely sectioned proventriculus of *S. spongiphila*. The long axis of the central lumen, l, is oriented in the dorso-ventral plane. *Large arrowheads* indicate nerve trunks. Both fibrillar and nonfibrillar *(asterisks)* regions of the radially oriented myoepithelial cells (m) are visible. The exterior surface shows that the cells are aligned in rows along the length of the organ and that the dorsal and ventral aspects of the proventriculus are grooved; one groove is indicated by the *small arrow*. × 240. (From Smith et al., ref. 41, with permission.)

capability of EGTA. Although extraction in the presence of Mg and EGTA (c) was not so complete as that in the presence of EGTA alone (a), these results taken together indicate that Ca is present in high concentrations in the crystalline material. This result agrees with the work of Ponsolle et al. (35) who used electron probe analysis on myoepithelial cells of *Syllis amica* and found that the cores contained high concentrations of calcium and phosphate. The interesting question of the function of the Ca-containing vesicles within the cores remains unanswered, however.

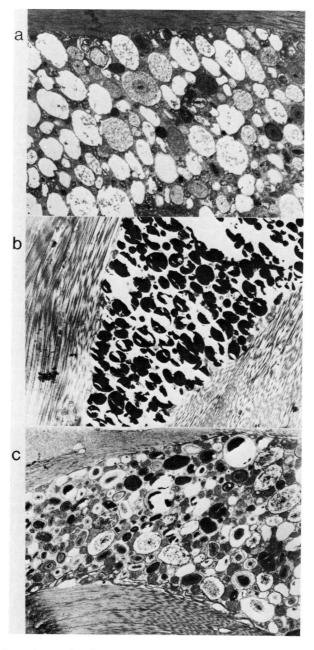

FIG. 3. Electron micrographs of transverse sections of the proventriculus showing the cores of myoepithelial cells treated with 0.01 M EGTA alone **(a)**, with 0.01 M EGTA and 0.2 M CaCl₂ **(b)**, and with 0.01 M EGTA and 0.2 M MgCl₂ **(c)**. Myofibrillar components flank the cores. Proventriculi were fixed in 2.5% glutaraldehyde in cacodylate buffer and 14% sucrose. They were then treated for 20 hr with EGTA alone or with EGTA and competing salt, dehydrated, infiltrated, and embedded in Epon by conventional procedures. Cores from which the crystalline material was not extracted were poorly infiltrated and difficult to section; this problem accounts for the torn regions visible in **b** and **c**. × 4,550.

INNERVATION

The myoepithelial cells are innervated by populations of both excitatory and inhibitory neurons (3). The neurons are confined to a layer on the surface facing the lumen, and they are separated from the membranes of the myoepithelial cells by a layer of collagen 0.3 to 1.0 μm in thickness (41).

Each myoepithelial cell appears to receive both excitatory and inhibitory input (3). Experimentally, excitatory and inhibitory junction potentials can be elicited indirectly in the myoepithelial cells by applying pulse stimuli to the anterior end of the animal. The elicited responses resemble spontaneous activity generated by the preparation (see 12, for records of spontaneous activity). The excitatory junction potentials are reversibly diminished by d-tubocurarine (10^{-5} to 5×10^{-4} M) but not by atropine (10^{-6} to 10^{-4} M), and they are mimicked by the iontophoretic application of 0.1 to 1.0 M acetylcholine or carbamylcholine (4). Thus, the excitatory neuromuscular system appears to be cholinergic with nicotinic postsynaptic receptors. The inhibitory transmitter has not yet been identified; applied to the bathing medium, both 10^{-4} M γ-aminobutyric acid (GABA) and 10^{-4} M 5-hydroxytryptamine (5-HT) elicited hyperpolarizations in the myoepithelial cells; however, at higher concentrations, they elicited both hyperpolarizations and depolarizations. Both of these substances have been shown either to be present in or to produce specific effects in various annelids (19); thus they seem likely candidates for the inhibitory transmitter(s).

CORRELATION OF MECHANICAL AND ELECTRICAL ACTIVITY

Contraction and relaxation of the myoepithelial cells are related to their electrical activity. The light micrographs of Fig. 4 were taken from a single preparation and show the proventriculus and its lumen during various phases of indirectly elicited electrical activity. In a, no electrical activity was elicited; the cells are at resting potential (approximately -60 mV), and the lumen of the proventriculus is slightly open. In b, elicited excitatory junction potentials summed to produce overshooting depolarizations associated with opening of the lumen. In c, elicited hyperpolarizing junctional activity was associated with closing of the lumen. Thus, it appears that the myoepithelial cells contract when depolarized (thereby pulling open the lumen) and relax while hyperpolarized (closing the lumen).

CALCIUM SPIKES

Overshooting, regenerative potentials can be elicited directly by intracellularly applied pulse stimuli. Such events, which are associated with contractions, are reversibly abolished by the removal of Ca^{2+} ions from the bathing medium or by the addition (to Ca-containing medium) of 20 mM Co^{2+} ions; the overshooting potentials are essentially irreversibly abolished by the addition of 10 mM La^{3+} ions (2). They are also reversibly abolished in the presence of the calcium antagonists D-600 or verapamil, each at a concentration of 1 mM. The spikes are unaffected

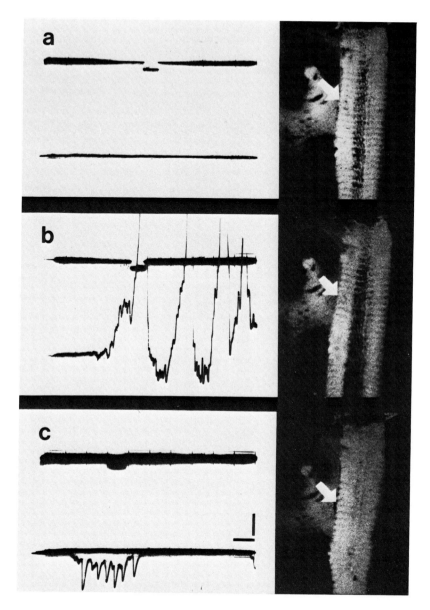

FIG. 4. Electrical activity associated with contraction and relaxation of the proventricular cells. The light micrographs on the **right** were taken at the time indicated by a downward deflection of the *top trace* in each of the three oscilloscope records at the **left**. The proventriculus was pinned with the raphes on *top* and *bottom*, and the lateral (longest) cells were brought into focus. The *arrows* indicate the position of the microelectrode that recorded the activity shown on the bottom traces. All records were obtained from the same cell. Indirect stimuli were applied to the anterior end of the animal with a suction electrode. At rest **(a)**, the lumen (central dark region) of the proventriculus was slightly open; during depolarization **(b)**, the lumen was opened wide; and during hyperpolarization **(c)**, the lumen was closed. Calibrations: Vertical, 10 mV; horizontal, 1.0 sec.

by 10^{-5} M tetrodotoxin or by replacing 95% of the Na ions in the bathing medium with sucrose; finally, spikes can be elicited in Ca-free solutions containing either 1 mM $BaCl_2$ or 10 mM $SrCl_2$ (2). All of these results support the conclusion that the overshooting depolarizations are pure calcium spikes.

MANGANOUS SPIKES

The Ca-spikes of the syllid myoepithelial cells are not abolished in the presence of Mn^{2+} ions, and in Ca-free solutions containing 5 to 50 mM Mn^{2+} ions, overshooting regenerative events can be elicited by direct stimulation (2). Figure 5 illustrates examples of spikes recorded in Ca-free solutions containing varying concentrations of Mn^{2+} ions. Like Ca-spikes, the Mn-spikes are propagated along the proventriculus, and the membranes generating Mn-spikes are refractory to further stimulation during a given response. The amplitude of the overshoot of Mn-spikes

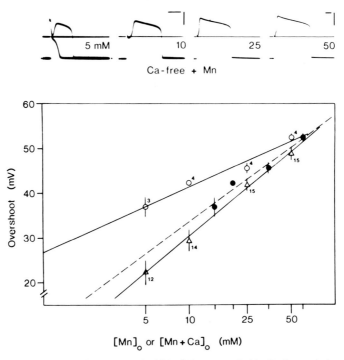

FIG. 5. Top: Examples of directly elicited Mn-spikes recorded in Ca-free solution containing 5, 10, 25, or 50 mM $MnCl_2$. The *top lines* indicate 3 mV below the zero reference level. Calibrations: Vertical, 40 mV and 2×10^{-6} A; horizontal, 1.0 sec (the two records at the left were taken at twice the sweep speed as the two on the right). **Bottom:** Overshoot amplitude plotted against either log10 $[Mn]_o$ or log10 $[Ca + Mn]$. *Triangles*, spikes recorded in Ca-free solution with varying $[Mn]_o$, plotted against $[Mn]_o$. *Open circles*, spikes recorded in Ca-containing solution with varying $[Mn]_o$, plotted against $[Mn]_o$. *Closed circles*, the same data as indicated by open circles, plotted against $[Ca + Mn]_o$. *Bars* indicate standard errors of the mean; *numbers* indicate the number of cells tested under each experimental condition.

increases with increasing $[Mn]_0$. The semilog10 graph of Fig. 5 shows that, in Ca-free solutions, the mean overshoot amplitude increases 27.5 mV for a 10-fold change in $[Mn]_0$ (pooled data from 10 preparations, triangles). This value is close to that predicted by the Nernst equation for a membrane permeable to a divalent cation. In the presence of 10 mM Ca^{2+} ions, the mean overshoot amplitude increases only about 15 mV for a 10-fold change in $[Mn]_0$ (pooled data from 4 preparations, open circles). This result suggests that both Ca^{2+} and Mn^{2+} ions compete to pass through the same channels. When the same mean amplitudes are replotted against the sum of $[Ca]_0$ and $[Mn]_0$, the resulting line indicates an increase in overshoot amplitude of 24 mV for a 10-fold change in $[Ca + Mn]_0$ (closed circles, interrupted line). This slope of 24 mV is within the 95% confidence interval of the slope of the line for Ca-free, Mn-containing solutions (22.6 to 32.4 mV), suggesting that the channels permit the passage of both Ca^{2+} and Mn^{2+} ions. Of the first order transition metal cations, Mn has the lowest energy of hydration (32). A possible explanation for the ability of Mn^{2+} ions to pass through the calcium channels is that these cations are able to lose sufficient waters of hydration to permit their passage.

UNCOUPLING OF DEPOLARIZATION AND CONTRACTION

Mn spikes are not associated with contractions. Figure 6 illustrates the appearance of a proventriculus along with simultaneously recorded electrical activity during immersion in Control ASW (artificial sea water), Ca-free ASW, and Ca-free ASW containing either Mn or Sr ions. Directly elicited spikes are associated with contraction in solutions containing Ca or Sr ions, but contraction does not occur in Ca-free solutions containing Mn ions. Neither contractions nor spikes could be elicited in Ca-free ASW. Thus, although Mn ions can replace Ca ions in regenerative depolarization of the cell, they cannot replace Ca ions in initiating contraction. Sr ions, on the other hand, can replace Ca ions both in the generation of spikes and in the initiation of contraction. These observations also strongly suggest that the initiation of contraction requires an influx of calcium ions from outside the cell.

An external source of Ca^{2+} ions correlates with the morphological observations of an elaborate pattern of invaginations of the plasma membrane (the highly branched transverse tubular system) and very little SR (41). Because the transverse tubular membrane enters all levels of the sarcomeres, the myofibrillar system appears to have access to the external medium. If the body fluid of *Syllis* contains about the same concentration of Ca ions as does its sea water environment (approximately 10 mM), then the extracellular medium would provide a rich source for the influx of Ca ions. The mechanism by which the sarcoplasmic concentration of free Ca ions is normally kept low is not known. In the absence of an elaborate SR, other organelles that could transport free Ca ions out of the sarcoplasm include the transverse tubular membranes, mitochondria, and/or the Ca-containing crystalline vesicles. If the vesicles are actively involved in the sequestration of Ca ions, it is possible that they are one-way sinks for excess Ca ions, somewhat like the storage excretion depots for other substances in insects (44). Alternatively, the vesicles

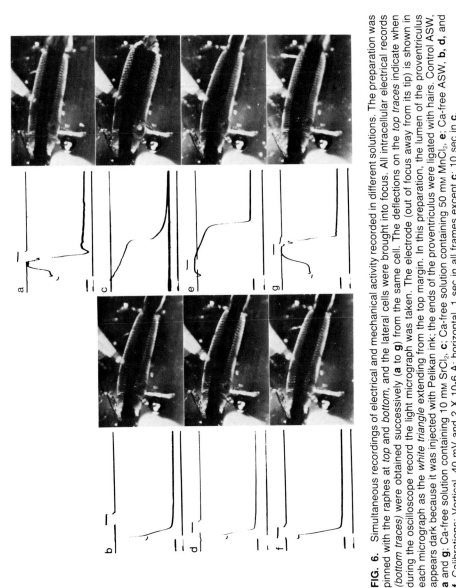

FIG. 6. Simultaneous recordings of electrical and mechanical activity recorded in different solutions. The preparation was pinned with the raphes at *top* and *bottom*, and the lateral cells were brought into focus. All intracellular electrical records (*bottom traces*) were obtained successively (**a** to **g**) from the same cell. The deflections on the *top traces* indicate when during the oscilloscope record the light micrograph was taken. The electrode (out of focus away from its tip) is shown in each micrograph as the *white triangle* extending from the top margin. In this preparation, the lumen of the proventriculus appears dark because it was injected with Pelikan ink; the ends of the proventriculus were ligated with hairs. Control ASW, **a** and **g**; Ca-free solution containing 10 mM SrCl₂, **c**; Ca-free solution containing 50 mM MnCl₂, **e**; Ca-free ASW, **b**, **d**, and **f**. Calibrations: Vertical, 40 mV and 2 X 10-6 A; horizontal, 1 sec in all frames except **c**; 10 sec in **c**.

could slowly exchange Ca ions and thereby serve as internal buffering organelles, a function similar to that suggested for the cytoplasmic granules in some protozoans (9,10). It seems unlikely that the vesicles could function as an SR, to sequester and release Ca, since the physiological data demonstrating a requirement for an influx of extracellular Ca ions do not support the idea of an intracellular source.

COMPARISON TO OTHER CONTRACTILE CELLS

The myoepithelial cells of *S. spongiphila* have some characteristics that are similar to other contractile cells, as well as certain unique properties. The cells are unique in consisting of only one or two unusually long sarcomeres. Sarcomeres of similar length (33 μm) occur in the notochordal muscle of *Branchiostoma lanceolatum* (amphioxus), but the fibers contain many sarcomeres (17). Prosser (36, and *this volume*) has pointed out that the syllid myoepithelial cells exhibit characteristics of both striated and smooth muscles. They resemble striated muscle cells because they possess myofilaments organized into sarcomeres, and they have transverse tubular invaginations; they resemble certain smooth muscle cells because they have a reduced SR and distances of only a few micra between the contractile elements and the transverse tubules or plasma membrane. As Prosser has noted, the syllid myoepithelial cells may more easily be described as "narrow-fibered" contractile cells than they can be described as either smooth or striated muscle cells.

As do several other muscle cells, the thick filaments of *Syllis* contain paramyosin. This protein has been found in muscle fibers of animals in many phyla, including molluscs, arthropods, and annelids (e.g., 5,6,13,14,17,27,30). Several investigators have proposed that paramyosin forms a core to which myosin molecules are attached (6,15,28,42,43). Ikemoto and Kawaguti (27) suggested that the presence of a paramyosin core permits a thick filament to attain greater length than would be permitted by polymerization only of myosin molecules, and Levine et al. (31) showed for several different animals that the thick filament length increased as the ratio of paramyosin to myosin increased. An additional functional role of paramyosin (e.g., providing a "catch" mechanism or thick filament shortening) in *S. spongiphila* is unknown.

Similarly, although unusual among vertebrate muscle cells, the organization of the myoepithelial cells of *Syllis* into fibrillar and nonfibrillar regions is not unique. For example, many coelenterate myoepithelial cells have myofilaments confined to their basal regions (36). Further, the somatic muscle cells of the nematode *Ascaris* are divided into three parts (11): the belly, which contains the nucleus; the arm, which extends to the nerve cord and receives synaptic input (39); and the fiber, or spindle, which is the obliquely striated (38) contractile part of the cell.

By their abundance, the crystalline inclusions of the syllid myoepithelial cells inspire speculation regarding their functional role; however, crystalline inclusions occur in a variety of cells in numerous animals, including protozoans (9,10), insects (40,44), and amphibians and mammals (16). Although the storage excretion of uric acid in insects can be understood as a mechanism for water conservation, the physiological significance and the mechanism(s) of production of many such inclusions remain unclear.

Finally, the generation of Ca-spikes is a common property of invertebrate muscle cells and of vertebrate smooth muscle cells; the plateau of the action potential of vertebrate heart muscle also appears to be dependent on an increase in the permeability of the membrane to Ca ions (for reviews of Ca-spikes, see 20–22,37). The function of the inward calcium current in all of these types of muscle cells appears to be the initiation of contraction (22). To understand the functional selectivity of calcium channels, it is of interest to consider the properties of those cations to which the calcium channels are permeable. A preparation that has calcium channels that are permeable to Mn^{2+} ions permits comparison of the actions of Ca ions both with those of cations in the same group as Ca (alkali earth metals) and with those of (transition metal) cations of the same period. Although many voltage-dependent calcium channels, such as those of the barnacle muscle fiber (23), appear to be blocked by Mn ions, others are not. In addition to the proventriculus of *S. spongiphila*, other muscle cells that have calcium channels that permit the passage of Mn ions include guinea pig heart papillary muscle cells (33,34), larval beetle muscle cells (18), and smooth muscle cells of sheep carotid arteries (29). Further, the calcium channels of frog atrial cells (7,8) appear to be blocked by Mn ions at low concentrations but to permit the passage of Mn ions at higher concentrations. Thus, the permeability of calcium channels to Mn ions may be a more general phenomenon than is currently believed.

ACKNOWLEDGMENTS

It is a pleasure to thank Dr. J. del Castillo for rediscovering *S. spongiphila* as a physiological preparation, Dr. R. T. Briggs for the electron microscopic examination of the tissues of Fig. 2, Ms. P. E. Lessie for technical assistance, Dr. R. F. Olivo for critically reading the manuscript, and Mr. F. McKenzie for supplying experimental animals. This work was supported by NIH grant NS12196.

REFERENCES

1. Amos, W. B., Routledge, L. M., Weis-Fogh, T., and Yew, F. F. (1976): The spasmoneme and calcium-dependent contraction in connection with specific calcium binding proteins. In: *Calcium in Biological Systems*, edited by the Society for Experimental Biology XXX, pp. 298–299. Cambridge University Press, Cambridge.
2. Anderson, M. (1979): Mn^{2+} ions pass through Ca^{2+} channels in myoepithelial cells. *J. Exp. Biol.*, 82:227–238.
3. Anderson, M., and del Castillo, J. (1976): Electrical activity of the proventriculus of the polychaete worm *Syllis spongiphila*. *J. Exp. Biol.*, 64:691–710.
4. Anderson, M., and Mrose, H. (1978): Chemical excitation of the proventriculus of the polychaete worm *Syllis spongiphila*. *J. Exp. Biol.*, 75:113–122.
5. Bailey, K. (1957): Invertebrate tropomyosin. *Biochim. Biophys. Acta*, 24:612–619.
6. Bullard, B., Luke, B., and Winkelman, L. (1973): The paramyosin of insect flight muscle. *J. Mol. Biol.*, 75:359–367.
7. Chapman, R. A., and Ellis. D. (1977): The effects of manganese ions on the contraction of the frog's heart. *J. Physiol. (Lond.)*, 272:331–354.

8. Chapman, R. A., and Ellis, D. (1977): Uptake and loss of manganese from perfused frog ventricles. *J. Physiol. (Lond.)*, 272:355–366.
9. Coleman, J. R., Nilsson, J. R., Warner, R. R., and Batt, P. (1972): Qualitative and quantitative electron probe analysis of cytoplasmic granules in *Tetrahymena pyriformis*. *Exp. Cell Res.*, 74:207–219.
10. Coleman, J. R., Nilsson, J. R., Warner, R. R., and Batt, P. (1973): Electron probe analysis of refractive bodies in *Amoeba proteus*. *Exp. Cell Res.*, 76:31–40.
11. DeBell, J. T., del Castillo, J., and Sanchez, V. (1963): Electrophysiology of the somatic muscle cells of *Ascaris lumbricoides*. *J. Cell. Comp. Physiol.*, 62:159–178.
12. del Castillo, J., Anderson, M., and Smith, D. S. (1972): Proventriculus of a marine annelid: Muscle preparation with the longest recorded sarcomere. *Proc. Natl. Acad. Sci.*, 69:1669–1672.
13. de Villafranca, G. W., and Leitner, V. E. (1967): Contractile proteins from horseshoe crab muscle. *J. Gen. Physiol.*, 50:2495–2496.
14. Elfin, M., Levine, R. J. C., and Dewey, M. M. (1976): Paramyosin in invertebrate muscles. I. Identification and localization. *J. Cell Biol.*, 71:261–272.
15. Elliott, A. (1974): The arrangement of myosin on the surface of paramyosin filaments in the white adductor muscle of *Crassostrea angulata*. *Proc. R. Soc. Lond. [Biol.]*, 186:53–66.
16. Fawcett, D. W. (1967): *The Cell. Its Organelles and Inclusions*. Saunders, Philadelphia.
17. Flood, P. R., Guthrie, D. M., and Banks, J. R. (1969): Paramyosin muscle in the notochord of Amphioxus. *Nature*, 222:87–88.
18. Fukuda, J., and Kawa, K. (1977): Permeation of manganese, cadmium, zinc, and beryllium through calcium channels of an insect muscle membrane. *Science*, 196:309–311.
19. Gerschenfeld, H. M. (1973): Chemical transmission in invertebrate central nervous systems and neuromuscular junctions. *Physiol. Rev.*, 53:1–119.
20. Hagiwara, S. (1973): Calcium spike. *Adv. Biophys.*, 4:71–102.
21. Hagiwara, S. (1975): Ca-dependent action potential. In: *Membranes. A Series of Advances. Vol. 3. Lipid Bilayers and Biological Membranes: Dynamic Properties*, edited by G. Eisenman, pp. 359–381. Marcel Dekker, New York.
22. Hagiwara, S., and Byerly, L. (1981): Calcium channel. *Annu. Rev. Neurosci.*, 4:69–125.
23. Hagiwara, S., and Nakajima, S. (1966): Differences in Na and Ca spikes as examined by application of tetrodotoxin, procaine and managanese ions. *J. Gen. Physiol.*, 49:793–806.
24. Hanson, J., Lowy, J., Huxley, H. E., Bailey, K., Kay, C. M., and Ruegg, J. C. (1957): Structure of molluscan tropomyosin. *Nature*, 180:1134–1135.
25. Haswell, W. A. (1890): A comparative study of striated muscle. *Q. J. Microsc. Sci.*, 30:31–50.
26. Haswell, W. A. (1921): The proboscis of the Syllidea. Part I. Structure. *Q. J. Microsc. Sci.*, 65:323–337.
27. Ikemoto, N., and Kawaguti, S. (1967): Elongating effect of Tropomyosin A on the thick myofilaments in the long-sarcomere muscle of the horseshoe crab. *Proc. Jap. Acad.*, 43:974–979.
28. Kahn, J. S., and Johnson, W. H. (1960): The localization of myosin and paramyosin in the myofilaments of the byssus retractor muscle of *Mytilus edulis*. *Arch. Biochem. Biophys.*, 86:138–143.
29. Keatinge, W. R. (1978): Mechanism of slow discharges of sheep carotid artery. *J. Physiol. (Lond.)*, 279:275–289.
30. Kominz, D. R., Saad, F., and Laki, K. (1957): Chemical characteristics of annelid, mollusc and arthropod tropomyosins. *Conference on the Chemistry of Muscular Contraction*, pp. 1–11. Igaku Shoin Ltd., Tokyo.
31. Levine, R. J. C., Elfin, M., Dewey, M. M., and Walcott, B. (1976): Paramyosin in invertebrate muscles. II. Content in relation to structure and function. *J. Cell Biol.*, 71:273–279.
32. Noyes, R. M. (1962): Thermodynamics of ion hydration as a measure of effective dielectric properties of water. *J. Am. Chem. Soc.*, 84:513–522.
33. Ochi, R. (1975): Manganese action potentials in mammalian cardiac muscle. *Experientia*, 31:1048–1049.
34. Ochi, R. (1976): Manganese-dependent propagated action potentials and their depression by electrical stimulation in guinea pig myocardium perfused by sodium-free media. *J. Physiol. (Lond.)*, 263:139–156.
35. Ponsolle, L., Wissocq, J.-C., and Galle, P. (1974): Étude des inclusions microcristallines dans le proventricule des Syllidiens (Annelida polychaeta). *Cytobiologie*, 9:169–179.
36. Prosser, C. L. (1980): Evolution and diversity of nonstriated muscles. In: *Handbook of Physiology, Section 2: The Cardiovascular System, Volume II*, edited by D. F. Bohr, A. P. Somlyo, and H. V. Sparks, Jr., pp. 635–670. American Physiological Society, Bethesda, Maryland.

37. Reuter, H. (1973): Divalent cations as charge carriers in excitable membranes. *Prog. Biophys. Mol. Biol.*, 26:1–43.
38. Rosenbluth, J. (1965): Ultrastructural organization of obliquely striated muscle fibers in *Ascaris lumbricoides*. *J. Cell Biol.*, 25:495–515.
39. Rosenbluth, J. (1965): Ultrastructure of somatic muscle cells in *Ascaris lumbricoides*. II. Intermuscular junctions, neuromuscular junctions, and glycogen stores. *J. Cell Biol.*, 26:579–591.
40. Smith, D. S. (1968): *Insect Cells*. Oliver and Boyd, Edinburgh.
41. Smith, D. S., del Castillo, J., and Anderson, M. (1973): Fine structure and innervation of an annelid muscle with the longest recorded sarcomere. *Tissue Cell*, 5:281–302.
42. Squire, J. M. (1971): General model for the structure of all myosin-containing filaments. *Nature*, 233:457–462.
43. Szent-Gyorgyi, A. G., Cohen, C., and Kendrick-Jones, J. (1971): Paramyosin and the filaments of molluscan "catch" muscles. II. Native filaments: Isolation and characterization. *J. Mol. Biol.*, 56:239–258.
44. Wigglesworth, V. B., (1972): *The Principles of Insect Physiology*, pp. 579–582. Chapman and Hall, London.
45. Wissocq, J.-C. (1974): Étude ultrastructurale d'un organe musculaire constitué de fibres possedant les plus longs sarcomeres du regne animal.: le proventricule des Syllidiens (Annelides Polychetes). I. Fibres non contractees. *J. Microsc.*, 19:285–306.

Basic Biology of Muscles: A Comparative
Approach, edited by B. M. Twarog,
R. J. C. Levine, and M. M. Dewey.
Raven Press, New York © 1982.

Giant Barnacle and *Peripatus* Muscles: Striated—Smooth Transitions

Graham Hoyle

Biology Department, University of Oregon, Eugene, Oregon 97403

This chapter concerns the question of assumed fundamental similarity of muscle molecular mechanisms. In 1956, Pantin wrote: "It has long been tacitly assumed that muscle is muscle wherever it is found" (8). Paraphrasing Gertrude Stein, anatomist Hess wrote: "A muscle is a muscle is a muscle" (2). This assumption of commonality has recently been criticized by Huxley (7) on the grounds that it was largely responsible for muscle scientists following a number of false leads about contractile mechanisms for more than half this century. He claims that the return to a solid foundation, on which he clearly earnestly believes muscle studies are now based, was achieved by paying intensive attention to a single specialized form, the skeletal cross-striated one of tetrapod chordates.

Muscle cells have a greater than 10^6-fold range in maximum speed capability, a 15 to 7,000% range in maximum lengths over which they can function, about a hundredfold range in maximum force, from 0.3 to 30 kg.cm^{-2}, and a range from a few seconds to indefinite over which they can sustain maximum tension without fatigue. To encompass such a wide range, they have evolved a bewildering seemingly limitless range of detailed ultrastructural, biophysical, and biochemical properties. No two muscles or even muscle cells, from different species, have yet been found that cannot rather easily be distinguished from each other. It is a major task for the comparative muscle physiologist to understand the significance of these variations on the muscle theme. The variations provide an extremely rich set of natural experiments on which a fundamental muscle scientist may test concepts. I'm sorry to say, however, that I am unable to name a single such scientist who has utilized this invaluable opportunity.

My specific purpose is to describe and compare the ultrastructural and physiological details of the characteristic component cells of two muscles, from animals with very different life-styles, one that is cross-striated and the other smooth in the classic sense. The animals are a sedentary crustacean, the giant barnacle *Balanus nubilus*, and the predatory onychophoran *Peripatus*. For more than a century, *Peripatus*, a soft-bodied, multi-legged, air-breathing terrestrial viviparous animal, has been considered to be a living fossil, a "missing link" between annelids and arthropods. The giant barnacle has a rock-hard unarticulated shell; the muscle cells of interest are the six retractors of equally hard scutal and tergal plates. They each contain about 40 giant cells up to 3 mm diameter and 6 cm in length, depending on the size of the specimen. The larger muscle cells of the body of a mature *Peripatus* range up to 40 μm in diameter, which alone separates them from conventional SM cells, and about 1 cm in length with tapered, often forked, ends. The average sarcomere length for barnacle cells is 6.2 μm, but Hoyle et al. (5) reported that the range in one cell is from 3.8 to 23 μm. Their maximum speed at 20°C is 1.1 cm/sec/cm. *Peripatus* cells are not striated: they have a maximum speed at 30°C of 1.1 cm/sec/cm.

PHYSIOLOGICAL FEATURES

At 0°C, the heat production of a single giant barnacle cell is very closely similar to that produced by a *Rana temporaria* whole sartorius; their shortening heats are virtually identical (Fig. 1) (Abbott and G. Hoyle, *unpublished results*). The maximum speed of contraction at their average habitat temperature, 11°C, is equal to that of frog sartorius at 0°C. Their maximum tension/unit area of cross-section at mean body length (l_0) is 5.8 kg.cm^{-2}, about double that of the frog. But there the similarity ends; the giant barnacle cell shortens reversibly under light load down to 1/6 l_0. Its maximum shortening velocity is unchanged, with decreasing length down to 75% l_0, whereas in the frog, shortening velocity falls precipitously with length below l_0. There is no shortening at all below .5 l_0 (Fig. 2). For some other frog muscle cells, shortening is even further limited down to about 70% l_0.

Very little force is required to extend the barnacle cell by up to about 20% above l_0, from which length the cell returns exactly to l_0 when released. After a *Peripatus* cell has been stretched 20%, which also requires very little force, it remains close to the extended length, as do most smooth muscle cells. The compliance is approximately seven times stiffer in frog sartorius cells (Fig. 3), as well as shorter. In barnacle cells, it is 14% rest length and entirely located within the contractile material.

When extension is continued beyond 120% l_0, the tension begins to rise steeply in the barnacle cell, but it falls progressively with time to a steady value. The length/steady tension curve during extension is not followed during return; there is a marked hysteresis and the cell returns to a length substantially greater than l_0. When repeated, the steady tension/length curve follows a flatter path than the first. It is progressively flatter, with each extension up to third or fourth, with less hysteresis each time, until there is none (Fig. 4). The shapes of these curves depend markedly on two additional factors: the rate of stretch and the maximum extent of stretch. The slower the stretch, the less the amount of hysteresis; the greater the final length reached at the faster stretch speeds, the greater the amount of hysteresis. The length assumed on unloaded release is proportional to the final length reached (Fig. 5). With extremely slow stretches beyond 160% l_0, at about 1 cm/hr, a final length of about 300% l_0 can be achieved before breaking. Breaking occurs at shorter

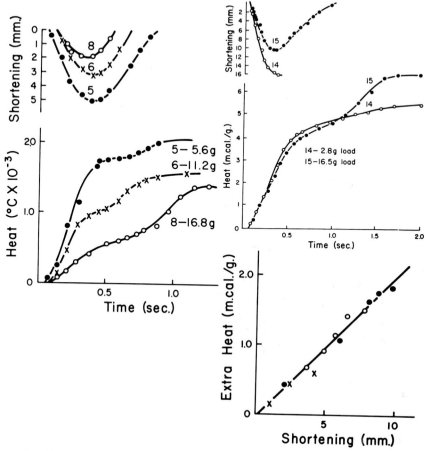

FIG. 1. Heat production by single giant barnacle retractor cell. **Left:** Heat production during isotonic twitches under three different loads. **Upper right:** Heat production under light and heavy loads in which the cell was prevented from relaxing at the peak of shortening under the light load. **Lower right:** Extra heat of shortening plotted as a function of extent of shortening. (From B. C. Abbott and G. Hoyle, *unpublished results.*)

lengths with faster stretch or if the cell is not in perfect condition. The slack length is maximally increased by about 25%.

If as little as a single strong electric shock is applied to the unloaded or lightly loaded stretched cell so that free shortening occurs, the shortened cell returns to its original length. Now, when stretched again, the tension/length curves follow very closely the first curve. Therefore, active shortening restores the original internal conditions, which are progressively altered by extensive slow stretches. Clusters of *Peripatus* cells behave in a similar manner to the barnacle cell except that the forces required for extension are much less, and the stretches do not need to be so slow. This will not be surprising to smooth muscle biophysicists because most vertebrate smooth muscles and some invertebrate obliquely striated muscles (that

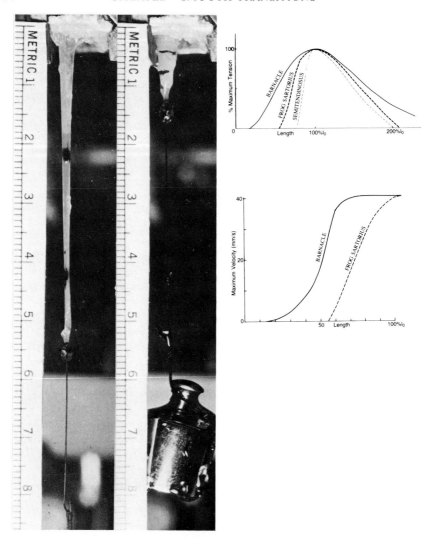

FIG. 2. Left: Giant barnacle *(Balanus nubilus)* single short cross-striated retractor cell under 20-g load marked at rest length at 1 cm intervals. **Right:** Third sequence of supercontraction under 20-g load evoked by a drop of isotonic KCl. Each shortening was maintained for about 1 min. **Upper right:** Maximum active tension/length curves for giant barnacle cell compared with those for frog sartorius and semitendinosus single cells (sarcomere length not servo-length-controlled). **Lower right:** Maximum velocity/length curves for barnacle giant cell compared with that for frog sartorius.

were for long treated as if smooth) have long been known to show similar features. The surprise is finding them in undoubted cross-striated cells of the barnacle.

During the entire slow stretching, no matter to what length, the barnacle and *Peripatus* cells develop active tension in response to either pulses of current applied through straddling parallel plate electrodes, or high K^+ saline. Maximum active

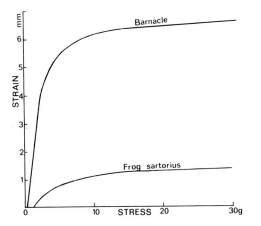

FIG. 3. Stress/strain curve of series elastic component of single giant barnacle depressor muscle cell compared with that of whole frog sartorius of similar mass. The barnacle attachments are inelastic.

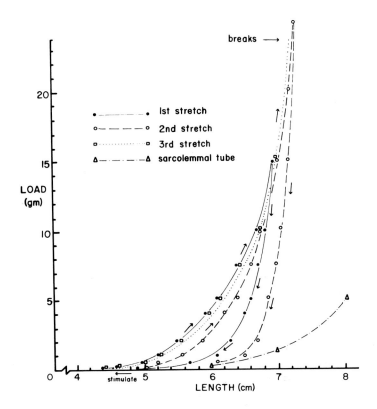

FIG. 4. Hysteresis in load/length curve for giant barnacle cell stretched relatively quickly. The cell was given a single, brief electric shock following the return after the second stretch. This restores the original length. The load/extension curve for the "sarcolemmal tube" of a neighboring cell of similar thickness after coagulation of its myoplasm is shown also.

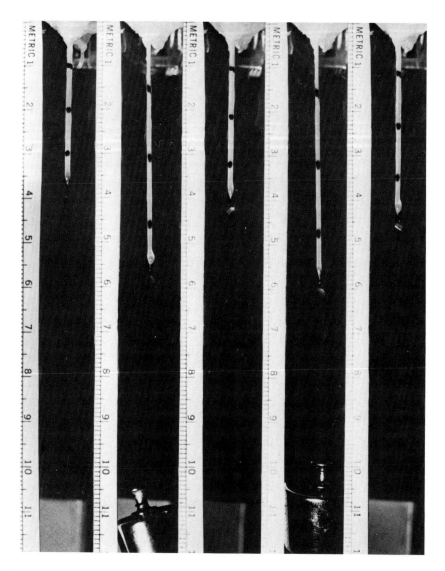

FIG. 5. Effects of stretch on unloaded length of barnacle giant cell. First load, 10 g, second load, 20 g. Load removed in third and fifth photographs.

tension declines with increasing length, but even in the barnacle it is not eliminated completely at any extent of stretch, provided the stretching is done very slowly.

Both barnacle and *Peripatus* cells give graded calcium spikes and neither is 100% activated by a single action potential even if it overshoots. The rate of rise of active state is less than one-tenth that in frog sartorius. Any degree of activation from zero to maximum can be achieved by varying duration and amplitude of a DC current pulse applied across the entire cell (Fig. 6). The change in state may be

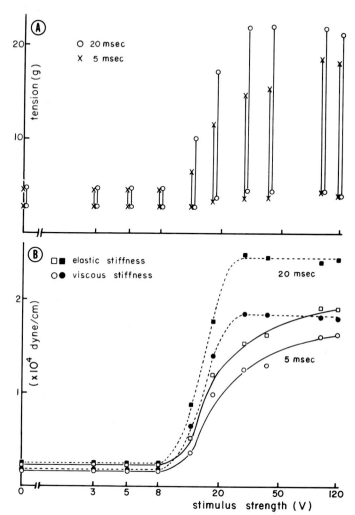

FIG. 6. A: Tension at two different stimulus durations and increasing strength (constant voltage) applied by long plate electrodes straddling single giant barnacle depressor cell. **B:** Elastic stiffness and viscous stiffness as determined by sinusoidal stretch results from same cell as A. (From M. Kobayashi, *unpublished results*.)

followed by sinusoidal stretch analysis. Barnacle giant cells tolerate repeated extension up to 25% l_0 at frequencies up to 5 Hz under maximal stimulation. Both elastic stiffness and viscosity rise steadily with increasing extent of activation, but with the viscosity increase lagging behind the stiffness increase.

ULTRASTRUCTURE

We must look to the ultrastructures to seek an explanation of these capabilities, which these muscles share in common with vertebrate smooth muscles. The ability of the barnacle cross-striated cells to shorten reversibility was explained by their having Z discs that are partly perforated and that become more opened up following activation, as reported by Hoyle et al. (4). The Z discs of glycerinated barnacle myofilament clusters actively expand following addition of high concentrations of ATP. This behavior of the Z permits the thick filaments to pass through it readily and to interdigitate across it (Fig. 7). This is nice because it is directly compatible with sliding filament theory. It does call for some angling away from the longitudinal axis of thin filaments, and of clustering of myofilaments as lateral thickening occurs, on cross-bridge theory. The thick filaments retain their original length, remaining stiff and without shortening, down to about 30% l_0, but below this extent of shortening they begin to coil loosely. It is still possible to imagine the coiling as being compatible with a cross-bridge mechanism. But this and indeed all other aspects of barnacle cell dynamics are more readily explained by a hypothesis of contractility in ultrathin filaments that are in parallel with both A and I filaments. The principal restoring force for relaxation may be radial elasticity. Alternatively, an active reversal of contraction is required.

Stretch beyond overlap reveals, as in all other cross-striated cells, the presence of the highly elastic ultrathin filaments in parallel with the ordinary thick (A) and thin (I) filaments. However, unlike the situation in stretched mammalian cells, the extensively stretched Z regions become highly distorted longitudinally (Fig. 8). This brings some I filaments from one sarcomere into contact with A filaments from neighboring ones after they have lost contact with their natural partners. This would not be possible if there were a lot of sarcoplasmic reticulum (SR) around myofilament clusters, but in barnacle cells there is relatively little, with numerous gaps.

Thus, no matter how much the barnacle muscle is stretched, there is the possibility of end-to-end continuity via cross-bridges. This kind of association has long been one of the suggestions put forward to explain the wide length range capability of smooth muscles. Nevertheless, this possibility has been denied for stretched frog sartorius cells. Complete loss of tension-generating mechanism at exactly the point of nonoverlap of I and A filaments is a cornerstone of cross-bridge theory, as reported by Gordon et al. (1).

In *Peripatus* the very long I filaments are attached to Z material that has no lateral alignment (Fig. 9.) Material is attached to the surface membrane, some to invaginated wide tubules, both radial and longitudinal, and some is located in the body of the cell. The latter is in the form of ovoid dense Z bodies, whose long axis is parallel to the long axis of the cell. These, and also Z bodies attached to the surface membrane, closely resemble the attachment sites for 6 nm diameter filaments in vertebrate smooth muscle, as determined by Hoyle and Williams (6).

In contrast to the latter, *Peripatus* cells contain large numbers of uniform 23 nm diameter thick filaments, no matter what the fixative or pH (Fig. 10).The overall ratio of ordinary thin-to-thick is 4:1. But no matter what the state of contraction at the time of fixation, there are very few orbits of thin around thick. The ones that do occur have 10 thin ones, compatible with the 4:1 ratio. The majority of thin

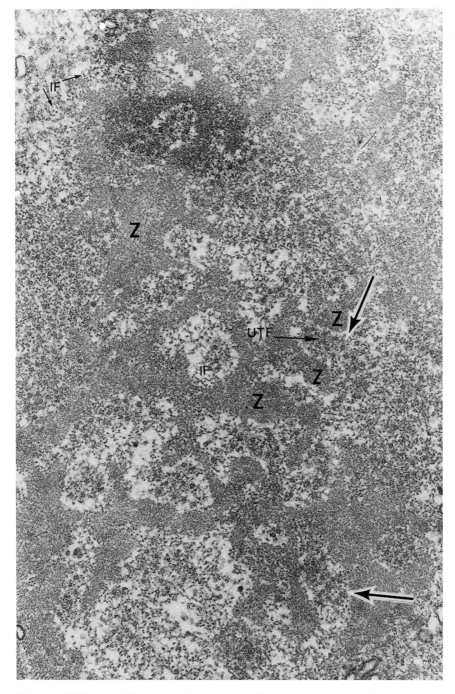

FIG. 7. EM TS through Z region of giant barnacle cell in state of supercontraction with isometric tension recording. Numerous thick filaments are seen penetrating perforations in the Z region. Each is accompanied by dense clusters of ultrathin filaments (UTF). Ordinary thin I filaments (IF) run independently.

FIG. 8. EM LS through single sarcomere of giant barnacle cell stretched about 250%. Parts of the regions are considerably displaced longitudinally and interaction between some displaced A regions with non-neighbor I regions has occurred *(arrows)*. The cell in general is holding together via longitudinal ultrathin (T) filaments.

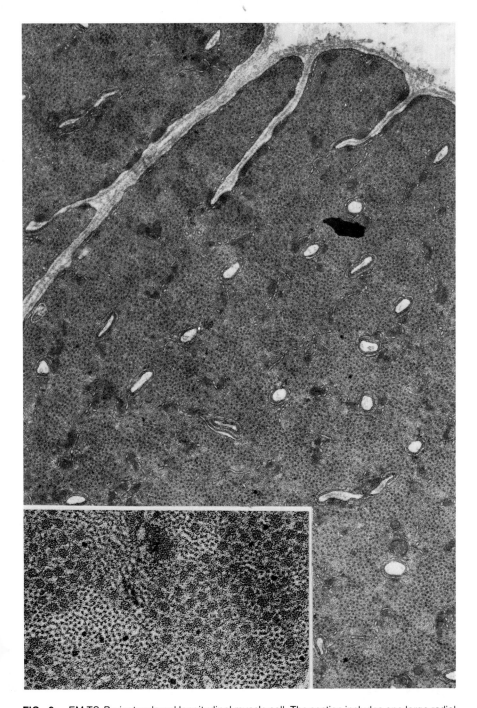

FIG. 9. EM TS *Peripatus* dorsal longitudinal muscle cell. The section includes one large radial wide invaginated tubule (WIT) and two short ones. These occur in addition to a few ordinary narrow ITs. Cut in cross-section are 22 longitudinal WITs. Each is surrounded by cysternal elements of SR (C). **Inset:** high magnification image of thick and ordinary thin filaments and a portion of a Z body. Ultrathin filaments surround the thick filaments but very few ordinary thin filaments. (From Hoyle and Williams, ref. 6.)

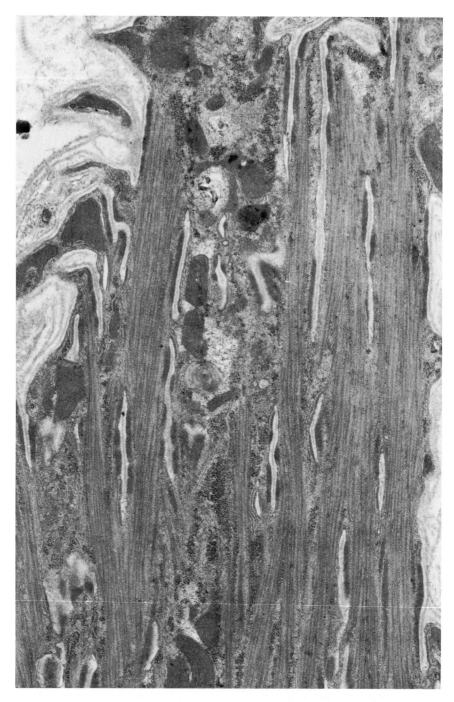

FIG. 10. EM LS *Peripatus* dorsal longitudinal muscle cell near its attachment showing origins of longitudinal WITs with material, including ultrafine collagen strands, invaginated from sarcolemmal complex that attach the cell to the body wall. Long dense Z bodies serve as attachments for ordinary thin filaments. (From Hoyle and Williams, ref. 6.)

and thick filaments form clusters with their own kind, about 30 thick ones in each thick filament cluster, lacking a markedly hexagonal array, and thin ones in clusters of 100 or so.

Despite the disarray of Z material and the lack of orbiting, an elemental Z-I-A-I-Z sequence is recognizable. A case could be made for calling this *disorderly striated muscle* to contrast it with cross-striated or obliquely striated. This disorderly array is clearly the aspect that permits the great length range over which the cells can operate, and the general physiological similarity to vertebrate smooth muscles.

PLASTICITY

Smooth muscles of vertebrates are much more plastic when fully activated than are their cross-striated partners. They do not show abrupt falls in tension when subjected to long quick stretches. Exactly the same is true of giant barnacle cross-striated cells. Large sinusoidal stretches up to 25% l_0, at frequencies up to 10 Hz, can be applied to them during or following maximal stimulation, yet they show normal twitch properties afterward. Their properties change in the manner of a viscoelastic body undergoing a steady rise in stiffness when activated, very much along lines once considered to explain vertebrate cross-striated cells and still used in analysis of vertebrate smooth muscle. In carrying out stretch experiments on giant barnacle cells, one is always impressed by their rubber-like elasticity, as have been most investigators of smooth muscles of all kinds.

CONCLUSIONS

When directly examined and compared, classic vertebrate cross-striated cells, such as those of frog sartorius, seem markedly different from smooth cells of taenia coli in both ultrastructure and physiological properties. But invertebrate skeletal muscles have an extremely wide range of ultrastructural features and biophysical/physiological properties. Although only two examples from within the vast available range were considered—one undoubtedly cross-striated cells, the other cells in which no order occurs—no discordancy exists between their properties, which ally each of them with aspects of both the vertebrate extremes.

It still seems to me that there is an underlying unity in muscle, as has long been supposed, but if we have to select an odd one, certainly from among the ones we have considered, it is clearly the frog sartorius. Since this is the special case, it may be that a great mistake is being made in using it as the standard model on which to base all our ideas about the fundamental mechanism of contraction.

There are major differences between the majority of invertebrate skeletal cells in general, and the special case, frog sartorius. First, regardless of intrinsic maximum speed or tension capabilities, the majority function over a severalfold greater length range. Second, the contractile machinery of the majority is infinitely gradedly activatable. Third, the viscosity and elastic stiffness of the majority are proportional to the extent of activation. Fourth, the majority are much more extensible and viscoelastically plastic at all extents of activation of their contractile machinery than is frog sartorius.

The properties of both muscles favor a model I proposed (3) in which the fundamental elastic structural/contractile elements are ultrathin T filaments that run from end to end of the cell where they attach to the surface membrane. If Z regions are present, they run freely through them. I filaments of cross-striated muscle and their equivalent, ordinary thin filaments of smooth muscle, and also thick filaments, run in parallel with T filaments. The purpose of I/A interaction in vertebrate cross-striated cells on this model is to regulate the energy supply for contraction in the T filaments.

REFERENCES

1. Gordon, A. M., Huxley, A. F., and Julian, F. T. (1966): The variation in isometric tension with sarcomere length in vertebrate muscle fibres. *J. Physiol. (Lond.)*, 184:170–192.
2. Hess, A. (1967): The structure of vertebrate slow and twitch muscle fibers. *Invest. Ophthalmol. Vis. Sci.*, 6:217–228.
3. Hoyle, G. (1968): Ultrastructure of striated muscle. *Symp. Biol. Hung.*, 8:34–48.
4. Hoyle, G., McAlear, J. H., and Selverston, A. (1965): Mechanism of supercontraction in a striated muscle. *J. Cell. Biol.*, 26:621–640.
5. Hoyle, G., McNeill, P. A., and Selverston, A. (1973): Ultrastructure of barnacle giant muscle fibers. *J. Cell. Biol.*, 56:74–91.
6. Hoyle, G., and Williams, M. (1980): The musculature of *Peripatus* and its innervation. *Philos. Trans. R. Soc. Lond. [Biol.]*, 288:481–510.
7. Huxley, A. F. (1980): *Reflections on Muscle*. Princeton University Press, Princeton.
8. Pantin, C. F. A. (1956): Comparative physiology of muscle. *Br. Med. Bull.*, 12:199–202.

Basic Biology of Muscles: A Comparative Approach, edited by B. M. Twarog, R. J. C. Levine, and M. M. Dewey. Raven Press, New York © 1982.

Crustacean Muscles: Atrophy and Regeneration During Molting

Donald L. Mykles and Dorothy M. Skinner

Biology Division, Oak Ridge National Laboratory, Oak Ridge, Tennessee 37830

The mechanical problem of pulling claw muscle from the old exoskeleton at the time of ecdysis of a crustacean was recognized by nineteenth-century biologists. Herrick (25) observed that muscle occupying a cross-sectional area of 882 mm^2 had to be pulled through an opening of 93 mm^2 in the American lobster, *Homarus americanus*. In 1860, Salter (58) noted in molting lobsters "the large soft hands being drawn by compression through the narrow joints, as a wire is drawn through the contracting holes of a draw plate."

Claw muscle atrophy facilitates withdrawal of the propus, the large, distal segment of the claw, through the narrow basiischial joint at ecdysis. Claw muscle atrophy was first described in the crab, *Cancer pagurus*, by Couch in 1837 (10). In animals about to molt, he "found the limbs shrunk to three-fourths of what the shell could well contain." In a later paper (11), Couch states that "for without much wasting, the muscular structure of the hand or fingered claw, could not be made to pass through the narrow joints at its insertion into the body."

The claw closer muscle of the Bermuda land crab *Gecarcinus lateralis* undergoes a sequential atrophy and restoration during each molting cycle. Muscle protein decreases 40% during proecdysis and is restored following ecdysis (63). Amino acid incorporation into protein of postecdysial muscle is five times greater than that in anecdysial muscle (75). Since the rates of protein synthesis in anecdysial and proecdysial muscles are the same (75), it appears that proecdysial muscle atrophy is caused primarily by an increase in protein degradation. Cathepsin D activity is twofold greater in proecdysial than in anecdysial muscle (76).

The ultrastructure of muscle atrophy has been extensively studied in numerous vertebrate systems (13,19,23,43,45,46,51,59,60). By contrast, atrophy of arthropod muscle has been the subject of few investigations. The effects of denervation have been examined in the muscles of moth larvae (53) and locusts (55), as have the effects of disuse on cockroach leg muscle (1). A process similar to that in *Gecarcinus lateralis* occurs in a small group of intersegmental muscle fibers in the blowfly, *Calliphora erythrocephala*, at metamorphosis, except that in the latter case all the contractile material is degraded (12).

337

All forms of atrophy share certain ultrastructural characteristics. The decrease in fiber width corresponds to a decline in myofibrillar cross-sectional area. As myofilament degradation progresses, interfibrillar space increases, there is increased organelle turnover (13,19), and the absolute amount of sarcoplasmic reticulum (SR) decreases (19).

In mammalian skeletal muscle, hydrolysis of myofilaments during atrophy is caused by a simultaneous decreased synthesis and increased degradation of myofibrillar proteins (21). Generally, proteolysis is nonspecific and both thick and thin filaments are degraded (43,51). However, preferential hydrolysis of thick filaments has been reported in degenerating insect muscle (2,6,12,34), dystrophic muscle (13), and pigeon pectoralis muscle after chronic exposure to dantrolene sodium (62).

The Ca^{2+}-dependent proteinases (CDP) (5) purified from the skeletal muscle of several vertebrates are believed to play an important role in myofibril protein turnover (3,14,15,22,29,41,54,68–70). In various vertebrate muscles maintained *in vitro*, increased intracellular levels of Ca^{2+} stimulate protein turnover (35,68) and are correlated with dissolution of myofilaments (17,18,52,68). Increased levels of CDP activity have been measured in dystrophic muscle (36,49) and in atrophying muscle caused by vitamin E deficiency (16).

We have examined the ultrastructural basis of atrophy of claw closer muscle of the land crab and the organization of myofibrils and SR during the hydrolysis of protein that occurs during proecdysis. Proecdysis was initiated by autotomy of seven or eight walking legs (64,65). In some ways, the ultrastructure of the accompanying claw muscle atrophy resembles that of other atrophies induced by denervation, disuse, and disease in that lysosomes do not appear to be involved in myofilament degradation. We have investigated the changes that occur in contractile proteins during claw muscle atrophy and the involvement of CDP in myofilament degradation. In contrast to other systems, there is a preferential degradation of thin filaments relative to thick filaments, resulting in an increase in thick-filament packing density.

The hydrolysis of actin, myosin, tropomyosin, and troponin is stimulated by Ca^{2+} and inhibited by EGTA and the proteinase inhibitors antipain and leupeptin. These data strongly suggest the presence of CDP capable of degrading contractile proteins in claw muscle fibers and that the CDP plays a significant role in claw muscle atrophy. To our knowledge, this is the first CDP that degrades the major contractile proteins actin and myosin.

FIG. 1.   Normal closer fibers from anecdysial crabs fixed with 2.5% glutaraldehyde. **A:** Longitudinal section of a fiber in which a Z-tubule (ZT) makes dyadic contact (D) with sarcoplasmic reticulum (SR) via a T-tubule (T). A-band (A); I-band (I); Z-line (Z); Bar = 1 μm. **B:** Transverse section of a fiber shows myofibril (Mf) surrounded by sarcoplasmic reticulum (SR) at level of A-band. Dyad (D); Z-line (Z); Bar = 1 μm. **C:** Stack of SR lamellae (SR) in transverse section. Bar = 0.5 μm. **D:** Nucleus (N) and mitochondria (M) in peripheral sarcoplasm. Basement membrane (BM); Myofibril (Mf); Bar = 1 μm. (From Mykles and Skinner, ref. 47.)

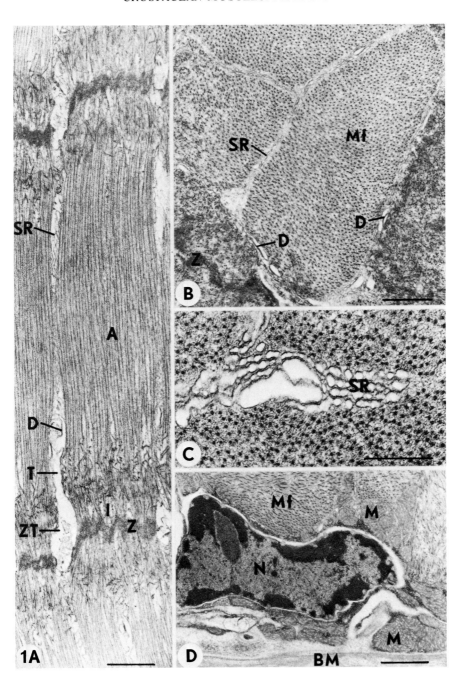

ULTRASTRUCTURE OF NORMAL MUSCLE: ANECDYSIAL FIBERS

The ultrastructure of fibers of claw closer muscle from *Gecarcinus lateralis* resembles that of slow, or tonic, muscles from other decapod crustaceans (8,20,26,27,30–33,57,61). The myofibrils are comprised of long sarcomeres (7 to 12 μm) with irregular Z-lines (Fig. 1A). Thick filaments are surrounded by orbits of 10 to 15 thin filaments (see below); the ratio of thin:thick myofilaments is about 9:1. Thick filaments are usually separated by one or two rows of thin filaments. The packing density of the thick filaments is 288 filaments per μm^2 (Table 1) and corresponds to an interfilament distance of 59 nm.

In cross-sections, the SR forms a well-defined boundary around each myofibril and comprises a fenestrated sleeve around the A band (Fig. 1B). Lamellae often occur in stacks of up to four layers (Fig. 1C). T-tubules, which are continuous with the cell membrane, make dyadic or triadic contact with the SR toward the end of the A-band near the A-I junction (Figs. 1A and B). Nuclei and mitocondria are located primarily in the peripheral sarcoplasm (Fig. 1D).

ULTRASTRUCTURE OF ATROPHIC MUSCLE: LATE PROECDYSIAL FIBERS

The ultrastructure of proecdysial muscle atrophy is similar in several characteristics to that of other striated muscle atrophies induced by disuse (1), denervation (19,43,46,51,55), and disease (13,45,60). The decrease in fiber width is accompanied by a fourfold decrease in myofibrillar cross-sectional area (Figs. 2B and 3). We find that fiber width is proportional to myofibril cross-sectional area rather than to numbers of myofibrils. The variability in the cross-sectional area of the myofibrils is greatest in anecdysial and early proecdysial muscles, and decreases as the fibers become more uniform in size following atrophy in late proecdysis. Clearly, large fibers undergo a more substantial atrophy and show a much greater reduction in cross-sectional area than small fibers (Fig. 3).

Interfibrillar space is dilated in atrophic fibers (Figs. 2A and B), causing the fraction of fiber area occupied by myofibrils to decrease from 0.912 to 0.80 (Table

TABLE 1. *Anecdysis and late proecdysis[a]*

Molt stage	Myofibril cross-sectional area (μm^2)	Length of A-band (μm)	Fraction of fiber occupied by myofibrils	SR surface density[b]	Thick filament packing density[c]
Anecdysis	6.60 ± 3.35 (16)	6.2 ± 0.9 (5)	0.912 ± 0.027 (16)	1.52 ± 0.44 (12)	288 ± 79 (12)
Late proecdysis	1.48 ± 0.56 (10)	6.8 ± 0.4 (3)	0.800 ± 0.054 (10)	1.49 ± 0.28 (8)	494 ± 101 (10)

From Mykles and Skinner (47).
[a]Mean ± 1 SD (number of fibers in parentheses).
[b]SR membrane surface area (μm^2) per cell volume (μm^3).
[c]Filaments per myofibril cross-sectional area (μm^2).

1) suggesting that dissolution of myofilaments at the periphery of the fibril accounts for the decline in myofibril diameter, as has been proposed in other atrophies (43,45,51,55).

Late proecdysial fibers are also characterized by filament-free areas within the fibrils (Fig. 2C). These areas of erosion are often observed to extend several micrometers (Fig. 2A). Erosion areas probably result from proteolysis in the center of myofibrils as in denervated rat muscle (43) and degenerating insect muscle (6,34), in which some myofilaments are lost from the central portions of the myofibrils and the regular alignment of myofilaments is interrupted. Despite the occurrence of erosion areas in late proecdysial claw muscle, the packing arrangement of its myofilaments is more regular than that of anecdysial muscle. The greater stability inherent in the regular packing may maintain the gaps in the myofilament lattice caused by protein degradation and would explain the delayed appearance of erosion areas until late in proecdysis.

There are no significant changes in the dimensions of the myofilaments during atrophy. Thick filaments measure 18 to 20 nm in diameter and about 6.5 μm in length in both anecdysial and proecdysial fibers (Table 1; cf., Figs. 1A and 2A). The diameter of the thin filaments remains at about 7 nm.

The packing density of thick filaments is 72% greater in atrophic than in anecdysial muscle and is correlated with a decrease in the filament ratio from 9:1 to 6:1 (Table 1, cf., Figs. 4A and B). The density of 494 μ^{-2} in atrophic fibers corresponds to an interfilament distance of 45 nm. Thick filaments are surrounded by orbits of 7 to 11 thin filaments. In addition, many thick filaments are no longer separated by rows of thin filaments (Fig. 4B, arrows).

Lysosomes appear primarily involved with the turnover of organelles, principally mitochondria and SR, but not in the dissolution of myofilaments. Putative lysosomes were usually found near mitochondria and nuclei at the periphery of the fiber at some distance from myofilament dissolution. Although lysosomes appear more active in atrophic fibers, as evidenced by the presence of secondary lysosomes (Fig. 5) and myelin figures (Fig. 6), and increased activity of lysosomal enzymes (76), filamentous material was never observed in their interiors.

Few changes in the internal membrane systems are observed. Although sarcolemmal clefts often become enlarged, no swelling of T- and Z-tubules and SR is seen. Both anecdysial and proecdysial fibers have an SR membrane surface density of approximately 1.5 $\mu^2/\mu m^3$ (Table 1). Although the surface density is unaltered, the SR appears more dispersed in late proecdysial fibers: stacks of two or more lamellae are uncommon and the lamellae consist of shorter segments (Fig. 2C). Atrophic fibers of cockroach leg muscle also showed a reduced number of lamellar stacks (1).

Many of the degenerative changes that appear in other atrophic systems do not occur in claw muscle atrophy. Organelles, such as nuclei and mitochondria, retain their normal appearance (Fig. 2D). The general organization of the sarcomere remains intact despite a substantial hydrolysis of myofilaments and the consequent

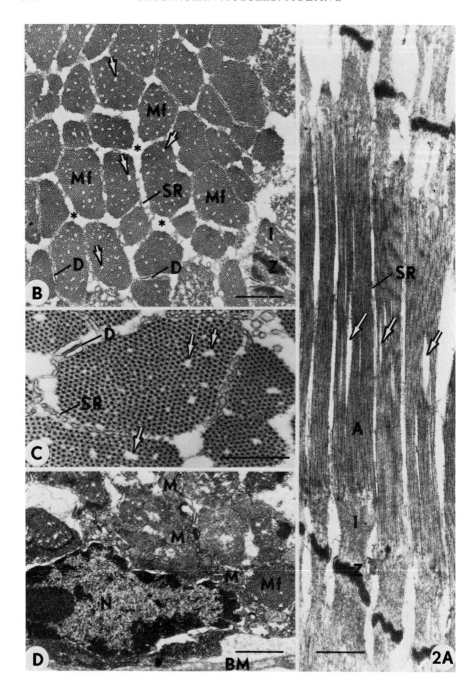

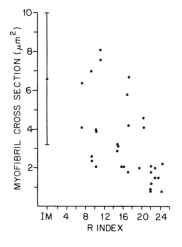

FIG. 3. Mean myofibrillar cross-sectional area (μm^2) in closer fibers from crabs in proecdysis as a function of limb regenerate length (R index). The length of the regenerating limbs, expressed as the R index (length of regenerate $\times 10^2 \times$ carapace width^{-1}), increased linearly during proecdysis to a maximum of 22 to 25 immediately before ecdysis and provided an external measurement of the progress of proecdysis. Crabs underwent ecdysis 54 to 90 days after limb autotomy. Mean cross-sectional area ($\pm$ 1 SD) in anecdysial (IM) fibers is provided for comparison. (From Mykles and Skinner, ref. 47.)

rearrangement of the myofilaments remaining in the myofibril, while structures associated with disruption of SR and T-tubules in both vertebrate and arthropod denervated fibers are never observed. The T-system and SR do not swell, and vesiculate (19,53,55,59) and labyrinthine structures, derived from T-tubules (19,23,43,51,55,59), and lamellar structures, derived from the SR (12,23,43), are never observed. The dyads and triads in claw fibers never become disoriented as in denervated frog and rat skeletal muscle (13,23,43,59), but rather maintain their normal positions during atrophy. This suggests that proecdysial atrophy is non-pathological and normal contraction-coupling of the muscle may be retained. We have observed that the claws of late proecdysial animals are still functional, although the tension developed by the closer muscles is much less than that of anecdysial crabs.

ULTRASTRUCTURE OF LEG MUSCLE

Atrophy occurs only in the claw during proecdysis, lending support to the nineteenth-century hypothesis (11) that reduction in muscle mass facilitates withdrawal of the broad distal segment through the narrow basiischial joint at ecdysis. The tonic muscles in the largest segment (merus) of walking legs were examined in anecdysial and late proecdysial crabs. The ultrastructure of leg fibers resembles

---

FIG. 2. Atrophic closer fibers from crabs in late proecdysis fixed with 2.5% glutaraldehyde. **A:** Longitudinal section of fiber in which areas of erosion *(arrows)* extend several micrometers. A-band (A); I-band (I); sarcoplasmic reticulum (SR); Z-line (Z); Bar = 1 μm. **B:** Transverse section of a fiber showing enlarged interfibrillar space *(asterisks)* and numerous erosion areas *(arrows)* at level of A-band. Dyad (D); I-band (I); myofibrils (MF); sarcoplasmic reticulum (SR); Z-line (Z); Bar = 1 μm. **C:** Transverse section of myofibrils with erosion areas *(arrows)*. Normal-looking sarcoplasmic reticulum (SR) occurs as single lamella. Dyad (D) appears normal, Bar = 0.5 μm. **D:** Transverse section of fiber showing nucleus (N) and mitochondria (M) in peripheral sarcoplasm. Basement membrane (BM); Myofibril (Mf); Bar = 1 μm. (From Mykles and Skinner, ref. 47.)

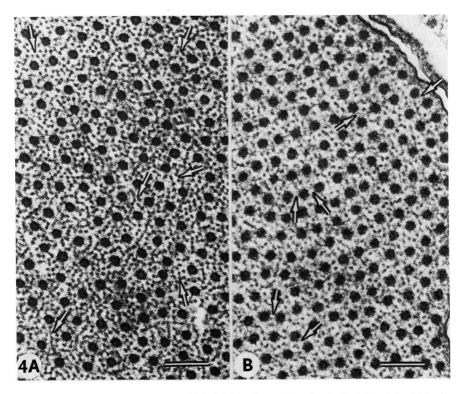

FIG. 4. Transverse sections of myofibrils in claw closer fibers fixed with 2.5% glutaraldehyde and 8% tannic acid. **A:** A myofibril from an anecdysial animal with thick filaments surrounded by orbits of 10 to 15 thin filaments; thin:thick filament ratio is about 9:1. *Arrows* show areas where thick filaments are separated by two rows of thin filaments. The average center-to-center distance between thick filaments is 51 nm. Bar = 0.1 μm. **B:** In this section of a late proecdysial myofibril, the thick filaments are surrounded by orbits of 7 to 11 thin filaments; thin:thick filament ratio is about 6:1. *Arrows* indicate areas where thick filaments are not separated by thin filaments. Average center-to-center distance between thick filaments is 45 nm. Bar = 0.1 μm. (From Mykles and Skinner, ref. 47.)

that of tonic muscles of the claw, except that the mean myofibrillar cross-sectional area is much smaller in leg muscle (Table 2). There are no apparent differences in the ultrastructure of leg muscle at the two molt stages (Figs 7A and B) and the average cross-sectional areas of the myofibrils is about 1.3 μm^2 in both stages (Table 2).

ATROPHY IN CLAW MUSCLE FROM CRABS REGENERATING ONE WALKING LEG

The degree of claw muscle atrophy depends on the metabolic demands placed on the animal. The decrease in myofibril cross-sectional area is much less in crabs regenerating one leg (52%) than that in crabs regenerating seven legs (78%) (Table 2).

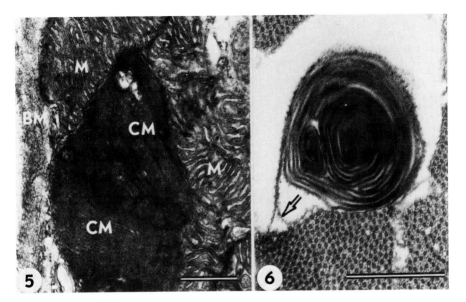

FIG. 5. Secondary lysosome in peripheral sarcoplasm of late proecdysial fiber containing condensed mitochondria (CM). Adjacent mitochondria (M) appear normal. Basement membrane (BM); Bar = 0.5 μm.

FIG. 6. Myelin figure in a late proecdysial fiber. *Arrow* indicates membrane continuity between myelin figure and sarcoplasmic reticulum. Bar = 0.5 μm.

TABLE 2. *Mean myofibrillar cross-sectional areas in claw and leg muscle from anecdysial and late proecdysial crabs[a]*

	Cross-sectional area (μm^2)
Leg	
Anecdysial crabs	1.25 ± 0.16 (9)
Proecdysial crabs	1.31 ± 0.29 (9)
Claw	
Anecdysial crabs	6.60 ± 3.35 (16)
Proecdysial crabs: 7 LA	1.48 ± 0.56 (10)
Proecdysial crabs: 1 LA	3.18 ± 0.99 (9)

[a]Mean ± 1 SD (n).

QUANTIFICATION OF MYOFIBRILLAR PROTEINS IN MUSCLE HOMOGENATES

Myosin heavy chain (HC; 200 K) and actin (A; 43 K) are the major muscle proteins in whole muscle homogenates (Fig. 8). Six hemolymph proteins ranging in molecular weight from 65 to 90 K (Fig. 8) are present in the muscle homogenates. Homogenates from late proecdysial crabs contained greater amounts of these proteins, reflecting the larger quantity of hemolymph in atrophic muscle (Fig. 8, lane b). Other proteins associated with the contractile apparatus were troponin-T (TNT;

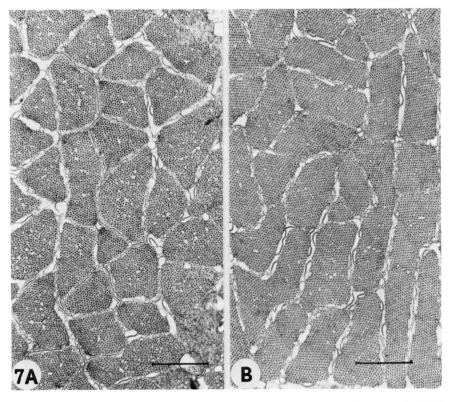

FIG. 7. Transverse sections of leg muscle fibers from anecdysial **(A)** and late proecdysial **(B)** crabs fixed with 2.5% glutaraldehyde. Bar = 1 μm.

47 K), tropomyosin (TM; 40 K), troponin-I (TNI; 29 K), myosin light chain (LC; 18 K), and an unidentified protein of 95 K. The 95 K protein may be α-actinin since it is similar in size to vertebrate α-actinins (24). Tropomyosin and the two troponin subunits were identified on the basis of electrophoretic mobilities similar to those of other arthropod troponins and tropomyosins (38,39,56). Two other unidentified proteins with molecular weights of 16 and 37 K were soluble in low-salt buffer.

In crab muscle atrophy, although all the contractile proteins are degraded, there is a preferential hydrolysis of actin relative to myosin. As described above, ultrastructural observations demonstrate a decrease in the ratio of thin to thick myofilaments from 9:1 to 6:1. Densitometric scans of muscle homogenates show a relative decrease of proteins that comprise the thin filament: Actin, tropomyosin, and troponin (Fig. 9).

To quantify the amounts of actin and myosin HC more precisely, homogenates were electrophoresed on polyacrylamide gels cross-linked with dialyltartardiamide, stained with Coomassie blue and solubilized in periodic acid. Absorbancies were compared with those of actin and myosin standards electrophoresed in parallel. We

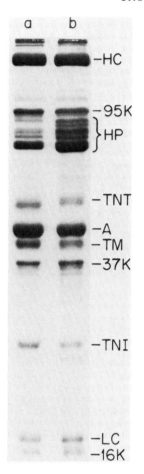

FIG. 8. SDS-PAGE of whole muscle homogenates from crab claw. Lane *a*: Whole muscle homogenate from anecdysial animal (25μg). Lane *b:* Whole muscle homogenate from late proecdysial animal (25 μg). Myosin heavy chain (HC) and actin (A) are major proteins in the contractile apparatus. Troponin-T (TNT), tropomyosin (TM), troponin-I (TNI), myosin light chain (LC), and one unidentified protein (95 *K*) are also associated with myofibrils. Two other unidentified proteins (37 and 16 *K*) are soluble in low-salt buffer. Hemolymph proteins (HP) are *bracketed*. Note that there is less actin relative to myosin HC in a late proecdysial homogenate (lane *b*) than in an anecdysial homogenate (lane *a*).

found that the molar ratios of actin to myosin HC decreased 31%, from 8.86 to 6.07, during muscle atrophy (48).

CALCIUM CONTENT OF MUSCLE HOMOGENATES

As the exoskeleton is degraded during proecdysis, large amounts of calcium are released. The concentration of calcium in hemolymph is about 17 mM and increases 36% in proecdysis (66). The calcium content of late proecdysial homogenates is double that of anecdysial homogenates (Table 3). It is not known what the relative contributions the hemolymph and tissue make to total calcium content. The volume of the hemolymph is greater in late proecdysial homogenates and may contribute to the increase in calcium. Since hemolymph protein concentrations increase three-fold during proecdysis (4,72), expressing calcium content on the basis of milligrams of protein should more than compensate for this increase in volume. Thus, it seems

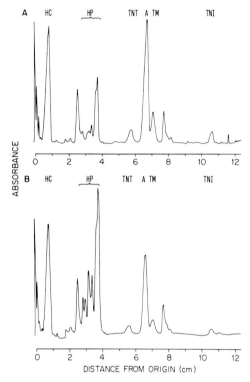

FIG. 9. Densitometer scans of lanes *a* and *b* from Fig. 8. **A:** Whole muscle homogenate from anecdysial animal. **B:** Whole muscle homogenate from late proecdysial animal. Note that there is less actin (A), tropomyosin (TM), troponin-T (TNT), and troponin-I (TNI) relative to myosin heavy chain (HC) in late proecdysial homogenate. Hemolymph proteins (HP). (From Mykles and Skinner, ref. 48.)

TABLE 3. *Calcium contents of anecdysial and late proecdysial muscle homogenates*[a]

	Calcium content (μg/mg protein)
Anecdysis	3.42 ± 0.31 (5)
Late proecdysis	6.83 ± 0.29 (4)

[a]Mean ± 1 SD (n).
From Mykles and Skinner (48).

likely that elevated intracellular calcium makes a substantial contribution to the increase in total calcium in late proecdysial homogenates.

CALCIUM-DEPENDENT PROTEOLYTIC ACTIVITY IN MUSCLE HOMOGENATES

A series of experiments was performed to characterize the enzyme(s) that degraded contractile proteins and whose activity increased in the proecdysial period.

Since late proecdysial homogenates have increased calcium levels and proteinases isolated from vertebrate muscles are activated by calcium, the first experiments tested the effects of Ca^{2+}.

Proteolysis in whole muscle homogenates is stimulated by 5 mM Ca^{2+} at neutral pH (Fig. 10).The proteolytic activity is equivalent to 6.31 nmol leucine released/mg protein/hr. EGTA inhibits most of the activity (0.75 nmol leucine/mg/hr). The proteinase(s) is not associated with the myofibrils since activity remains in 6,000 × g supernatants. Washing myofibrils three times with buffer reduces activity sevenfold, from 2.42 to 0.35 nmol leucine/mg/hr.

Pepstatin, leupeptin, and antipain are group-specific proteinase inhibitors that have been used to determine the role of several different types of muscle proteinases in protein degradation. Both leupeptin and antipain are inhibitors of cysteine proteinases and inhibit vertebrate CDPs (3,28,37,69,71). Pepstatin inhibits aspartic proteinases such as lysosomal cathepsin D (73).

Table 4 shows the effects of EGTA and proteinase inhibitors on CDP activity. EGTA is more effective than either leupeptin or antipain in inhibiting proteolysis. The depressed activity (77%) observed with pepstatin is probably the result of having 4% dimethylsulfoxide (DMSO), in which the pepstatin was dissolved, in the reaction mixture. The activity with 4% DMSO alone is the same as that when pepstatin is present.

We have measured CDP activity in anecdysial and late proecdysial claw muscle homogenates and compared these to CDP activity in limb regenerates (Table 5), in which a rapid net synthesis of muscle protein is taking place. Ca^{2+}-dependent proteinases activity in late proecdysial homogenates is 47% greater than that in anecdysial homogenates; CDP activity is also greater in homogenates of limb re-

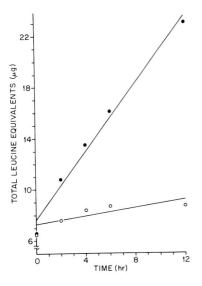

FIG. 10.   Release of TCA-soluble material (μg leucine equivalents) from an anecdysial muscle homogenate incubated with 5 mM $CaCl_2$ (*closed circles*, correlated coefficient of 0.992) or 10 mM EGTA and 5 mM $CaCl_2$ (*open circles*, correlation coefficient of 0.795). Reaction mixtures contained 80 mM KCl, 80 mM Hepes-NaOH, pH 7.5, 4 mM $MgCl_2$, 4 mM NaN_3, 0.8 mM dithiothreitol, and 40 μg/ml gentamicin; incubated at 37°C.

TABLE 4. *Effects of inhibitors on CDP activity in muscle homogenates*

Condition[a]	Activity[b]	Percent activity remaining
Control	2.01 ± 0.46 (5)	100
EGTA, 10mM	0.46 ± 0.05 (5)	23
Leupeptin, 400 μM	0.93 ± 0.14 (5)	46
Antipain, 400 μM	0.65 ± 0.16 (5)	32
Pepstatin, 400 μM	1.54 ± 0.30 (5)	77
DMSO, 4%	1.61 ± 0.40 (5)	80

[a]Reaction conditions same as in Fig. 10 with 5 mM Ca^{2+}; incubated at 37°C, 48 hr.
[b]nmol leucine released/mg protein/hr ± 1 SD (number of animals in parentheses).

TABLE 5. *CDP activities in claw muscle and limb regenerates[a]*

Tissue	Molt stage	Activity[b]
Claw muscle	Anecdysis	4.85 ± 1.04 (5)
Claw muscle	Late proecdysis	7.11 ± 1.06 (5)
Limb regenerates	Late proecdysis	6.28 ± 0.37 (3)

[a]Reaction conditions same as in Fig. 10 with 5 mM Ca^{2+} ± 10 mM EGTA; incubated at 37°C, 12 hr.
[b]nmoles/leucine released/mg protein/hr ± 1 SD (number of animals in parentheses).

generates. In vertebrates, anabolic increases in protein degradation have been observed in young, growing animals and in induced hypertrophy (see 44, for review).

ANALYSIS OF MYOFIBRILLAR PROTEIN DEGRADATION

There is a specific hydrolysis of myofibrillar proteins in homogenates containing 5 mM $CaCl_2$ at neutral pH. Actin, myosin HC and LC, tropomyosin, troponin-T, troponin-I, and two unidentified proteins (95 *K* and 37 *K*) are degraded: there is no detectable degradation of hemolymph proteins contained in the muscle homogenates (Fig. 11). Thus, hemolymph proteins serve as an internal control and attest to the specificity of the proteinase(s) for muscle proteins.

In anecdysial homogenates, as the single myosin HC band was degraded, it separates into two distinct bands (Fig. 11A, arrowhead; compare 0 hr, lane a and 24 hr, lane g). In addition, a doublet of approximately 140 *K* appears and a band at 105 *K* becomes more prominent as incubation continues. After 12 hr there are no further changes in the appearance of the 140 and 105 *K* bands, suggesting that a steady state has been reached between accumulation and further hydrolysis to peptides and amino acids. The amounts of actin, myosin LC (not shown), tropomyosin, 95 *K* protein, troponin-T, 37 *K* protein, and troponin-I (not shown) decrease

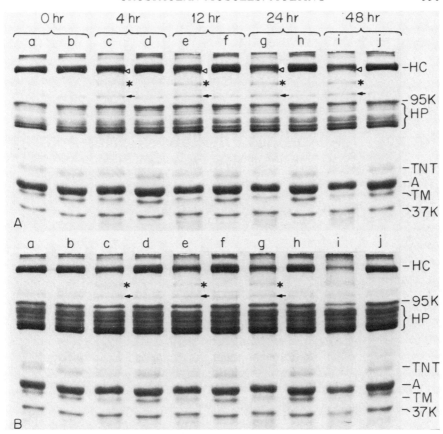

FIG. 11. Proteolysis in whole muscle homogenates. Homogenates contained either 5 mM CaCl₂ alone (lanes *a, c, e, g,* and *i*) or 5 mM CaCl₂ and 10 mM EGTA (lanes *b,d,f,h,* and *j*). Reactions were terminated at 0 (lanes *a* and *b*), 4 (lanes *c* and *d*), 12 (lanes *e* and *f*), 24 (lanes *g* and *h*), and 48 (lanes *i* and *j*) hr by addition of SDS sample buffer. **A:** Homogenate of muscle from anecdysial animal. In the presence of Ca²⁺, the single myosin HC (HC) band separates into two bands *(arrowhead)* as two additional bands appear at 140 K *(asterisk)* and 105 *K (arrow).* These two bands become more prominent during the first 12 hr of incubation. Actin (A), tropomyosin (TM), troponin-T (TNT), troponin-I (not shown), myosin LC (not shown), and 37 *K* protein are degraded. Hemolymph proteins (HP) remain undegraded throughout the 48-hr incubation period. **B:** Homogenate of muscle from late proecdysial animal. Proteolysis is similar to, but more rapid than, that of the anecdysial homogenate. With Ca²⁺ present, bands appear at 140 and 105 *K,* but are completely degraded at 48 hr. Much of the actin (A), myosin HC (HC), myosin LC (not shown), and 37 *K* protein is degraded. Tropomyosin (TM), troponin-T (TNT), and troponin-I (not shown) are completely hydrolyzed at the end of 48-hr incubation. Hemolymph proteins (HP) remain undegraded. Composition of reaction mixture same as in Fig. 10, except that 80 mM Tris-HCl, pH 7.5, replaced Hepes. (From Mykles and Skinner, ref. 48.)

during a 48-hr incubation. Proteolysis appears to be completely inhibited by 10 mM EGTA (Fig. 11A, lanes b, d, f, h, and j).

Although the pattern of proteolytic activity in late proecdysial homogenates is qualitatively the same as that in anecdysial homogenates, the activity is much greater

(Fig. 11B). The same proteins are degraded as those in anecdysial preparations. Most of the actin, myosin HC and LC, 95 K protein is degraded; tropomyosin, troponin-T, and troponin I (not shown) are completely hydrolyzed by the end of a 48-hr incubation (Fig. 11B, lane i) as are the 140 and 105 K degradation products of myosin HC. Even in the increased activity characteristic of proecdysial muscle, hemolymph proteins are not hydrolyzed. Proteolysis is inhibited by EGTA (Fig. 11B, lanes b, d, f, h, and j).

Both antipain and leupeptin inhibit the degradation of myosin HC and LC, actin, tropomyosin, 95 K protein, troponin-T, 37 K protein, and troponin-I; 140 K protein and 105 K protein are less prominent (Fig. 12, lanes e and f). Pepstatin A and 4% DMSO, in which pepstatin was dissolved, failed to inhibit proteolysis (Fig. 12, lanes g and h).

CONCLUSIONS

Atrophy in the crustacean claw muscle promises to be a useful model in the study of protein metabolism and its manifestation in the structure of the sarcomere. Despite a substantial hydrolysis of protein during proecdysis, the organization of the sarcomere remains intact. The fibers do not degenerate, but rather remain viable through atrophy. This distinguishes claw muscle atrophy from many other systems in which atrophy is induced by trauma and often result in the eventual degeneration of entire fibers.

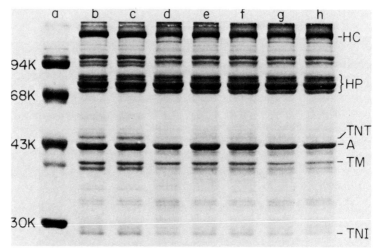

FIG. 12. Effects of proteinase inhibitors (400 μM) on proteolysis in whole muscle homogenate from anecdysial animal with 5 mM CaCl$_2$. Reactions terminated after 0 hr (lane *b*) and 24 hr (lanes *c* to *h*) by addition of SDS sample buffer. *a*, Protein standards; *b*, Ca^{2+} alone at 0 hr; *c*, Ca^{2+} and EGTA at 24 hr; *d*, Ca^{2+} alone at 24 hr; *e*, Ca^{2+} and leupeptin at 24 hr; *f*, Ca^{2+} and antipain at 24 hr; *g*, Ca^{2+} and pepstatin A in 4% DMSO at 24 hr; *h*, Ca^{2+} and 4% DMSO at 24 hr. Both leupeptin (lane *e*) and antipain (lane *f*) inhibit degradation of myosin HC (HC), troponin-T (TNT), actin (A), tropomyosin (TM), troponin-I (TNI), and myosin LC (not shown). Hemolymph proteins (HP) are not degraded. Neither pepstatin A (lane *g*) nor DMSO (lane *h*) inhibit proteolysis. Composition of reaction mixtures same as in Fig. 10. (From Mykles and Skinner, ref. 48.)

We have demonstrated a preferential hydrolysis of thin relative to thick myofilaments and the consequent rearrangement of myofilaments in the sarcomere. This phenomenon is not unique to *Gecarcinus* since other crustaceans are capable of even greater changes in myofilament packing during development. In the snapping shrimp, Mellon and Stephens (42) show that the thin:thick filament ratio increases from 3:1 to 7:1 as the pincer claw differentiates, through several ecdyses, into a snapper claw. Although not examined at the ultrastructural level, it is clear that fiber transformation occurs in lobster claw muscle as claws differentiate into the cutter and crusher. Histochemistry (50) and measurements of sarcomere length (9) show that transformation from fast to slow fibers (crusher) and from slow to fast fibers (cutter) takes place. Since the myofilament ratio is greater in slow (6:1) than in fast (3:1) fibers, fiber transformation in lobster claw necessitates a rearrangement in myofilament packing.

Proteolysis in whole muscle homogenates is stimulated by Ca^{2+} at neutral pH and inhibited by EGTA, antipain, and leupeptin. Although at the concentrations used the proteinase inhibitors were not as effective as EGTA in inhibiting protein degradation, the qualitative effects of EGTA, antipain, and leupeptin on proteolysis were identical. These results suggest that a CDP, which degrades myosin HC and LC, actin, and other myofibrillar proteins, is present in crab claw muscle.

Ca^{2+}-dependent proteinases in the skeletal muscles of several vertebrates including rabbit, cow, pig, and chicken are believed important in the turnover of myofibril proteins. Purified preparations of these proteinases degrade several myofibrillar proteins, including tropomyosin and troponin (3,22,29,70), and also release α-actinin from Z-lines (41,54). However, unlike the CDP(s) of crustacean muscle, those from vertebrate skeletal muscle hydrolyze neither actin nor myosin (see 7, for review). Although leupeptin decreased total protein degradation in intact fetal mouse heart, it had no effect on the degradation of myosin (74).

In vertebrate muscle, the degradative pathway of myofibrillar proteins remains unclear (see 40, for review). Either lysosomal or extralysosomal mechanisms may be involved. Since vertebrate CDPs hydrolyze neither actin nor myosin but have been shown to degrade Z-lines, it has been hypothesized that myofibril disassembly is initiated by extralysosomal CDPs with subsequent proteolysis by lysosomes (13,44). In crab claw muscle, the CDP appears to play a more direct role in myofibril protein turnover. Our results strongly suggest that there is present in crab claw muscle one or more CDPs that degrade the major (all?) myofibrillar proteins, including actin and myosin. The fact that Ca^{2+}-activated proteolytic activity is greater in atrophic muscle suggests that the proteinase plays an important role in myofibrillar protein turnover and in the loss of 40% of the muscle mass in crustacean claw that occurs during proecdysis.

ACKNOWLEDGMENTS

This research was sponsored by the Office of Health and Environmental Research, U.S. Department of Energy, under contract W-7405-eng-26 with the Union Carbide

Corporation. D.L.M. is a postdoctoral fellow supported initially by NIH-AG 0028-03 and subsequently by the Muscular Dystrophy Association through Oak Ridge Associated Universities.

REFERENCES

1. Anderson, M. (1979): Ultrastructural evidence for disuse atrophy in insect skeletal muscle. *Cell Tissue Res.*, 203:503–506.
2. Auber-Thomay, M., and Srihari, T. (1973): Évolution ultrastructural de fibres musculaires inter-segmentaires chez *Pieris brassicae* (L), pendant le dernier stage larvaire et la nymphose. *J. Microsc. (Paris)*, 17:27–36.
3. Azanza, J.-L., Raymond, J., Robin, J.-M., Coffin, P., and Ducastaing, A. (1979): Purification and some physico-chemical and enzymic properties of a calcium ion-activated neutral proteinase from rabbit skeletal muscle. *Biochem. J.*, 183:339–347.
4. Barlow, J., and Ridgway, G. J. (1969): Changes in serum protein during the molt and reproductive cycles of the American lobster *(Homarus americanus)*. *J. Fish. Res. Bd. Can.*, 26:2101–2109.
5. Barrett, A. J., and McDonald, J. K., editors (1980): *Mammalian Proteases: A Glossary and Bibliography, Vol. I, Endopeptidases.* Academic Press, London.
6. Beaulaton, J., and Lockshin, R. A. (1977): Ultrastructural study of the normal degeneration of the intersegmental muscles of *Antheraea polyphemus* and *Manduca sexta* (Insecta, Lepidoptera) with particular reference to cellular autophagy. *J. Morphol.*, 154:39–58.
7. Bird, J. W. C., Carter, J. H., Triemer, R. E., Brooks, R. M., and Spanier, A. M. (1980): Proteinases in cardiac and skeletal muscle. *Fed. Proc.*, 39:20–25.
8. Brandt, P. W., Reuben, J. P., Girardier, L., and Grundfest, H. (1965): Correlated morphological and physiological studies on isolated single muscle fibers. I. Fine structure of the crayfish muscle fiber. *J. Cell Biol.*, 25:233–260.
9. Costello, W. J., and Lang, F. (1979): Development of the dimorphic claw closer muscles of the lobster *Homarus americanus*. IV. Changes in functional morphology during growth. *Biol. Bull.*, 156:179–195.
10. Couch, J. (1837): Observations on the process of exuviation in the common crab *(Cancer pagurus*, Linn.). *Mag. Zool. Bot.*, 1:341–344.
11. Couch, J. (1843): On the process of exuviation and growth in crabs and lobsters, and other British species of stalk-eyed crustacean animals. *Annu. Rep. R. Cornwall Polytechnic Soc.*, 11:1–15.
12. Crossley, A. C. (1972): Ultrastructural changes during transition of larval to adult intersegmental muscle at metamorphosis in the blowfly *Calliphora erythrocephala*. I. Dedifferentiation and myoblast fusion. *J. Embryol. Exp. Morphol.*, 27:43–74.
13. Cullen, M. J., Appleyard, S. T., and Bindoff, L. (1979): Morphologic aspects of muscle breakdown and lysosomal activation. *Ann. NY Acad. Sci.*, 317:440–464.
14. Dayton, W. R., Goll, D. E., Zeece, M. G., Robson, R. M., and Reville, W. J. (1976): A Ca^{2+}-activated protease possibly involved in myofibrillar protein turnover. Purification from porcine muscle. *Biochemistry*, 15:2150–2158.
15. Dayton, W. R., Reville, W. J., Goll, D. E., and Stromer, M. H. (1976): A Ca^{2+}-activated protease possibly involved in myofibrillar protein turnover. Partial characterization of the purified enzyme. *Biochemistry*, 15:2159–2167.
16. Dayton, W. R., Schollmeyer, J. V., Chan, A. C., and Allen, C. E. (1979): Elevated levels of a calcium-activated muscle protease in rapidly atrophying muscles from vitamin E-deficient rabbits. *Biochim. Biophys. Acta*, 584:216–230.
17. Duncan, C. J., and Smith, J. L. (1978): The action of caffeine in promoting ultrastructure damage in frog skeletal muscle fibres. Evidence for the involvement of the calcium-induced release of calcium from the saroplasmic reticulum. *Naunyn Schmiedeberg's Arch. Exp. Pathol. Pharmacol*, 305:159–166.
18. Duncan, C. J., and Smith, J. L. (1980): Action of caffeine in initiating myofilament degradation and subdivision of mitochondria in mammalian skeletal muscle. *Comp. Biochem. Physiol.*, 65C:143–145.
19. Engel, A. G., and Stonnington, H. H. (1974): Morphological effects of denervation of muscle. A quantitative ultrastructural study. *Ann. NY Acad. Sci.*, 228:68–88.
20. Franzini-Armstrong, C. (1970): Natural variability in the length of thin and thick filaments in single fibres from a crab, *Portunus depurator*. *J. Cell Sci.*, 6:559–592.

21. Goldberg, A. L. (1972): Mechanisms of growth and atrophy of skeletal muscle. In: *Muscle Biology*, edited by R. G. Cassens, pp. 89–118. Marcel Dekker, New York.
22. Goll, D. E., Okitani, A., Dayton, W. R., and Reville, W. J. (1978): A Ca^{2+}-activated muscle protease in myofibrillar protein turnover. In: *Protein Turnover and Lysosome Function*, edited by H. L. Segal and D. J. Doyle, pp. 587–588. Academic Press, New York.
23. Gori, Z. (1972): Proliferations of the sarcoplasmic reticulum and the T system in denervated muscle fibers. *Virchows Arch. [Cell Pathol.]*, 11:147–160.
24. Harrington, W. F. (1979): Contractile proteins of muscle. In: *The Proteins, Vol. IV*, edited by H. Neurath and R. L. Hill, pp. 245–409. Academic Press, New York.
25. Herrick, F. H. (1895): The American lobster. A study of its habits and development. *Bull. US Fish. Comm.*, 15:1–252.
26. Hoyle, G. (1969): Comparative aspects of muscle. *Annu. Rev. Physiol.*, 31:43–84.
27. Hoyle, G., and McNeill, P. A. (1968): Correlated physiological and ultrastructural studies on specialized muscles. Ib. Ultrastructure of white and pink fibers of the levator of the eyestalk of *Podophthalmus vigil* (Weber). *J. Exp. Zool.*, 167:487–522.
28. Ishiura, S., Murofushi, H., Suzuki, K., and Imahori, K. (1978): Studies of a calcium-activated neutral protease from chicken skeletal muscle. I. Purification and characterization. *J. Biochem. (Tokyo)*, 84:225–230.
29. Ishiura, S., Sugita, H., Suzuki, K., and Imahori, K. (1979): Studies of a calcium-activated protease from chicken skeletal muscle. II. Substrate specificity. *J. Biochem. (Tokyo)*, 86:579–581.
30. Jahromi, S. S., and Atwood, H. L. (1967): Ultrastructural features of crayfish phasic and tonic muscle fibers. *Can. J. Zool.*, 45:601–606.
31. Jahromi, S. S., and Atwood, H. L. (1969): Correlation of structure, speed of contraction, and total tension in fast and slow abdominal muscle fibers of the lobster *(Homarus americanus)*. *J. Exp. Zool.*, 171:25–38.
32. Jahromi, S. S., and Atwood, H. L. (1971): Structural and contractile properties of lobster leg-muscle fibers. *J. Exp. Zool.*, 176:475–486.
33. Jahromi, S. S., and Govind, C. K. (1976): Ultrastructural diversity in motor units of crustacean stomach muscles. *Cell Tissue Res.*, 166:159–166.
34. Johnson, B. (1980): An electron microscopic study of flight muscle breakdown in an aphid *Megoura viciae*. *Tissue Cell*, 12:529–539.
35. Kameyama, T., and Etlinger, J. D. (1979): Calcium-dependent regulation of protein synthesis and degradation in muscle. *Nature*, 279:344–346.
36. Kar, N. C., and Pearson, C. M. (1976): A calcium activated neutral protease in normal and dystrophic human muscle. *Clin. Chim. Acta*, 73:293–297.
37. Kubota, S., Suzuki, K., and Imahori, K. (1981): A new method for the preparation of a calcium activated neutral protease highly sensitive to calcium ions. *Biochem. Biophys. Res. Commun.*, 100:1189–1194.
38. Lehman, W., Regenstein, J. M., and Ransom, A. L. (1976): The stoichiometry of the components of arthropod thin filaments. *Biochim. Biophys. Acta*, 434:215–222.
39. Lehman, W., and Szent-Gyorgyi, A. G. (1975): Regulation of muscular contraction. Distribution of actin control and myosin control in the animal kingdom. *J. Gen. Physiol.*, 66:1–30.
40. Libby, P., and Goldberg, A. L. (1980): Effects of chymostatin and other proteinase inhibitors on protein breakdown and proteolytic activities in muscle. *Biochem. J.*, 188:213–220.
41. Martin, A. F., Reddy, M. K., Zak, R., Dowell, R. T., and Rabinowitz, M. (1974): Protein metabolism in hypertrophied heart muscle. *Circ. Res.*, 35:32–40.
42. Mellon, D. F., Jr., and Stephens, P. J. (1980): Modifications in the arrangement of thick and thin filaments in transforming shrimp muscle. *J. Exp. Zool.*, 213:173–179.
43. Miledi, R., and Slater, C. R. (1969): Electron-microscopic structure of denervated skeletal muscle. *Proc. R. Soc. Lond. [Biol.]*, 174:253–269.
44. Millward, D. J. (1980): Protein degradation in muscle and liver. In: *Comprehensive Biochemistry, Vol. 19B*, edited by M. Florkin, pp. 153–232. Elsevier, Amsterdam.
45. Monckton, G., and Marusyk, H. (1975): ^{3}H leucine incorporation into myofibrils of normal and dystrophic mouse skeletal muscle. *Can. J. Neurol. Sci.*, 2:1–4.
46. Muscatello, U., Margreth, A., and Aloisi, M. (1965): On the differential response of sarcoplasm and myoplasm to denervation in frog muscle. *J. Cell Biol.*, 27:1–24.
47. Mykles, D. L., and Skinner, D. M. (1981): Preferential loss of thin filaments during molt-induced atrophy in crab claw muscle. *J. Ultrastruct. Res.*, 75:314–325.

48. Mykles, D. L., and Skinner, D. M. (1982): Molt-cycle associated changes in calcium-dependent proteinase activity that degrades actin and myosin in crustacean muscle. *Dev. Biol.*, 92.
49. Neerunjun, J. S., and Dubowitz, V. (1979): Increased calcium-activated neutral protease activity in muscles of dystrophic hamsters and mice. *J. Neurol. Sci.*, 40:105–111.
50. Ogonowski, M. M., Lang, F., and Govind, C. K. (1980): Histochemistry of lobster claw-closer muscles during development. *J. Exp. Zool.*, 213:359–367.
51. Pellegrino, C., and Frazini, C. (1963): An electron microscope study of denervation atrophy in red and white skeletal muscle fibers. *J. Cell Biol.*, 17:327–349.
52. Publicover, S. J., Duncan, C. J., and Smith, J. L. (1978): The use of A23187 to demonstrate the role of intracellular calcium in causing ultrastructural damage in mammalian muscle. *J. Neuropathol. Exp. Neurol.*, 37:544–557.
53. Randall, W. C. (1970): Ultrastructural changes in the proleg retractor muscles of *Galleria mellonella* after denervation. *J. Insect Physiol.*, 16:1927–1943.
54. Reddy, M. K., Etlinger, J. D., Rabinowitz, M., Fischman, D. A., and Zak, R. (1975): Removal of Z-lines and α-actinin from isolated myofibrils by a calcium-activated neutral protease. *J. Biol. Chem.*, 250:4278–4284.
55. Rees, D., and Usherwood, P. N. R. (1972): Effects of denervation on the ultrastructure of insect muscle. *J. Cell Sci.*, 10:667–682.
56. Regenstein, J. M., and Szent-Gyorgyi, A. G. (1975): Regulatory proteins of lobster striated muscle. *Biochemistry*, 14:917–925.
57. Reger, J. F. (1967): A comparative study on striated muscle fibers of the first antenna and the claw muscle of the crab *Pinnixia* sp. *J. Ultrastruct. Res.*, 20:72–82.
58. Salter, S. J. A. (1860): On the moulting of the common lobster *(Homarus vulgaris)* and the shore crab *Carcinus maenas). J. Proc. Linn. Soc. (Lond.)*, 4:30–35.
59. Schiaffino, S., and Settembrini, P. (1970): Studies on the effect of denervation in developing muscle. I. Differentiation of the sarcotubular system. *Virchows Arch. [Cell Pathol.]*, 4:345–356.
60. Shafiq, S. A., Askanas, V., Asiedu, S. A., and Milhorat, A. T. (1972): Structural changes in human and chicken muscular dystrophy. In: *Muscle Biology, Vol. 1*, edited by R. G. Cassens, pp. 255–272. Marcel Dekker, New York.
61. Sherman, R. G., and Atwood, H. L. (1971): Structure and neuromuscular physiology of a newly discovered muscle in the walking legs of the lobster *Homarus americanus. J. Exp. Zool.*, 176:461–474.
62. Silverman, H., Hikida, R. S., and Staron, R. S. (1979): Loss of thick filaments from fast-twitch glycolytic muscle fibers of the pigeon pectoralis after chronic administration of Dantrolene sodium. *Am. J. Anat.*, 155:69–82.
63. Skinner, D. M. (1966): Breakdown and reformation of somatic muscle during the molt cycle of the land crab, *Gecarcinus lateralis. J. Exp. Zool.*, 163:115–124.
64. Skinner, D. M., and Graham, D. E. (1970): Molting in land crabs: Stimulation by leg removal. *Science*, 169:383–385.
65. Skinner, D. M., and Graham, D. E. (1972): Loss of limbs as a stimulus to ecdysis in Brachyura (true crabs). *Biol. Bull.*, 143:222–233.
66. Skinner, D. M., Marsh, D. J., and Cook, J. S. (1965): Physiological salt solution for the land crab, *Gecarcinus lateralis. Biol. Bull.*, 129:355–365.
67. Statham, H. E., Duncan, C. J., and Smith, J. L. (1976): The effect of the ionophore A23187 on the ultrastructure and electrophysiological properties of frog skeletal muscle. *Cell Tissue Res.*, 173:193–209.
68. Sugden, P. H. (1980): The effects of calcium ions, ionophore A23187 and inhibition of energy metabolism on protein degradation in the rat diaphragm and epitrochlearis muscles *in vitro.* Biochem. J., 190:593–603.
69. Suzuki, K., Ishiura, S., Tsuji, S., Katamoto, T., Sugita, H., and Imahori, K. (1979): Calcium activated neutral protease from human skeletal muscle. *FEBS Lett.*, 104:355–358.
70. Toyo-Oka, T., and Masaki, T. (1979): Calcium-activated neutral protease from bovine ventricular muscle: isolation and some of its properties. *J. Mol. Cell. Cardiol.*, 11:769–786.
71. Toyo-Oka, T., Shimizu, T., and Masaki, T. (1978): Inhibition of proteolytic activity of calcium activated neutral protease by leupeptin and antipain. *Biochem. Biophys. Res. Commun.*, 82:484–491.
72. Truchot, J.-P. (1978): Variations de la concentration sanguine d'hémocyanine fonctionelle au cours du cycle d'intermue chez le crabe *Carcinus maenas* (L). *Arch. Zool. Exp. Gén.*, 119:265–262.

73. Umezawa, H. (1976): Structures and activities of protease inhibitors of microbial orgin. *Methods Enzymol.*, 45:678–695.
74. Wildenthal, K., Wakeland, J. R., Ord, J. M., and Stull, J. T. (1980): Interference with lysosomal proteolysis fails to reduce cardiac myosin degradation. *Biochem. Biophys. Res. Commun.*, 96:793–798.
75. Yamaoka, L. H., and Skinner, D. M. (1974): Synthesis of structural proteins in regenerating and non-regenerating crustacean muscle in the molt cycle. *J. Cell Biol.*, 63:381a.
76. Yamaoka, L. H., and Skinner, D. M. (1975): Cytolytic enzymes in relation to the breakdown of the chelae muscle of the land crab, *Gecarcinus lateralis*. *Comp. Biochem. Physiol.*, 52B:499–502.

Basic Biology of Muscles: A Comparative Approach, edited by B. M. Twarog, R. J. C. Levine, and M. M. Dewey. Raven Press, New York © 1982.

Physiological and Ultrastructural Studies on the Intracellular Calcium Translocation During Contraction in Invertebrate Smooth Muscles

Haruo Sugi and Suechika Suzuki

Department of Physiology, School of Medicine, Teikyo University, Itabashi-ku, Tokyo 173, Japan

It has been well established that the contraction-relaxation cycle in all kinds of muscle is controlled by the change in myoplasmic free Ca ion concentration (4). In vertebrate fast striated muscle, the Ca ions activating the contractile mechanism (activator Ca) are stored in the lumen of the sarcoplasmic reticulum (SR) by the active Ca transport mechanism, and released into the myoplasm to cause contraction. In various kinds of smooth muscle, on the other hand, the sources of activator Ca in physiological contraction still remain to be investigated. In general, smooth muscles exhibit a considerable variety with respect to their structure and function, so that the experimental results obtained from one type of muscle may not be directly applied to another. Much more experimental work from the standpoint of comparative physiology is desired.

This chapter is concerned with a series of correlated physiological and ultrastructural studies on the intracellular localization of activator Ca and its translocation during contraction in some invertebrate somatic smooth muscles.

MATERIALS AND METHODS

The materials used were the anterior byssal retractor muscle (ABRM) of a mussel *Mytilus edulis*, the longitudinal body wall muscle (LBWM) of an opisthobranch mollusc *Dolabella auricularia*, and the longitudinal retractor muscle (LRM) of a sea-cucumber *Stichopus japonicus*. The animals were collected at the Misaki Marine Biological Station, and kept in aerated sea water at 16°C.

Physiological Experiments

All experiments were performed with small muscle strips in the isometric condition, the isometric tension being recorded with a strain gauge (compliance, 2 μm/g; resonance frequency, 150 Hz). The standard experimental solution (artificial sea water) had the following composition (mM): NaCl, 497; KCl, 10; CaCl$_2$, 10; MgCl$_2$, 50 (pH adjusted to 7.2 by NaHCO$_3$). For recording the electrical activity, the muscle strip preparation was mounted in a three-compartment chamber with Vaseline-lined removable partitions. The central compartment was filled with paraffin oil or isotonic sucrose solution, whereas, the two end compartments were filled with the experimental solution and isotonic KCl solution respectively; thus, the demarcation potential between the KCl-depolarized and the experimental solution segments of the preparation was recorded with a pair of Ag-AgCl electrodes (29). The electrical and mechanical responses were recorded on an ink-writing oscillograph. All experiments were made at room temperature (18–27°C).

Electron Microscopy

For conventional electron microscopy, the preparation was prefixed with a 6% glutaraldehyde solution containing 2 mM CaCl$_2$ (pH 7.2 by 0.1 M cacodylate buffer), postfixed in 2% OsO$_4$ (unbuffered). For determining the intracellular localization of Ca, the preparation at rest or during the mechanical activity was fixed in a 1% OsO$_4$ solution (pH 6.2 to 6.8 by 0.01 N acetic acid) containing 2% K pyroantimonate (K$_2$H$_2$Sb$_2$O$_7$·4H$_2$O), which is known to penetrate intact cell membrane in the presence of Os to produce electron-opaque precipitate with intracellular cations (15,16). This isometric tension in the preparation was measured during the course of fixation. The fixed tissue was then dehydrated with ethanol, embedded in Epon 812. Sections were cut on a Porter-Blum MT-1 ultramicrotome, and examined with a Hitachi HU-12AS electron microscope unstained or stained with uranyl acetate and lead citrate.

X-Ray Microanalysis

For chemical identification of the precipitate of pyroantimonate salts, the sections (unstained, about 2,000 Å in thickness) were analyzed with an energy dispersive X-ray microanalyzer (Edax 707B or 711) attached to a Hitachi HHS-2R scanning microscope, or JEOL 100CX analytical electron microscope, which was used as a scanning transmission electron microscope (STEM). The spot analysis of the pyroantimonate precipitate was performed on the STEM image of the muscle fibers with an accelerating voltage of 20 KV and a sample current of 20 nA. The X-ray emission was collected over a detecting time of 200 or 400 sec. The quantitative analysis of the precipitate was further made with a computerized EDIT system.

RESULTS

Physiological Evidence for the Presence of Intracellularly Stored Activator Ca

Mytilus *ABRM*

In *Mytilus* ABRM, ACh (10^{-6}-10^{-3} M) produces contractures that are little affected by a reduction of [Ca]$_0$, Mn ions and low pH, i.e., the factors known to inhibit Ca-influx (e.g., 8,31), indicating that ACh may activate the contractile mechanism mainly by the release of intracellulary stored activator Ca (25).

More direct evidence for the presence of intracellularly stored activator Ca has been obtained from the electrical and mechanical responses of the ABRM to the removal of external divalent cations (25). As shown in Fig. 1A, the ABRM fibers exhibited repetitive spike-like electrical activity superimposed on a gradual decline of membrane potential when they were soaked in a O-Ca, O-Mg solution, and each spike-like potential change was followed by a distinct increment of isometric tension, so that a large mechanical response, as large as the maximum ACh- or K-induced contractures, was built up as a result of summation of each tension increment. This result indicates that the ABRM fibers actually contains intracellularly stored activator Ca enough to activate the contractile mechanism fully. The one-to-one relation between the spike-like potential change and the tension increment strongly suggests that the sites of activator Ca localization is the plasma membrane per se or its close vicinity. The ABRM showed, however, little tension development in response to

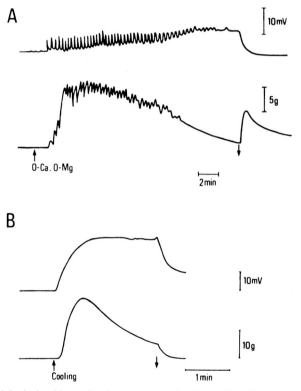

FIG. 1. Physiological evidence for the presence of activator Ca within invertebrate smooth muscle fibers. **A:** Electrical *(upper trace)* and mechanical *(lower trace)* responses of *Mytilus* ABRM to a O-Ca, O-Mg solution (25). **B:** Electrical *(upper trace)* and mechanical *(lower trace)* responses of *Dolabella* LBWM to rapid cooling. (Modified from Sugi and Suzuki, ref. 23.)

rapid cooling of the external medium, which is known to produce a marked mechanical response in various types of striated and smooth muscles (20).

Dolabella *LBWM*

The mechanical response to the removal of external divalent cations was not observed in *Dolabella* LBWM. Instead, it could be made to contract maximally by a rapid cooling of the surrounding medium from 20 to 3 or 5°C; the mechanical response to rapid cooling took place even when the LBWM fibers were depolarized in a high-K solution or in a Ca-free solution (Fig. 1B), indicating that they also contain intracellularly stored activator Ca enough to activate the contractile mechanism fully (23). The mechanical response to rapid cooling has been explained to be due to a reduced Ca-binding capacity and/or Ca-uptake in the SR (17,20).

Stichopus *LRM*

The mechanical response to a O-Ca, O-Mg solution was also observed in *Stichopus* LRM, though its magnitude was smaller than that of the maximum ACh- or K-induced contractures (24). The LRM could also be made to contract in hypertonic solutions, having an osmolarity about two times that of the standard solution even in the absence of external Ca ions, although the magnitude of the mechanical response to hypertonic solutions was smaller than the maximum ACh- or K-contractures. These results indicate that the LRM fibers also contain intracellularly stored activator Ca, although its amount may not be enough to activate the contractile mechanism fully. Like the ABRM fibers, the LRM fibers showed the mechanical response to O-Ca, O-Mg solution.

Intracellular Localization and Translocation of Activator Ca as Studied by the Pyroantimonate Method

Mytilus *ABRM*

As expected from the physiological experiments, the ABRM fibers fixed in the relaxed state exhibited distinct electron opaque pyroantimonate precipitate at the inner surface of the plasma membrane, and at the vesicles and the mitochondria in close apposition to the plasma membrane (Figs. 2A and B).

FIG. 2. Intracellular localization and translocation of electron-opaque pyroantimonate precipitate in *Mytilus* ABRM fixed in the relaxed state. **A:** Cross-section of the fibers showing the precipitate along the plasma membrane (P), at the vesicles (V) and at the mitochondria (M). Unstained. Calibration 1 μm. **B:** High-magnification view around the plasma membrane showing the localization of the precipitate along its inner surface and at the vesicles (V) closely apposed to it. *Arrows* indicate the outer surface of the plasma membrane. Stained. Calibration 0.1 μm. **C:** Cross-section of the fibers fixed in the contracted state. Note diffuse distribution of the precipitate in the myoplasm in the form of small particles. Unstained. Calibration 1 μm. **D:** Cross-section of the fibers fixed during the catch state. Note that the precipitate is again localized at the peripheral part of the fibers. Vesicles (V). Unstained. Calibration 1 μm. (Modified from Atsumi and Sugi, ref. 1.)

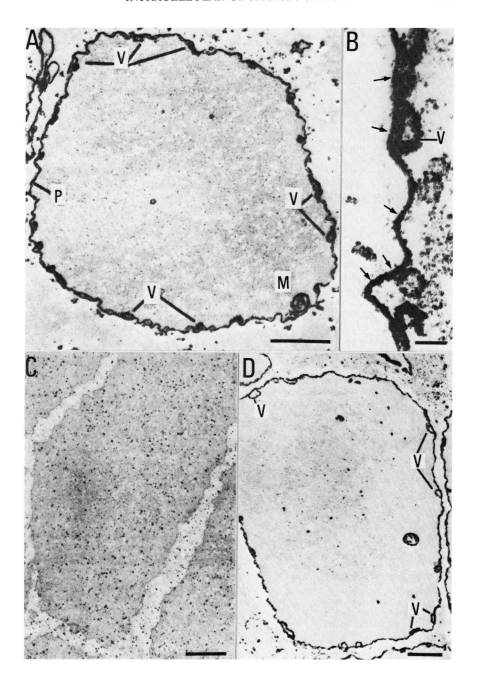

When the ABRM fibers were fixed at the peak of ACh-contractures or the mechanical response to the removal of external divalent cations, the precipitate was observed to distribute diffusely in the myoplasm in the form of numerous particles (Fig. 2C) (1).

An incidental observation was that, when the ABRM fibers were fixed during the catch state (e.g., 30), the precipitate almost returned to its resting sites of localization (Fig. 2D), suggesting that the catch state is established while the myoplasmic free Ca ion concentration is decreasing after active contraction.

Dolabella *LBWM*

The LBWM fibers had a fairly well-developed intracellular membranous structure compared to the ABRM and the LRM. The vesicular elements of the SR were closely apposed to the plasma membrane and to the surface tubules, i.e., tubular invaginations of the plasma membrane with many branches (Fig. 3A) (23). In the relaxed fibers, the precipitate was also observed at the inner surface of the plasma membrane and the surface tubules and at the SR (Figs. 3B and C) (26).

In the LBWM, fibers fixed during contraction also showed the diffuse distribution of the precipitate in the myoplasm (Fig. 3D) (26).

Stichopus *LRM*

In the LRM, the only intracellular membranous structure underneath the plasma membrane was the flattened surface vesicles (Fig. 4A) (24), and the precipitate was seen at the surface vesicles as well as at the inner surface of the plasma membrane (Fig. 4B) (27).

The diffuse distribution of the precipitate in the myoplasm was also observed in the fibers fixed during ACh- or K-contractures (Fig. 4C) (27).

X-Ray Microanalysis of the Pyroantimonate Precipitate

As shown in Fig. 5, the X-ray spectrum of the precipitate in both related and contracted fibers always showed the most distinct peak at 3,620 eV, which was explained to be due to a combination of Sb-Lα emission (at 3,600 eV) and Ca-Kα emission (at 3,690 eV) (18,26). Table 1 is an example of quantitative analysis of the elemental concentration ratios in the precipitate (27). The precipitate contained not only Ca but also other cations, such as Na and K. Considering that the amount

---

FIG. 3. Intracellular localization and translocation of pyroantimonate precipitate in *Dolabella* LBWM. **A:** Conventional electron micrograph showing the surface tubules (ST) and the vesicular elements of the SR (SR). Stained. Calibration 0.5 μm. (23). **B:** Localization of pyroantimonate precipitate along the inner surface of the plasma membrane and at the vesicular elements of SR (SR) in close apposition to it. *Arrows* indicate the outer surface of the plasma membrane. Stained. Calibration 0.5 μm. **C:** High-magnification view of the surface tubules showing the precipitate along its inner surface. Stained. Calibration 0.5 μm. **D:** Cross-section of the fibers fixed in the contracted state showing the diffuse distribution of the precipitate in the myoplasm. Stained. Calibration 2 μm. (Modified from Suzuki and Sugi, ref. 26.)

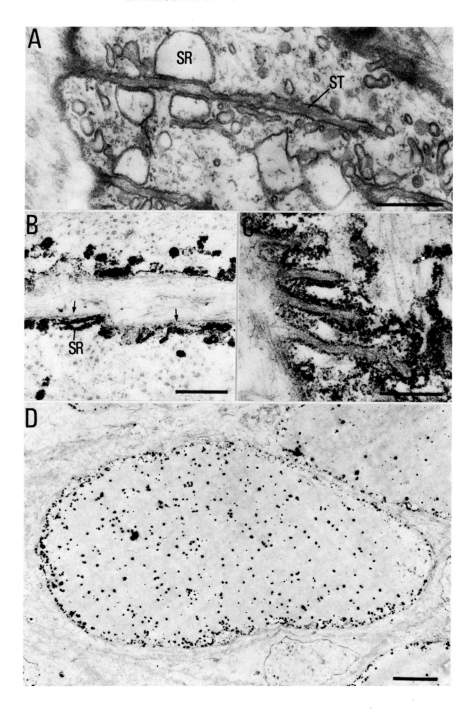

of Ca ions involved in the activation of the contractile mechanism is far smaller than those of Na and K in the myoplasm (4), these results may be taken to indicate that the pyroantimonate precipitate serves as a valid measure of Ca localization.

Thus, the marked change in distribution of the precipitate from the peripheral structures into the myoplasm during mechanical activity may reflect the intracellular translocation of activator Ca in these invertebrate smooth muscle fibers. Pyroantimonate combines with Ca most readily under the physiological condition in the myoplasm (13), but it also readily combines with Na (e.g., 14). In ethanol, pyroantimonate combines with K to form the precipitate (9). It may be that, when the fibers are exposed to the pyroantimonate-osmium solution, the formation of Ca-pyroantimonate precipitate first takes place reflecting the intracellular distribution of activator Ca, and followed by the further precipitation of Na pyroantimonate around the Ca pyroantimonate precipitate already present. In the subsequent dehydration procedure with ethanol, K pyroantimonate further precipitates around each pyroantimonate deposit. On this basis, the intracellular Na and K ions may serve to amplify the intracellular distribution of activator Ca, thus visualizing the intracellular translocation of activator Ca under the electron microscope.

DISCUSSION

The pyroantimonate method described in this chapter has also proved to be effective in determining the intracellular Ca translocation in some vertebrate visceral and vascular smooth muscles (7,22). In accordance with the physiological evidence for the presence of intracellularly stored activator Ca, all types of smooth muscle fibers studied exhibited distinct Ca localization at the peripheral structures, i.e., the inner surface of the plasma membrane, and the membranous structures in close apposition to the plasma membrane when they were fixed in the relaxed state. Similar results concerning the intracellular Ca localization have been obtained on various types of smooth muscle by use of oxalate (19,33), [45]Ca autoradiography (12), pyroantimonate (3), or Ca loading (11).

Our physiological experiments on dog coronary artery smooth muscle fibers indicate that they are only activated by the inwardly moving external Ca, and the fibers fixed at the relaxed state show no distinct Ca localization at the peripheral structures, whereas the fibers fixed during contraction exhibit the diffuse distribution of the precipitate in the myoplasm (H. Sugi, S. Suzuki, and N. Fujieda, *unpublished data*). Thus, the pyroantimonate method gives consistent results not only on smooth

---

FIG. 4. Intracellular localization and translocation of Ca in *Stichopus* LRM. **A:** Convertional electron micrograph of cross-section of the fibers showing the surface vesicles *(arrows)* underneath the plasma membrane. Stained. Calibration 1 μm (24). **B:** Localization of the precipitate at the inner surface of the plasma membrane and at the surface tubules. Stained. Calibration 0.2 μm. **C:** Cross-section of the fibers fixed in the contracted state showing the diffuse distribution of the precipitate in the myoplasm. Stained. Calibration, 1 μm. (Modified from Suzuki and Sugi, ref. 27.)

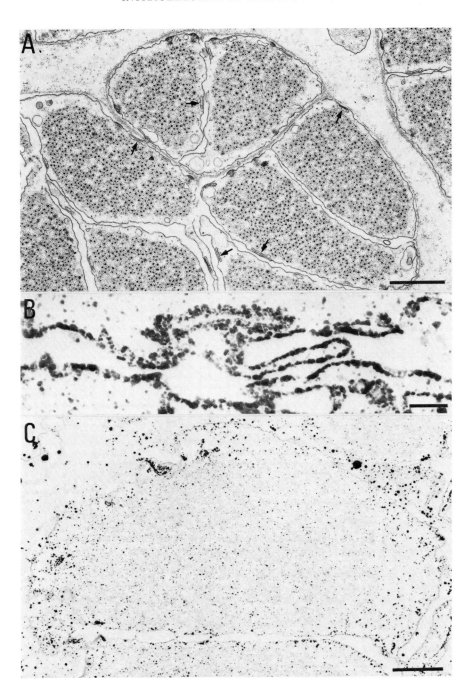

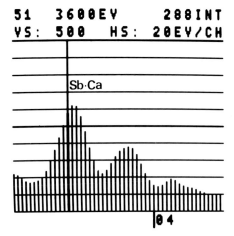

FIG. 5. Typical X-ray spectrum of the intracellular pyroantimonate precipitate. The *ordinate* gives the number of X-ray events; the *abscissa* is the individual energies of the X-ray in KeV (20 eV/CH). Note the most distinct peak at 3,620 eV. *Vertical line* indicates the position of Sb-Lα emission at 3,600 eV. (Modified from Suzuki and Sugi, ref. 26.)

TABLE 1. *Quantitative analysis of elemental concentration ratios in the pyroantimonate precipitate in the LRM fibres of a sea cucumber[a]*

	Mean ratios			
	Resting fibres		Contracted fibres	
Element-line	Plasma membrane	Subsarcolemmal vesicle	ACh-induced Myoplasm	K-induced Myoplasm
Sb-Lα	1.0000	1.0000	1.0000	1.0000
Ca-Kα	0.2540 ± 0.0520	0.1944 ± 0.0566	0.2845 ± 0.1197	0.2319 ± 0.0887
K-Kα	0.3365 ± 0.1169	0.4113 ± 0.1091	0.4036 ± 0.1201	0.3335 ± 0.1576
Mg-Kα	0.1406 ± 0.1166	0.0676 ± 0.0279	0.1231 ± 0.0621	0.2547 ± 0.1493
Na-Kα	0.2401 ± 0.1536	0.1449 ± 0.0981	0.0893 ± 0.0734	0.1234 ± 0.1380

[a]Values are mean ± SD (n = 10).
Modified from Suzuki and Sugi (27).

muscles with intracellularly stored activator Ca, but also on those with activator Ca of extracellular origin.

The validity of the pyroantimonate method in determining the intracellular Ca localization is further supported by our recent X-ray microanalysis of freeze-dried cryosections of the resting muscle fibers; the intracellular concentration of Ca is definitely higher around the plasma membrane than in the myoplasm distant from the plasma membrane (S. Suzuki and H. Sugi, *unpublished data*), indicating that the presence of the Ca-containing precipitate along the inner surface of the plasma membrane may not be an artifact arising from the histochemical procedures.

The intracellular membranous structures, i.e., the vesicles, the SR, and the mitochondria, are known to accumulate Ca in their lumen by the active Ca transport mechanism (for *Mytilus* ABRM, 10,28). In the case of the plasma membrane, its ability to accumulate Ca may be associated with the Ca-binding sites along its inner surface, and it seems possible that the Ca-binding capacity of the plasma membrane is influenced by membrane potential changes by ACh. In this connection, it would be of interest to examine whether some Ca-binding proteins, such as calsequestrin and calmodulin, exist along the inner surface of plasma membrane.

In *Mytilus* ABRM and *Dolabella* LBWM, there are bridge-like structures in the gap between the plasma membrane and the vesicles or the SR (1,23). Since analogous structures are known to be present in the triadic junction of vertebrate striated muscle (21,32), the Ca release from the vesicles or the SR may be controlled through these structures by membrane potential changes.

It may be argued that the Ca storage sites do not need to be close to the plasma membrane if a Ca-induced Ca-release mechanism is taken into consideration. However, the Ca-induced Ca-release mechanism, which is observed in skinned striated muscle fibers (5,6), is regenerative in nature, and seems to be inconsistent with the physiological contraction that is graded and under continuous control of membrane potential (2).

Much more experimental work is needed to solve the problems of excitation-contraction coupling in smooth muscles.

REFERENCES

1. Atsumi, S., and Sugi, H. (1976): Localization of calcium-accumulating structures in the anterior byssal retractor muscle of *Mytilus edulis* and their role in regulation of active and catch contractions. *J. Physiol. (Lond.)*, 257:549–560.
2. Costantin, L. L. (1975): Contractile activation in skeletal muscle. *Prog. Biophys. Mol. Biol.*, 29:197–224.
3. Debbas, G., Hoffman, L., Landon, E. J., and Hurwitz, L. (1975): Electron microscopic localization of calcium in vascular smooth muscle. *Anat. Rec.*, 182:447–472.
4. Ebashi, S., and Endo, M. (1968): Calcium ion and muscle contraction. *Prog. Biophys. Mol. Biol.*, 18:123–183.
5. Endo, M., Tanaka, M., and Ogawa, Y. (1970): Calcium-induced release of calcium from the sarcoplasmic reticulum of skinned skeletal muscle fibres. *Nature*, 228:34–36.
6. Ford, L. E., and Podolsky, R. J. (1970): Regenerative calcium release within muscle cells. *Science*, 167:58–59.
7. Fukuoka, H., Takagi, T., Nagai, H., Hotta, K., Suzuki, S., and Sugi, H. (1980): Localization and translocation of intracellular calcium in smooth muscle cells of bovine cerebral artery. *J. Electron Microsc. (Tokyo)*, 29:266–269.
8. Hagiwara, E., and Nakajima, T. (1966): Difference in Na and Ca spikes as examined by application of tetrodotoxin, procaine, and manganese ions. *J. Gen. Physiol.*, 49:793–806.
9. Hayat, M. A. (1975): *Positive Staining for Electron Microscopy.* Van Nostrand Reinhold, New York.
10. Humann, H.-G. (1969): Calciumakkumulierende Strukturen in einem glatten Wirbellosenmuskel. *Protoplasma*, 67:111–115.
11. Hurwitz, L. (1975): Some characteristics of the excitation-contraction coupling process in smooth muscle. In: *Concepts of Membranes in Regulation and Excitation*, edited by M. Rocha e Silva and G. Suarez-Kurtz, pp. 55–72. Raven Press, New York.
12. Jonas, L., and Zelck, U. (1974): The subcellular calcium distribution in the smooth muscle cells of the pig coronary artery. *Exp. Cell Res.*, 20:65–78.
13. Klein, R. L., Yen, S.-S., and Thureson-Klein, Å. (1972): Critique on the K-pyroantimonate method for semiquantitative estimation of cations in conjunction with electron microscopy. *J. Histochem. Cytochem.*, 20:65–78.
14. Komnick, H. (1962): Electronenmikroskopische Lokalisation von Na$^+$ and Cl$^-$ in Zellen und Geweben. *Protoplasma*, 55:414–418.

15. Komnick, H., and Komnick, U. (1963). Electronenmikroskopische Untersuchungen zur funktionellen Morphologie des Ionentransportes in der Sarzdrüse von *Larus argentatus*. *Z. Zellforsch. Mikrosk. Anat.*, 60:163–203.
16. Legato, M. J., and Langer, G. A. (1969): The subcellular localization of calcium ion in mammalian myocardium. *J. Cell Biol.*, 41:401–423.
17. Lüttgau, H. C., and Oetliker, H. (1968). The action of caffeine on the activation of the contractile mechanism in striated muscle fibres. *J. Physiol. (Lond.)*, 194:51–74.
18. Mizuhira, V. (1976): Elemental analysis of biological specimens by electron probe X-ray microanalysis. *Acta Histochem. Cytochem.*, 9:69–87.
19. Popescu, L. M., Diculescu, I., Zelck, U., and Ionescu, N. (1974): Ultrastructural distribution of calcium in smooth muscle cells and quantitative study. *Cell Tissue Res.*, 154:357–378.
20. Sakai, T. (1965): The effects of temperature and caffeine on activation of the contractile mechanism in the striated muscle fibers. *Jikeikai. Med. J.*, 12:88–102.
21. Somlyo, A. V. (1969): Bridging structures spanning the junctional gap at the triad of skeletal muscle. *J. Cell Biol.*, 80:743–750.
22. Sugi, H., and Daimon, T. (1977): Translocation of intracellularly stored calcium during the contraction-relaxation cycle in guinea-pig taenia coli. *Nature*, 269:436–438.
23. Sugi, H., and Suzuki, S. (1978): Ultrastructural and physiological studies on the longitudinal body wall muscle of *Dolabella auricularia*. I. Mechanical response and ultrastructure. *J. Cell Biol.*, 79:454–466.
24. Sugi, H., Suzuki, S., Tsuchiya, T., Gomi, S., and Fujieda, N. (1982): Physiological and ultrastructural studies on the longitudinal retractor muscle of a sea cucumber *Stichopus japonicus*. I. Factors influencing the mechanical response. *J. Exp. Biol.*, 97:101–111.
25. Sugi, H., and Yamaguchi, T. (1976): Activation of the contractile mechanism in the anterior byssal retractor muscle of *Mytilus edulis*. *J. Physiol. (Lond.)*, 257:531–547.
26. Suzuki, S., and Sugi, H. (1978): Ultrastructural and physiological studies on the longitudinal body wall muscle of *Dolabella auricularia*. II. Localization of intracellular calcium and its translocation during mechanical activity. *J. Cell Biol.*, 79:467–478.
27. Suzuki, S., and Sugi, H. (1982): Physiological and ultrastructural studies on the longitudinal retractor muscle of a sea cucumber *Stichopus japonicus*. II. Intracellular localization and translocation of activator calcium during mechanical activity. *J. Exp. Biol.*, 97:113–119.
28. Stössel, W., and Zebe, E. (1968): Zur intracellulären Regulation der Kontraktionsaktivität. *Pfluegers Arch.*, 302:38–56.
29. Twarog, B. M. (1954): Response of a molluscan smooth muscle to acetylcholine and 5-hydroxytryptamine. *J. Cell Comp. Physiol.*, 44:141–163.
30. Twarog, B. M. (1967): The regulation of catch in molluscan muscle. *J. Gen. Physiol.*, 50:157–169.
31. Van Breemen, C., Farinas, B. R., Casteels, R., Gerba, P., Wuytack, F., and Deth, R. (1973): Factors controlling cytoplasmic Ca^{2+} concentration. *Philos. Trans. R. Soc. Lond. [Biol.]*, 265:57–71.
32. Walker, S. M., and Schrodt, G. R. (1966): T-system connections with the sarcolemma and sarcoplasmic reticulum. *Nature*, 211:935–938.
33. Zelck, U., Jonas, L., and Wiegershausen, B. (1972): Ultrahistochemischer Nachweis von Calcium in glatten Muskelzellen der Arteria coronaria sinistra des Schweis. *Acta Histochem.*, 44:180–182.

Basic Biology of Muscles: A Comparative
Approach, edited by B. M. Twarog,
R. J. C. Levine, and M. M. Dewey.
Raven Press, New York © 1982.

Control of Activity of Cardiac, Skeletal, and Insect Fibrillar Muscle by Positive Feedback

Emil Bozler

The Ohio State University, Department of Physiology, Columbus, Ohio 43210

Almost 60 years ago, Fenn discovered that shortening of a muscle during a twitch markedly increased energy release (8). Later work showed that, at the same time, there are large changes in the chemistry of contraction. This shows that events during a twitch are not rigidly programmed, and that they are altered by movement. Under normal conditions, this control increases energy release when a load is lifted, a striking example of positive feedback. Such a response can be expected to be reflected also in the mechanics of muscle. This is the topic of this chapter.

Work on this subject was undertaken when it was found that, during a contracture of cardiac muscle, movement produced mechanical responses (3,4). When it was observed, furthermore, that such responses could be obtained in muscles depolarized in potassium solutions, it became clear that no membrane changes are involved and that we are dealing here with an internal control mechanism. A study of these responses led to the conclusion that shortening increases, lengthening decreases activity. When applied to normal muscle, such a mechanism would increase the energy liberation when lifting a load, and speed up relaxation, thus acting as positive feedback. This would increase the capacity to perform work. However, these conclusions are in apparent conflict with results of other investigators.

EVIDENCE FOR LENGTH FEEDBACK

The duration of an isotonic twitch can be markedly changed by passive changes in length produced by changes in load. In each panel of Fig. 1, A and B, three twitches were superimposed, one during which the load was diminished (A) or increased (B), and controls at the smaller and larger loads. Passive shortening prolonged the twitch; lengthening caused an early onset of relaxation and abbreviation of the twitch. Thus, shortening activates, lengthening deactivates, both effects representing positive feedback.

This result is confirmed by another experiment in which the load was changed twice, first increased of decreased, later returned to the original value near the peak (Fig. 1, C and D). Therefore, the load was the same during relaxation in all three twitches of each panel. Control experiments showed that a change in load during

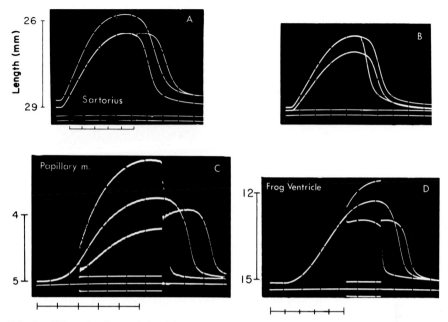

FIG. 1. Effect of a change in load during isotonic twitch. In **A** and **B** (from same muscle), load changed only once, was diminished in **A** and increased in **B**. Passive length change 0.4%; load change by 16% P_0. In each panel also control twitches at the greater and smaller lengths. In **C** and **D**: effect of transient length change. In each panel a transient increase and decrease and a control. Load the same during relaxation in all three twitches. Load changed by 8% P_0 in **C** and by 15% P_0 in **D**. Lowest tracings indicate changes in load. Temperature: 3°C in **A**, **B**, and **D**, and 25°C in **C**. Time marks: 0.2 sec in **A** and **B**, 0.1 sec in **C**, and 1 sec in **D**.

the early and middle part of the rising phase of a twitch by itself had no effect on the onset and time course of relaxation. Therefore, the effects observed were due to the second change. This experiment again shows that extension has a relaxing effect, and that passive shortening prolongs activity. From similar experiments on the effect of transient shortening, other investigators have concluded that, on the contrary, shortening has a relaxing action (6,9). However, they overlooked that, in such an experiment, two length changes are imposed on the muscle and did not inquire which phase of the passive movement caused the physiological effect. As shown above, the effect observed depends on the second change.

The length changes imposed on the muscle in these experiments were small. The large size of the responses indicated that we are dealing here with a very powerful control mechanism that can be expected to influence greatly the time course of normal relaxation. In one of the experiments confirming this conclusion, twitches were isotonic first, but a stop prevented shortening below a certain length (Fig. 2). The final isotonic phase of the twitch began much later than in a normal twitch and, therefore, would have been expected to be fast. Actually, relaxation was at first even slower than in a normal twitch, gradually accelerated, and then was parallel with normal relaxation. This indicates that normal velocity depends on the

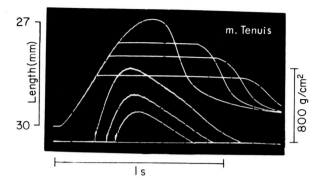

FIG. 2. Isotonic-isometric twitch. Shortening was stopped below a certain length. *Upper tracings:* length; *lower tracings:* tension.

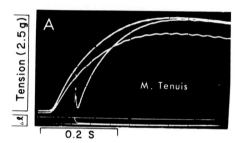

FIG. 3. Effect of release during isometric twitch of frog skeletal muscle. Two control twitches at the greater and smaller resting lengths and another twitch during which muscle shortened rapidly 1%. Temperature: 1°C.

deactivation by previous lengthening. It is also significant that the time course of relaxation is time independent.

These observations raise some questions about the nature of relaxation, particularly the role of Ca. Neither the time independence nor the rapidity of relaxation can easily be explained by the assumption that absorption of Ca is an important factor. The rapidity of relaxation is particularly remarkable in frog cardiac muscle, because of the poor development of the sarcoplastic reticulum.

How length changes influence the rising phase has been studied by several methods. In one of these, the muscle was released during the rising phase of an isometric twitch. Because tension recovered incompletely, Edman (7) and other investigators (for references, see 5) concluded that shortening diminished activity. However, this can be accounted for by the loss of mechanical energy without assuming any change in energy output. That release of energy is increased is suggested by several observations. It is well known that in skeletal muscle, after release during a tetanus, tension rises more rapidly than in controls. This was found to be true also in twitches, most strikingly in cardiac muscle. In some frog skeletal muscles, tension rose after the release higher than in controls at the short length (Fig. 3). Still more convincingly, J. A. Rall and E. Bozler *(unpublished results)* found that, in this type of experiment, heat production of the frog sartorius was increased, and total output of energy rose by 12% after shortening by 2%. Thus, it is clear that shortening has an activating action and that the muscle is not completely activated under isometric conditions.

Is the increased output of work and heat due to increased release of Ca? There is good evidence that in skeletal muscles the amount of Ca released in a twitch supersaturates the contractile elements (2). The experimental results, therefore imply that the increase in energy output is not due to increased release of Ca.

SIGNIFICANCE OF FEEDBACK

This question can be discussed by comparing the time course of an isotonic and isometric twitch (Fig. 4). It might be said that such a comparison is meaningless because of the large difference in mechanical conditions. However, the following features of these responses are significant. At 0°C, the velocity of shortening becomes maximal less than 40 msec after stimulation, probably before activation by Ca is complete. In Fig. 4, velocity and, therefore, output of mechanical power remained maximal up to the time when in an isometric twitch 98% of peak tension was reached and when muscle length had dropped to a level at which, in an isometric twitch peak, tension was only about 50% of the largest tension reached in an isometric twitch. Relaxation, on the contrary, was much faster in isotonic than isometric relaxation. The rapid onset and long duration of shortening completely agree with the findings presented above and are the result of positive feedback. Similarly, the rapid isotonic relaxation is fully explained by the relaxing effect of lengthening demonstrated above experimentally. What would happen under the influence of negative feedback, the accepted view on feedback in striated muscles? This would weaken and abbreviate the rising phase and would make relaxation slower. Both of these effects would diminish the ability to perform work; they are also contrary to elementary facts.

DIFFERENCES BETWEEN CARDIAC AND SKELETAL MUSCLE

The length responses discussed so far are very similar in cardiac and skeletal muscles, but large quantitative differences were found in other experiments. In Fig.

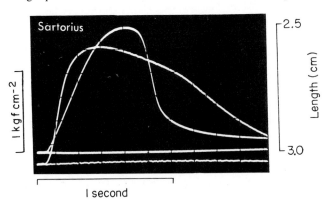

FIG. 4. Isotonic and isometric twitches of the frog sartorius at the same resting length 1.1 L_{max}. Load: 0.1 P_0. Temperature: 1°C.

5, the effect of a rapid change in load during the rising phase of an isotonic twitch was studied. In cardiac muscle, the rate of shortening was increased after a diminution of the load (Fig. 5A), compared to controls at the same load, decreased after an increase in load (Fig. 5B). This agrees with the results described above, that passive shortening activates and lengthening deactivates. However, under the same conditions, the rate of shortening was exactly parallel with controls in skeletal muscles (Fig. 5C and D). The difference between the two types of muscles can be

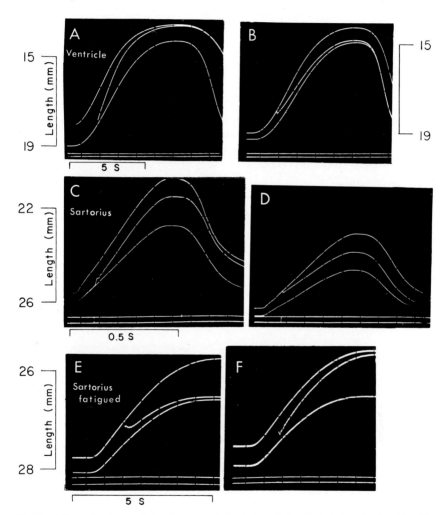

FIG. 5. Effect of a change in load on rate of shortening during isotonic twitch of a strip of frog ventricle and frog sartorius. Each panel depicts an experimental twitch, during which load was changed, and twitches with the old and new loads. Same muscle in **A** and **B**, and same muscle and **C**, **D**, **E**, and **F**. Load, indicated by *lower horizontal lines*, changed from 0.35 to 0.22 P_0 in **A**, from 0.35 to 0.5 P_0 in **B**, from 0.2 to 0.14 P_0 in **C**, from 0.2 to 0.26 P_0 in **D**, from 0.2 to 0.3 P_0 in **E**, and from 0.2 to 0.12 P_0 in **F**. Temperature: 1°C.

explained by assuming that skeletal muscles are fully activated after a short period of shortening, but cardiac muscle is not. That there is no principle difference between these types of muscle is also suggested by the observation that fatigued skeletal muscle behaves mechanically like cardiac muscle (Fig. 5E and F).

The difference between the two types of muscle is also illustrated by comparing freeloaded and afterloaded twitches. If shortening is essential for full activation, it must be expected that if the first part of a twitch is isometric, as is true in an afterload twitch, shortening at first is slower than in a freeloaded twitch at the same time. In each panel of Fig. 6, three twitches were recorded (upper series)—the first freeloaded, the others afterloaded, but different afterloads. Two other twitches (lower series) were freeloaded with the same loads as in the afterloaded twitches. In cardiac muscle, shortening was at first slow in afterloaded twitches, whereas, in skeletal muscle, shortening was exactly parallel in both types of twitches.

This difference can also be shown to be quantitative. If the onset of shortening of skeletal muscle was examined with instruments with small inertia, it was found that shortening did not become instantaneously maximal. There was a lag of about 10 msec in frog muscles and 20 msec in turtle muscles. The lag increased in fatigue and could reach more than 50 msec.

The results illustrate that important characteristics of muscles may depend on quantitative differences in the length responses.

INSECT FIBRILLAR MUSCLE

As shown by Pringle (10), length feedback plays a unique role in insect fibrillar muscle. The oscillations of these muscles have been explained by assuming that shortening has a relaxing action and that lengthening increases activity. Therefore, the feedback is assumed to be opposite in direction from that suggested by the experiments on vertebrate muscles described above.

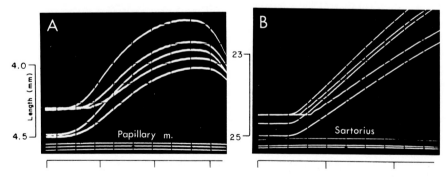

FIG. 6. Comparison of afterload and freeloaded twitches. Of the *upper groups* of twitches, the first is freeloaded; the next two are afterloaded with increasing loads. The *lower groups* are two freeloaded twitches with loads equal to those of the two afterloaded twitches. The loads, indicated by the three *lower horizontal lines*, were 90, 135, and 180 g/cm² in **A** and 400, 500, and 800 g/cm² in **B**. Calibration on left: length in millimeters. Time marks every 0.2 sec in **A** and every 0.05 sec in **B**.

What is the experimental basis for this view of the nature of feedback in insect muscles? Their mechanical properties have been studied by two methods. Sinusoidal analysis has shown that in a Ca-activated state the muscles produce more force while shortening than lengthening. This would actually suggest positive feedback. However, in such a system negative feedback can produce oscillations, if it acts with a delay. Therefore, the question cannot be decided by this method.

However, the problem was also studied by a more direct approach. It was found that, in a Ca-activated muscle, stretch caused an instant rise in tension, then a slower drop and finally a still slower rise to a sustained level. The first two phases are considered viscoelastic effects, and the last phase an active response, called stretch activation.

However, an entirely different interpretation of these results is possible. This has been suggested by experiments on cardiac muscle and is illustrated by the following experiment (Fig. 7). A contracture was produced in a strip of frog ventricle by an isosmotic K solution. Stretch induced a response like that of insect muscles; this has been considered as evidence of stretch activation (1,11,13,14). However, this interpretation is erroneous because, after stretch, tension is increased by two other factors. It is increased by the parallel elastic component. To determine this force, the muscle was stretched also during relaxation. It must be assumed then that, after stretch of the active muscle, the baseline is elevated by this force. It was seen then that tension always dropped briefly below the corrected baseline. Often tension even dropped briefly below original baseline. Thus, the first effect of stretch was relaxation. Later there was a further slow rise in tension, but this was also not directly due to stretch. The tension developed during a contracture was found to increase with length, like twitch tension. After the muscle had recovered in Ringer solution, a higher tension was induced in a second contracture at the new length. This tension was close to that reached after stretch during the first contracture. Therefore, the entire effect of the stretch was a rapid deactivation. What has been considered stretch activation was the recovery from this effect.

Naturally, the same control experiments are necessary in experiments on insect muscle. White et al. (15) have already gone one step in this direction. They found that, when the tension produced by stretching the relaxed muscle was subtracted from the stretch response of the Ca-activated muscle, tension actually dropped below the corrected baseline, indicating a deactivating response to stretch. The further rise probably has the same origin as in cardiac muscle. Also, in insect muscle, the sustained tension reached after Ca activation has been shown to increase sharply with length (12). This interpretation is supported by the fact that the tension produced by stretch is sustained. It is puzzling that a single stimulus, here stretch, produces a long lasting state of tension. This paradox disappears if the final tension is that which would have been produced without mechanical manipulation at the new length. Thus, stretch induces deactivation also in insect muscle. This conclusion applies also to other muscles for which stretch activation has been assumed. These considerations led to the conclusion that there is no essential difference between insect fibrillar and other striated muscles as far as feedback is concerned. The main

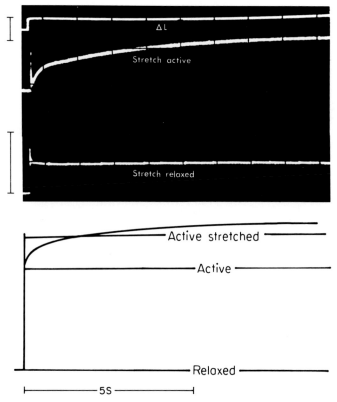

FIG. 7. Effect of stretch during a contracture of a strip of frog ventricle.*Upper tracing:* Length change. *Middle tracing:* Tension change caused by stretching during a contracture. *Lower tracing:* Effect of stretching of relaxed muscle. The *lower diagram* gives the baseline tension of the relaxed muscle, that of the active muscle before and after stretch, in relaxation to the response of the active muscle. Depolarizing solution contained in millimole per liter: KCl 116, Ca 1.5, phosphate 1 (pH 7.1), glucose 2. Temperature 0°C. Calibration on left: upper 10% length change, lower 100 g/cm².

distinguishing feature of insect muscle, which makes possible the oscillations, is the ability to raise internal Ca concentration to a level where a sustained state of activity is produced.

In summary, muscular activity is controlled not only by changes in Ca concentration, but also by a mechanism that increases activity above the isometric level during shortening and has a relaxing effect during lengthening. In this way, shortening under a load and relaxation are accelerated. Both effects increase the capacity to perform work. This control is present in all types of muscles tested, specifically skeletal, cardiac, and insect fibrillar muscle.

ACKNOWLEDGMENTS

This work was aided by grant 1 TO1 AM20048-01 of the National Institutes of Health.

REFERENCES

1. Abbot, R. H., and Steiger, G. H. (1977): Temperature and amplitude dependence of tension transients in glycerinated skeletal and insect fibrillar muscle. *J. Physiol. (Lond.)*, 266:13–42.
2. Blinks, J. R., Rudel, R., and Taylor, S. R. (1978): Calcium transients in isolated amphibian muscle fibers: Detection with aequorin. *J. Physiol. (Lond.)*, 277:291–323.
3. Bozler, E. (1972): Feedback in the contractile mechanism of the frog heart. *J. Gen. Physiol.*, 60:239–247.
4. Bozler, E. (1975): Mechanical control of the time course of contraction of the frog heart. *J. Gen. Physiol.*, 1965:329–344.
5. Bozler, E. (1977): Mechanical control of the rising phase of contraction of frog skeletal and cardiac muscle. *J. Gen. Physiol.*, 70:697–705.
6. Brutsaert, D. L. (1974): The force-velocity-time interaction of cardiac muscle. Physiological basis of Starling's law of the heart, pp. 155–174. *Ciba Foundation. (Lond.)*.
7. Edman, K. A. P. (1975): Mechanical deactivation induced by active shortening in isolated muscle fibers of the frog. *J. Physiol. (Lond.)*, 143:513–540.
8. Fenn, W. O. (1923): A quantitative comparison between the energy liberated and the work performed by the isolated sartorius muscle of the frog. *J. Physiol. (Lond.)*, 58:175–204.
9. Jewell, B. R., and Wilkie, D. R. (1960): The mechanical properties of relaxing muscle. *J. Physiol. (Lond.)*, 152:30–47.
10. Pringle, J. W. S. (1949): The excitation and contraction of the flight muscles of insects. *J. Physiol. (Lond.)*, 108:226–232.
11. Pringle, J. W. S. (1978): Stretch activation of muscle: Function and mechanism. *Proc. R. Soc. Lond. [Biol.]*, 201B:107–130.
12. Ruegg, J. C., and Stumpf, H. (1969): Activation of the myofibrillar ATPase activity by extension of glycerol extracted insect fibrillar muscle. *Pfluegers Arch.*, 305:34–46.
13. Steiger, G. J. (1977): Stretch activation and tension transients in cardiac, skeletal and insect flight muscle. In: *Insect Flight Muscle*, edited by R. T. Tregear, pp. 221–273. North-Holland, Amsterdam.
14. Steiger, G. J., Brady, A. J., and Tan, S. T. (1978): Intrinsic regulatory properties of contractility in the myocardium. *Circ. Res.*, 42:339–350.
15. White, D. C. S., Wilson, M. G. A., and Thorson, J. (1978): What does relaxed insect flight muscle tell us about the mechanism of active contraction? In: *Cross-bridge Mechanism in Muscle Contraction*, edited by Haruo Sugi and G. H. Pollack. University Park Press, Baltimore, Maryland.

Basic Biology of Muscles: A Comparative
Approach, edited by B. M. Twarog,
R. J. C. Levine, and M. M. Dewey.
Raven Press, New York © 1982.

Diversity of Narrow-Fibered and Wide-Fibered Muscles

C. Ladd Prosser

Department of Physiology and Biophysics, University of Illinois, Urbana, Illinois

In a volume like the present one we often become so concerned with our specialties and with experimental details that we lose sight of broad biological implications. This chapter attempts to put some of the facts of comparative muscle physiology into broad biological perspective.

This chapter will illustrate origins and diversity of muscles by brief statements regarding contractile proteins, morphology and physiology of myoepithelia and myocytes, adaptive properties of invertebrate nonstriated postural muscles, some aspects of vertebrate smooth muscle biology, and diversity of modulation of calcium regulation of actin-myosin interaction. It is proposed that transverse alignment of thick and thin filaments (striations) has evolved many times and that for a functional classification of muscles, cross-striations are less important than whether coupling between plasma membranes and contractile filaments is direct over short distances between plasma membrane and contractile filaments or indirect via T-tubules. It is proposed that the designations narrow-fibered or wide-fibered muscle are more appropriate than smooth or striated muscle. The two kinds of muscle differ in many ways, and many muscles with transversely aligned filaments are functionally narrow-fibered muscles (e.g., frog heart). Indeed, evidence (Fay et al., *this volume*) suggests that filaments of vertebrate smooth muscle are organized as sarcomeres with the dense bodies analogous to Z-lines.

CONTRACTILE PROTEINS

Several kinds of proteins for motility have evolved independently. Tubulins function in mitotic movements of chromosomes, ciliary beat, some kinds of protoplasmic flow (as in neurons). Flagellins power the movement of bacterial flagella. Spasmoneme is a contractile protein of ciliate protozoa sensitive to Ca^{2+} but not to ATP. In all muscles, tension is developed by interaction of actin and myosin (12).

Actins are more widely distributed than myosins, are highly conserved as to primary structure, and may have preceded myosins in evolution (17). Actins are relatively small proteins (43,000 to 45,000 daltons). They occur in all muscles, in many nonmuscle motile cells—myoepithelia, fibroblasts, amoeboid cells, blood platelets, in the acrosomes of some sperm—and in many nonmotile cells such as brain, liver, kidney, and red blood cells. In nonmuscle cells, actins may move by polymerization-depolymerization. Actin can bind to other proteins than myosin, e.g., to spectrin of red blood cells.

Myosins are large proteins (approximately 450,000 daltons) that occur in muscles and in some nonmuscle motile cells. There are many isotypes of myosin, e.g., in phasic, tonic, or embryonic striated muscles of vertebrates. Differences in actin-to-myosin ratios are reflected in relative numbers of thin and thick myofilaments.

Until recently it was believed that myosins and actins were proteins that occurred in eukaryotes only and that they may have evolved with muscle. Recent evidence indicates that actin and myosin may occur in bacteria, *Escherichia coli*, according to electrophoretic properties of extracted proteins, immunological identification, and capacity to combine with rabbit muscle counterparts (16). Actin and myosin have also been found in yeast, where they may function in budding (30) and recently they have been found in tomatoes, where they may function in streaming (28). Actin occurs in mycelia of Neurospora (2). Actin-myosins function in the formation of cleavage furrows of dividing cells (e.g., fertilized eggs) and are made visible by antibody staining. Cells that divide by interposition of a cell plate, as in higher plants, may not use actin-myosin in division. Actin-myosins function in cytoplasmic movement (cytoplasmic streaming), and this may be their function in plant cells. Actin-myosins also function in amoeboid movement, in slime molds (10). Actin-myosins function in prokaryotes (e.g., bacteria) is not known. It is concluded that actins probably preceded myosins: as interacting proteins of cell movement actins and myosins evolved in cytoplasmic streaming and cell division. The use of actin-myosin interaction as the basis of all muscle movement was secondary.

MYOEPITHELIA AND MYOCYTES AS PRECURSORS OF MUSCLES

Epithelial cells with contractile filaments are present in many kinds of animals. In diploblastic animals, such as coelenterates, myoepithelia occur in either ectoderm or endoderm. Contractions of coelenterate myoepithelia can be triggered by motor impulses in a nerve net or by impulses conducted through an epithelium by intercellular spread (15). Some jelly-fish myoepithelia have thick and thin filaments transversely aligned as striations (Fig. 1); in the same jellyfish other myoepithelial cells have bundles of nonaligned filaments (Fig. 2) (24). The striated regions contract more rapidly than the nonstriated.

Many myoepithelia have bundles of striated filaments along one side of a cell (epithelia of trematodes and hearts of ascidians). Flat myoepithelial cells occur in the pharynx of a polychaete worm (Scylla) and thick and thin filaments are clearly aligned, but in the flattened axis the cell membranes are not more than a few micrometers from filaments (25). Contractions of these cells are nerve activated and the cells show calcium action potentials (3). There are myoepithelia in some human sweat glands. In summary, myoepithelia are thin contractile cells, some with transversely aligned filaments and others with nonaligned filaments.

FIG. 1. Ectodermal myoepithelial cells of medusa *Aglantha* with bundles of striated myofilaments. Functions in swimming. Calibration 1 μm. (From Singla, ref. 24, with permission.)

Living sponges are simple metazoans without nervous systems; they have flagellated cells for fluid propulsion in canals and contractile myocytes arranged radially and circumferentially in the oscula. Myocytes are small (2 μm diameter), are widely separated by collagen, and contain thick and thin filaments (4). Mechanical stimulation of an osculum elicited contraction but the cells were not excitable to electric shocks and electrical responses were not obtained. Responses to mechanical stimulation occurred in high potassium (400 mM). In Ca-free sea water, contractions did not occur but Sr or Mg could be substituted for Ca; a univalent cation was also required but it could be Na, K, or Li. It is postulated that myocytes of sponges can be excited mechanically without the membrane events normal to other animals (18). It is concluded that myocytes and myoepithelia are primitive contractile cells and the transition from them to muscle is readily pictured.

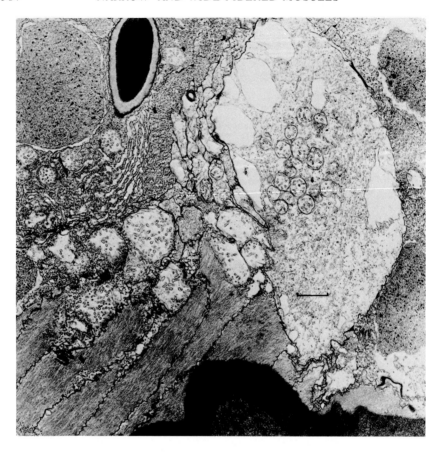

FIG. 2. Myoepithelium of tentacle of *Aglantha* showing bundles of nonstriated filaments. *Dark line at bottom* is mesogloea; large axon with vesicles at *right*. Calibration 1 μm. (From Singla, ref. 24, with permission.)

MUSCLES OF HOLLOW-BODIED ANIMALS AND POSTURAL NONSTRIATED MUSCLES OF INVERTEBRATES

Animals with endoskeletons (vertebrates) or exoskeletons (arthropods) have nerve-activated striated muscles for locomotion. These are of wide diameter (100 to 1,000 + μm) and have membrane invaginations, T-tubules that transmit excitatory signals inward from plasma membrane to contractile system. The muscles are multinucleate, have Z-bands separating sarcomeres and actin potentials are usually due to increased Na^+, sometimes to both Na^+ and Ca^{2+} conductance. Contractions of these striated muscles are largely isometric (tension developing). Most nonarthropod invertebrates are soft and hollow-bodied and the skeleton is hydrostatic. In these animals, the postural muscles are nonstriated or diagonally striated. In hollow-bodied animals and in visceral organs of vertebrates, the contractions are mostly isotonic (shortening) and the muscles extended six to eight times rest length can contract. The muscle fibers of hollow-body or hollow-organ animals are non-striated, small (2 to 4 μm in at least one dimension from membrane to contractile filaments), and the cells are uninucleate, lack membrane invaginations (T-tubules); in vertebrate viscera (but not invertebrates), there are caveolae of membranes. The postural nonstriated muscles of invertebrates are activated by nerves; many are dually innervated, hence multiunitary. The postural muscles may give both tonic and phasic contractions. Their action potentials are due to calcium currents, rarely to calcium plus Na (22).

Examples of nonstriated nerve-activated postural muscles are the five longitudinal retractor muscles of holothurians (sea cucumbers). Fibers of these retractors occur in bundles separated by connective tissue; individual fibers are small, spindle-shaped, and not connected by nexuses (Figs. 3 and 4) (14,23). Innervation is cholinergic and responses can be blocked by d-tubocurarine and enhanced by physostigmine. The retractor muscles can be excited by stretch and responses to stretch are mediated by nerve terminals. In some holothurians, the retractor muscles are rhythmically active but the rhythmicity, like the response to stretch, can be abolished by nerve-blocking drugs (23).

Gastropod molluscs have radula protractors and retractors that consist of narrow nonstriated fibers that are nerve-activated. In at least one species of snail, *Rapana*, the protractor is excited (depolarized) by acetylcholine and the retractor by glutamate; responses of each muscle are modulated by octopamine and by serotonin (5-HT) (11). In another snail, *Helix*, radula protractors are excited by acetylcholine, relaxed by serotonin. Nonstriated muscle fibers of the gill of *Aplysia* are innervated from three identified neurons, two of which are cholinergic, the third probably glutaminergic. Many molluscan postural muscles show both phasic and tonic contractions; in some species these may result from dual innervation, in others phasic contractions correlate with single spikes and tonic contractions with multiple spikes in a burst. Limited evidence indicates calcium and/or chloride conductance responses in molluscan postural muscles.

An adaptation for maintaining tension for long times (minutes or hours) with little expenditure of energy is the "catch" mechanism of adductors of bivalve molluscs, byssus retractor of the clam *Mytilus*, body wall muscles of gordius worms (27). Catch muscles have narrow nonstriated fibers that receive excitatory cholinergic innervation, which elicits a fast contraction. Catch muscles also receive relaxing nerves, usually serotonergic; once contracted, the muscles relax only slowly or not at all until the relaxing mechanism is activated. Contraction is triggered by Ca spikes. The thick filaments of molluscan catch muscles have a core of paramyosin, with myosin on the outside. The mechanism of catch is discussed in other chapters of this volume; there may be a delay in Ca sequestration and/or change in state of paramyosin once it is shortened by myosin contraction. Paramyosin is also found in noncatch muscles of some other animals (e.g., striated muscles of limulus) where it may be associated with very long myosin molecules. Paramyosin does not occur in vertebrate muscles.

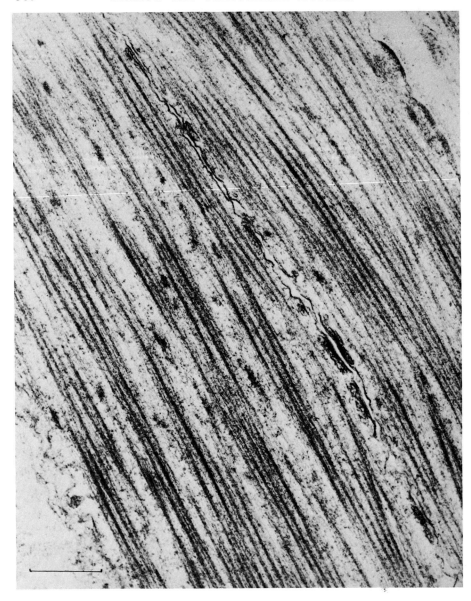

FIG. 3. Longitudinal section of two fibers of a holothurian longitudinal retractor muscle. Calibration 0.5 μm. (From Levin and Elfvin, ref. 14, with permission.)

An adaptation for isotonic contractions over a wide range of muscle lengths (six to eight times the rest length) occurs in the nonstriated, narrow-fibered proboscis retractors of sipunculid worms; the retractors pull the proboscis against hydrostatic pressure in the body cavity. The muscles are richly innervated and give both phasic and tonic contractions, both probably mediated by acetylcholine. In this one feature,

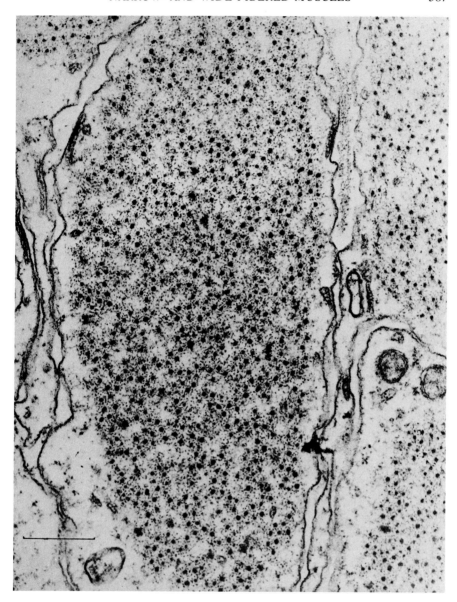

FIG. 4. Cross-section of one fiber of longitudinal retractor of a holothurian. Calibration 0.5 μm. (From Levin and Elfvin, ref. 14, with permission.)

proboscis retractors resemble striated muscles of crustaceans in which the same transmitter (glutamate) can elicit fast or slow contractions. When a sipunculid retractor contracts, the fibers fold in accordion fashion (Fig. 5) such that across the entire muscle the folds of very many fibers are aligned (19). The folds are 20 to 30 μm apart and folding is reflected in birefringent bands. A tension-length curve

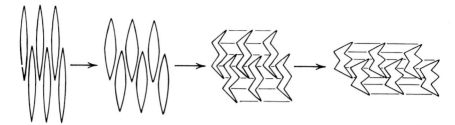

FIG. 5. Diagram representing transition of a proboscis retractor of a sipunculid worm from relaxed state (at *left*) to fully contracted (at *right*).

shows two peaks of optimal length, one where the pleats are extended and the other presumably for optimal overlap of thick and thin filaments.

Another adaptation for isotonic contractions of highly extensible muscles is diagonal striation. This occurs in the body wall (locomotor) muscles of many annelids, in longitudinal muscles of the roundworm *Ascaris* (13), and, with modification, in the prochordate *Branchiostoma* (amphioxus). Individual fibers of most muscles with diagonal striations are oval- or ribbon-shaped in cross-section and the distance from upper and lower membranes to contractile filaments is short (~2 μm). These diagonally striated muscles are well innervated and acetylcholine is usually the excitatory transmitter; some annelids show inhibition mediated by GABA. Excitation is by increases in conductance of Na^+ and Ca^{2+} in variable amounts; inhibition is by increase in K^+ conductance (earthworm) (8).

When diagonally striated muscles contract, the angle of the filaments with respect to the long axis of a fiber increases, often from about 5 to 15° when extended to 45° when shortened; this results in a change in length by shearing. In light microscopy of the ribbon-shaped fibers of earthworm, the myofilaments are seen as running diagonally in one direction near the upper face and in opposite direction in the lower. In longitudinal sections of a polychaete muscle, electron micrographs reveal prominent Z-lines in diagonal rows and the distance between Zs of adjacent rows is less in short than in extended fibers (Figs. 6 and 7) (9). The measured angle is the same for a given fiber length whether the length is reached by contraction or by passive stretch, hence the change in angle is postulated to result from fiber elasticity (9). Fiber diameter increases with shortening and maintains the volume constant. In contraction, since the striations are diagonal, sliding between thick and thin filaments causes shortening by shearing and considerable changes in length can occur (Fig. 8). The Z-lines are not discs, as in cross-striated muscles, but are narrow bands.

All of the postural nonstriated and diagonally striated muscles are neurally activated and all of them consist of narrow fibers without invaginating T-tubules. In many molluscs and annelids, the muscle of hearts and digestive organs is nonstriated and neurally activated. Arthropod heart and visceral muscles are usually striated and are neurally activated to beat rhythmically. Very few of the narrow-fibered nonstriated postural muscles of invertebrates show rhythmicity, and when it does occur it is due to neural activation.

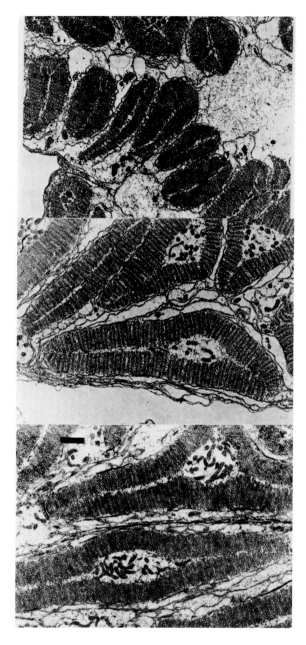

FIG. 6. Cross-sections of diagonally striated muscle fibers of a polychaete worm. **Top:** Fully extended muscle. **Middle:** At rest length. **Lower:** Contracted muscle. Calibration 1 μm. (From Iwamoto and Takahashi, ref. 9, with permission.)

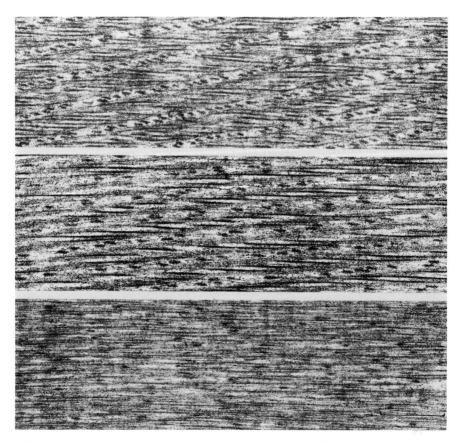

FIG. 7. Longitudinal sections of diagonally striated muscle fiber at different lengths. **Top:** Muscle at 0.4 L_0. **Middle:** 1.2 L_0. **Bottom:** 4.1 L_0. Note increasing separations of dark Z-lines at longer lengths. Calibration 1 μm. (From Iwamoto and Takahashi, ref. 9, with permission.)

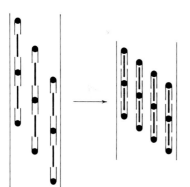

FIG. 8. Diagram of diagonally striated muscle fiber showing decrease in angle of diagonal striations with respect to long axis of fiber on shortening. Z-lines as *solid circles*, thin and thick filaments in sarcomeres.

A few invertebrate nonstriated muscles contract rhythmically and myogenically, for example, the hearts of many molluscs. The myogenic rhythmicity of these hearts resembles that of unitary smooth muscles of vertebrates, but the molluscan hearts cannot be precursors of vertebrate visceral muscle. Fibers of the heart of bivalve molluscs are nonstriated and may be connected by nexuses. The beat is myogenic but the hearts are neurally modulated. The hearts of most bivalves and snails are inhibited by cholinergic nerves; the ACh receptors are not strictly nicotinic or muscarinic. Whether acetylcholine depolarizes or hyperpolarizes the muscle depends on the relation of the resting potential to E_K and E_{Cl} (22). Recently, cardioaccelerator polypeptides have been identified in bivalves and gastropods. Another nonstriated muscle that appears to show myogenic rhythmicity is the stomach of squid.

VERTEBRATE NARROW-FIBERED NONSTRIATED MUSCLES

Relatively few nerve-activated, i.e., multiunitary, smooth muscles have been identified in vertebrates. Examples are: the muscle portions of nictitating membrane, piloerectors, some small blood vessels, muscles of lower esophagus, and some sphincters. Several mammalian muscles, vas deferens, bladder, and certain large blood vessels have properties of both multiunitary and unitary muscles. The nerve-activated muscles do not show spontaneous rhythmicity; they show no conduction from muscle fiber to fiber and are not stimulated by stretch. Nerve-activated muscles resemble but are not homolgous to the postural nonstriated muscles of hollow-bodied animals.

Most visceral smooth muscles of vertebrates are unitary. They are found in stomach, small bowel, colon, ureter, uterus, and many blood vessels (e.g., portal vein). Most of them show myogenic rhythmicity, conduct excitatory waves from cell to cell, and many of the muscles are very sensitive to stretch. They show functional similarities to molluscan heart but visceral smooth muscle appears to have evolved separately in the vertebrates (20). The muscle fibers are spindle-shaped, uninucleate, usually 3 to 5 μm in diameter and 200 to 400 μm long. The fibers are connected by nexuses of low electrical resistance, hence a sheet of visceral muscle is a functional syncytium. Thick and thin filaments are present and the ratio of actin to myosin is much greater than in striated muscle; intermediate filaments (100 Å) may have a structural function. Shallow invaginations of plasma membrane form rows of caveolae that may have pinocytotic function but there are no deep invaginations or T-tubules (7). Peripheral caveolae are not found in invertebrate nonstriated muscles. Vertebrate smooth muscles contain dense bodies to which thin filaments may attach and which contain some of the proteins of Z-bands of striated muscle. These dense bodies are also found in some invertebrate nonstriated muscles. Myogenic rhythmicity is modulated by nerves, and transmitters are liberated from varicosities; many muscle fibers may lie in the area between axons, hence trans-mitters must diffuse the distance of many cell widths. The innervation of unitary visceral muscles differs from that of multiunitary postural nonstriated muscles in which each fiber is activated by an axon. Vertebrate visceral muscles are regulated also by hormones. When a fiber of visceral smooth muscle contracts, there is no sliding between fibers but a shortening and increase in diameter of individual fibers (Fig. 9). Tension is transmitted through a sheet of fibers via a mesh of connective tissue.

Myogenic rhythmicity of visceral muscles of vertebrates differs from muscle rhythmicity of invertebrates, which is neurally activated. Several electrical patterns have been noted in gastrointestinal muscles (21): (a) pacemaker potentials, which may trigger spikes and which represent Ca-conductance change, have been observed in circular muscle of mammalian small intestine and in *Taenia coli*; (b) Ca-dependent "slow spikes" of stomach of amphibians and fish; (c) rhythmic fast spikes during maintained stretch in dogfish intestinal retractors; (d) slow waves that represent oscillatory sodium pump activity, occurring in longitudinal muscle of intestine of mammals (but not in guinea pig) and sensitive to metabolic inhibitors as well as to inhibitors of (Na-K)-ATPase; frequency of slow waves determined by unknown Ca-requiring processes; (e) minute rhythms on which slow waves are superimposed; (f) bursts of spikes at intervals of many minutes (approximately hourly), representing a rhythm of neurohormonal control; and (g) nonphysiologic rhythmic oscillations of large amplitude in muscles treated with Ca-free, EGTA-containing medium. Local reflex regulation associated with distension and stimulation of mechano-receptors may contribute to rhythmicity. This summary indicates that unitary visceral muscle of vertebrates is a specialization that appears not to be directly derived from invertebrate muscles.

REGULATION OF MYOSIN-ACTIN INTERACTION BY CALCIUM

All movement systems in which myosin and actin interact derive energy from ATP and require calcium ions. In muscles of all kinds, myosin binds to actin by cross-bridges; myosin ATPase is then activated, and sliding between thick and thin filaments (Fig. 10) ensues. The interaction between myosin and actin is precisely regulated by calcium-dependent processes. The timing of actin-myosin interactions makes possible the sequence of energy liberation in contraction and calcium sequestration in relaxation. Calcium ions act in one of several ways to remove the block on actin-myosin interaction in the resting state. The diverse modes of Ca-regulation may be considered in light of the evolution of types of muscles. There are three methods of Ca-regulation: (a) release of actin-myosin from block by a sequence of proteins—tropomyosin and three troponins, one of which is Ca-sen-

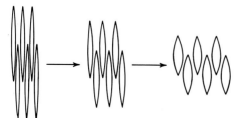

FIG. 9. Diagram of a vertebrate visceral nonstriated muscle from extended to shortened (contracted) state.

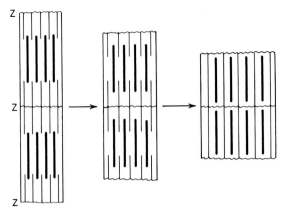

FIG. 10. Diagram of portion of a vertebrate striated muscle fiber showing sliding between thick and thin filaments as sarcomeres shorten in contraction.

sitive; (b) reaction of one myosin head with Ca^{2+}; and (c) phosphorylation of myosin by calcium-sensitive myosin kinase.

The first method, troponin-tropomyosin regulation, is best known in vertebrate striated muscles. Troponin-C is absent from nonstriated muscles—both vertebrate and invertebrate. Tropomyosin is present in smooth muscle of vertebrates, where it may have structural functions. Evidence has been presented that the striated muscles of vertebrates and arthropods evolved separately (22) and that the Ca-binding proteins (troponins) are clearly different.

The second method, Ca-regulation (use of myosin light chains that are directly activated by Ca^{2+}), is known in molluscs and some related animals (26).

The third method, phosphorylation of myosin by a Ca-activated kinase, is very widespread and may be more primitive than other modes of Ca regulation. It occurs in vertebrate visceral smooth muscles (5,6) and probably in many invertebrates. Most, perhaps all, cells contain kinases that trigger various enzyme reactions, especially those of protein synthesis. Protein kinases occur in many bacteria (29); cyclic AMP activates a protein kinase in *E coli*. In mammalian smooth muscles, a protein kinase is activated by cyclic AMP, synthesis of which is triggered by norepinephrine (1). It is easy to imagine the use of preexisting kinases by muscles early in their evolution and the appearance of a myosin-specific kinase that required Ca^{2+}. More information regarding myosin kinases of invertebrate nonstriated muscles is needed.

CONCLUSIONS

Actin and myosin are present in prokaryotes (a few bacteria) where their function is unknown. Actin and myosin occur in primitive eukaryotes, possibly in higher plants. These proteins probably evolved in eukaryotes in protoplasmic streaming and in formation of cleavage furrows of cell division. Actin is more widely distributed and may have preceded myosin in evolution. Actin and myosin occur in all muscles, in many nonmuscle motile cells, and in some nonmotile cells. The proportion of actin to myosin is greater in nonstriated than in striated muscles.

Myocytes and myoepithelia are contractile cells that may have been precursors of muscle. Myoepithelia derive from either ectoderm or endoderm. Myoepithelial cells may have transversely aligned thick and thin filaments (striations) or may have irregularly organized filaments (nonstriated bundles). Myoepithelia are flattened cells activated by nerves or by epithelial conduction; they probably occur in most phyla and are best known in coelenterates.

Hollow-bodied animals use nonstriated or diagonally striated muscles for locomotion. These muscle cells are uninucleate, measure approximately 2 μm from plasma membrane to myofilaments, and are activated by motor nerves. These muscles are very extensible and contraction is mainly isotonic. Adaptations for isotonicity in postural muscles of hollow-bodied animals are:

1. Contraction over several (four to eight) times rest length.
2. Responsiveness to stretch, usually via stimulation of motor nerve endings.
3. Folds (pleats) that are aligned in many fibers may cross an entire muscle.
4. Diagonal striations of filaments change angle of orientation on extension or contraction and thus implement considerable shortening or lengthening.
5. Maintenance of contracted state (catch) until release on signal from relaxing nerve fibers. Catch muscles have very thick paramyosin filaments with a cortex of myosin.
6. Calcium spikes permit a direct coupling between plasma membrane and contractile filaments.
7. Multiple innervation—phasic, tonic, relaxing. Multiunitary nonstriated muscles lack nexuses and invaginating T-tubules; hence there is no interfiber conduction or indirect excitation-contraction coupling. Nerve-activated nonstriated muscles are much less often found in vertebrates than in invertebrates. Some vertebrate visceral muscles are both nerve-activated and myogenically rhythmic.

Myogenic rhythmicity (with nexal conduction) occurs in molluscan hearts, possibly in mollusc and echinoderm digestive organs. Myogenic rhythmicity may have evolved several times in response to distension of organs. Rhythmicity of vertebrate visceral muscle is modulated by nerves and hormones. Vertebrate visceral fibers are narrow, and have caveolae, nexuses, and no T-tubules. They have Ca^{2+} spikes. Vertebrate visceral rhythmic muscles have no direct precursors in invertebrates.

Striated muscles contract and relax faster than nonstriated, even when the striated muscles lack T-tubules. Those with T-tubules are of much larger diameter than nonstriated fibers, and multinucleate. Striated fibers are primarily isometric and are attached to endoskeleton or exoskeleton.

Three general modes of calcium regulation of myosin-actin interaction have evolved:

1. Regulation by tropomyosin-troponin. Tropomyosins may have additional structural functions; troponins of arthropods differ from troponins of vertebrates. Tropomyosin-troponin regulation is general in striated muscles.

2. Ca-binding light chains of myosin, best known in molluscs.

3. Phosphorylation of myosin by a Ca-requiring kinase, best known in vertebrate visceral muscle and probably present in most invertebrate nonstriated postural muscles. Protein kinases (often Ca sensitive) occur widely, originated in prokaryotes; this may have been the earliest form of Ca-regulation. Ca-sensitive myosin light chains and troponin-tropomyosin may have evolved secondarily.

A general conclusion is that transverse alignment of thick and thin filaments has evolved many times. Striations occur in myoepithelia of coelenterates and a polychaete. There are diagonally striated flat muscle cells in annelids, nematodes, and molluscs. More significant than cross-striation functionally is the presence of direct coupling between plasma membrane and myofilaments (narrow-fibered muscles) or of indirect coupling via T-tubules (wide-fibered muscles).

In narrow-fibered muscles and in all myoepithelia, the distance from plasma membrane to contractile filaments is short (1 to 3 μm); in wide-fibered muscles, it may be 50 to 500 μm. Narrow fibers have no or limited T-tubules, and electromechanical coupling between plasma membrane and contractile filaments is direct. In wide-fibered muscles, long invaginations, T-tubules, and extensive sarcoplasmic reticular tubules are interposed between cell membrane and myofilaments. Narrow-fibered muscles have calcium action potentials and Ca^{2+}, which enters the cells, may also activate the contractile system. Wide-fibered muscles are usually uninucleate and have higher surface:volume ratio than multinucleate wide-fibered muscles. In narrow-fibered muscles, modulation of calcium control of actin-myosin interaction is by a Ca-sensitive kinase, which phosphorylates the myosin light chain,

TABLE 1. *A functional classification of narrow- and wide-fibered muscles*

Narrow-fibered muscles	Wide-fibered muscles
Random thick and thin filaments or striations present transversely or diagonally	Transverse alignment of thick and thin filaments
Fiber diameters 2–10 μm in cylindrical (spindle) fibers or from top to bottom in ribbon fibers	Fiber diameters 50–2,000 μm
Uninucleate	Usually multinucleate
Surface: volume ratio high	Surface: volume ratio low
No T-tubules, sparse SR; membrane caveolae in vertebrates	Membrane invaginations, large or as T-tubules; extensive SR
Ca^{2+} action potentials predominant, sometimes Ca^{2+} plus Na^+	Na^+ action potentials or junction potentials, rarely Ca^{2+}
Dense bodies for thin filament attachment, narrow Z-lines in the diagonally striated	Z-bands as transverse discs
Ca control by phosphorylation of myosin by Ca-dependent myosin kinase. Ca sensitive light chain on some myosins	Tropomyosin-troponin modulation of Ca control of actin-myosin action. Some myosin phosphorylation of uncertain function
Activation: nerve activated or myogenically rhythmic	All nerve activated
Electrotonic coupling by nexuses in myogenically rhythmic, not in neurally activated	No electrotonic coupling

or by a Ca-sensitive site on a myosin light chain, whereas most wide-fibered muscles have a troponin-tropomyosin sequence for Ca control. Narrow-fibered muscles are either activated by motor nerves or are endogenously rhythmic, and the activity of these spontaneously active muscles is modulated by nerves and hormones. All wide-fibered muscles are activated by motor nerves, some of them by both excitatory and inhibitory axons. Wide-fibered muscles usually have Na^+ conductance changes in junction potentials or spikes; these indirectly lead to release of Ca^{2+} from sarcoplasmic reticulum.

It is timely to deemphasize the terminology of nonstriated (smooth) and striated muscles and to use a functional classification based on the differing properties of narrow-fibered and wide-fibered muscles (Table 1).

ACKNOWLEDGMENTS

Support from grant PHS AM 27794-02 is acknowledged. Portions of this chapter were presented in a publication on the same subject in 4th International Symposium on Vascular Neuroeffector Mechanisms, Kyoto, Japan 1981.

REFERENCES

1. Adelstein, R., and Eisenberg, E. (1980): Regulation and kinetics of the actin-myosin ATP interaction. *Annu. Rev. Biochem.*, 49:921–956.
2. Allen, E. D., and Sussman, A. S. (1978): Presence of an actin-like protein in mycelium of Neurospora crossa. *J. Bacteriol*, 135:713–716.
3. Anderson, M., and del Castillo, J. (1976): Electrical activity of the proventriculus of the polychaete worm *Syllis spongiphila*. *J. Exp. Biol.*, 64:691–710.
4. Bagby, R. M. (1956): The fine structure of myocytes in sponges *Microciona prolifera* and *Tedania ignis*. *J. Morphol.*, 118:167–181.
5. de Laneolle, R. (1981): Characterization of antibodies to smooth muscle myosin kinase and their use in localizing myosin kinase in nonmuscle cells. *Proc. Natl. Acad. Sci. USA*, 78:4738–4742.
6. Driska, S. P., Askoy, M. O., and Murphy, R. A. (1981): Myosin light chain phosphorylation associated with contraction in arterial smooth muscle. *Am. J. Physiol.*, 240:C222–233.
7. Gabella, G. (1976): Quantitative morphological study of smooth muscle cells of guinea-pig *taenia coli*. *Cell Tissue Res.*, 170:161–186.
8. Ito, L., Kuriyama, H., and Tashiro, N. (1969): Miniature excitatory junction potentials in somatic muscles of earthworm. *J. Exp. Biol.*, 50:107–118.
9. Iwamoto, H., and Takahashi, H. (1982): Structural properties of diagnoally striated muscle of a polychaete worm. *(in press.)*
10. Kamiya, N. (1981): Physical and chemical basis of cytoplasmic streaming. *Annu. Rev. Plant Physiol.*, 32:205–236.
11. Kobayashi, M., Muneoka, Y., and Fujiwara, M. (1980): Modulatory actions of the possible neurotransmitters in the molluscan radular muscles. *Adv. Physiol. Sci.*, 22:319–336.
12. Korn, E. D. (1978): Biochemistry of actomyosin-dependent cell motility. *Proc. Natl. Acad. Sci. USA*, 75:588–599.
13. Lanzavecchi, G. (1977): Morphological variations in helical muscles (*Aschelminthes* and *Annelida*). *Int. Rev. Cytol.*, 51:133–186.
14. Levine, R. J. C., and Elfvin, M. (1982): Structure of a holothurian muscle. *(in press.)*
15. Mackie, G. O. (1975): Neurobiology of *Stomotoca*. II Pacemakers and conduction pathways. *J. Neurobiol.*, 6:357–378.
16. Nakamura, K., and Watanabe, S. (1978): Myosin-like protein and actin-like protein from *E. coli*. *J. Biochem. (Japan)*, 83:1459–1470.
17. Pollard, T. D., and Weiking, P. R. (1974): Actin and myosin in cell movements. *CRC Crit. Rev. Biochem.*, 2:1–65.

18. Prosser, C. L. (1967): Ionic analyses and effects of ions on contractions of sponge tissues. *Z. Vergl. Physiol.*, 54:109–120.
19. Prosser, C. L. (1967): Problems in the comparative physiology of non-striated muscle. In: *Invertebrate Nervous Systems*, edited by C. A. G. Wiersma, pp. 133–150. University of Chicago Press, Chicago.
20. Prosser, C. L. (1974): Smooth muscle. *Annu. Rev. Physiol.*, 35:503–535.
21. Prosser, C. L. (1978): Rhythmic potentials in intestinal muscle. *Fed. Proc.*, 37:2153–2157.
22. Prosser, C. L. (1979): Evolution and diversity of nonstriated muscles. In: *Handbook of Physiology, Cardiovascular System II*, edited by D. Bohr, pp. 635–670. American Physiology Society, Bethesda, Maryland.
23. Prosser, C. L., and Mackie, G. O. (1980): Contractions of holothurian muscles. *J. Comp. Physiol.*, 136:103–112.
24. Singla, C. L. (1978): Fine structure of neuromuscular system of *Polyorchis penicillatus (Hydromedusae, cnidaria). Cell Tissue Res.*, 193:163–174.
25. Smith, D. S., del Castillo, J., and Anderson, M. (1973): Fine structure and innervation of an annelid muscle with the longest recorded sarcomere. *Tissue Cell*, 5:281–502.
26. Szent-Gyorgyi, A. G. (1975): Calcium regulation and muscle contraction. *Biophys. J.*, 15:707–723.
27. Twarog, B. M. (1976): Aspects of smooth muscle function in molluscan catch muscle. *Physiol. Rev.*, 56:829–838.
28. Vakey, M., and Scordilis, S. (1980): Contractile proteins from the tomato. *Can. J. Bot.*, 58:797–801.
29. Wang, J. Y. J., and Koshland, D. E. (1981): Identification of distinct protein kinases and phosphatases in the prokaryote *Salmonella typhidimurium. J. Biol. Chem.*, 256:4640.
30. Water, R. D., Pringle, J., and Kleinsmith, L. J. (1980): Actin-like protein and its messenger RNA in Saccharomyces. *J. Bacteriol.*, 144:1145–1151.

Subject Index